Eye Tracking and Visual Analytics

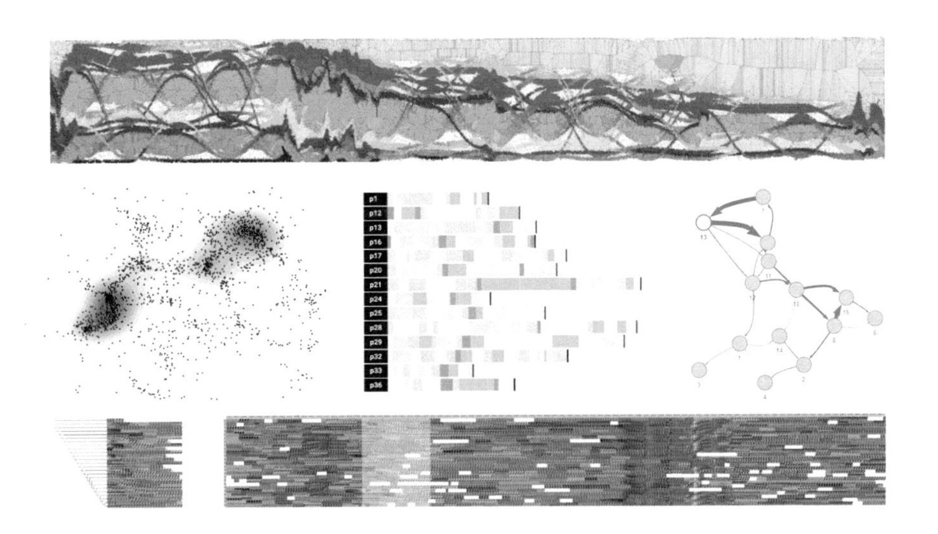

RIVER PUBLISHERS SERIES IN INFORMATION SCIENCE AND TECHNOLOGY

The "River Publishers Series in Information Science and Technology" covers research which ushers the 21st Century into an Internet and multimedia era. Multimedia means the theory and application of filtering, coding, estimating, analyzing, detecting and recognizing, synthesizing, classifying, recording, and reproducing signals by digital and/or analog devices or techniques, while the scope of "signal" includes audio, video, speech, image, musical, multimedia, data/content, geophysical, sonar/radar, bio/medical, sensation, etc. Networking suggests transportation of such multimedia contents among nodes in communication and/or computer networks, to facilitate the ultimate Internet.

Theory, technologies, protocols and standards, applications/services, practice and implementation of wired/wireless networking are all within the scope of this series. Based on network and communication science, we further extend the scope for 21st Century life through the knowledge in robotics, machine learning, embedded systems, cognitive science, pattern recognition, quantum/biological/molecular computation and information processing, biology, ecology, social science and economics, user behaviors and interface, and applications to health and society advance.

Books published in the series include research monographs, edited volumes, handbooks and textbooks. The books provide professionals, researchers, educators, and advanced students in the field with an invaluable insight into the latest research and developments.

Topics covered in the series include, but are by no means restricted to the following:

- Communication/Computer Networking Technologies and Applications
- Queuing Theory
- Optimization
- Operation Research
- Stochastic Processes
- Information Theory
- Multimedia/Speech/Video Processing
- Computation and Information Processing
- Machine Intelligence
- Cognitive Science and Brian Science
- Embedded Systems
- Computer Architectures
- Reconfigurable Computing
- Cyber Security

For a list of other books in this series, visit www.riverpublishers.com

Eye Tracking and Visual Analytics

Michael Burch

University of Applied Sciences, Chur, Switzerland

River Publishers

Published, sold and distributed by:
River Publishers
Alsbjergvej 10
9260 Gistrup
Denmark

www.riverpublishers.com

ISBN: 978-87-7022-433-8 (Hardback)
978-87-7022-432-1 (Ebook)

Contents

Preface xi

List of Figures xiii

List of Tables xxxi

List of Abbreviations xxxiii

1 Introduction 1
1.1 Tasks, Hypotheses, and Human Observers 3
1.2 Synergy Effects 7
1.3 Dynamic Visual Analytics 11

2 Visualization 17
2.1 Motivating Examples 19
2.2 Historical Background 27
2.2.1 Early Forms of Visualizations 28
2.2.2 The Age of Cartographic Maps 30
2.2.3 Visualization During Industrialization 32
2.2.4 After the Invention of the Computer 34
2.2.5 Visualization Today 36
2.3 Data Types and Visual Encodings 38
2.3.1 Primitive Data 39
2.3.2 Complex Data 42
2.3.3 Mixture of Data 48
2.3.4 Dynamic Data 50
2.3.5 Metadata 52
2.4 Interaction Techniques 53
2.4.1 Interaction Categories 54
2.4.2 Physical Devices 58
2.4.3 Users-in-the-Loop 61

2.5 Design Principles 62
2.5.1 Visual Enhancements and Decorations 63
2.5.2 Visual Structuring and Organization 65
2.5.3 General Design Flaws 66
2.5.4 Gestalt Laws 68
2.5.5 Optical Illusions 71

3 Visual Analytics **75**
3.1 Key Concepts 77
3.1.1 Origin and First Stages 78
3.1.2 Data Handling and Management 79
3.1.3 System Ingredients Around the Data 86
3.1.4 Involved Research Fields and Future Perspectives . . 88
3.2 Visual Analytics Pipeline 91
3.2.1 Data Basis and Runtimes 91
3.2.2 Patterns, Correlations, and Rules 93
3.2.3 Tasks and Hypotheses 97
3.2.4 Refinements and Adaptations 102
3.2.5 Insights and Knowledge 104
3.3 Challenges of Algorithmic Concepts 105
3.3.1 Algorithm Classes 106
3.3.2 Parameter Specifications 110
3.3.3 Algorithmic Runtime Complexities 111
3.3.4 Performance Evaluation 112
3.3.5 Insights into the Running Algorithm 114
3.4 Applications 116
3.4.1 Dynamic Graphs 117
3.4.2 Digital and Computational Pathology 118
3.4.3 Malware Analysis 119
3.4.4 Video Data Analysis 120
3.4.5 Eye Movement Data 122

4 User Evaluation **125**
4.1 Study Types 127
4.1.1 Pilot vs. Real Study 128
4.1.2 Quantitative vs. Qualitative 129
4.1.3 Controlled vs. Uncontrolled 130
4.1.4 Expert vs. Non-Expert 132
4.1.5 Short-term vs. Longitudinal 134

4.1.6 Limited-number Population vs. Crowdsourcing . . . 135
4.1.7 Field vs. Lab 136
4.1.8 With vs. Without Eye Tracking 138
4.2 Human Users 138
4.2.1 Level of Expertise 139
4.2.2 Age Groups 141
4.2.3 Cultural Differences 142
4.2.4 Vision Deficiencies 144
4.2.5 Ethical Guidelines 145
4.3 Study Design and Ingredients 147
4.3.1 Hypotheses and Research Questions 148
4.3.2 Visual Stimuli 149
4.3.3 Tasks 151
4.3.4 Independent and Dependent Variables 153
4.3.5 Experimenter 157
4.4 Statistical Evaluation and Visual Results 158
4.4.1 Data Preparation and Descriptive Statistics 160
4.4.2 Statistical Tests and Inferential Statistics 161
4.4.3 Visual Representation of the Study Results 163
4.5 Example User Studies Without Eye Tracking 167
4.5.1 Hierarchy Visualization Studies 168
4.5.2 Graph Visualization Studies 169
4.5.3 Interaction Technique Studies 171
4.5.4 Visual Analytics Studies 172

5 Eye Tracking **175**
5.1 The Eye 177
5.1.1 Eye Anatomy 178
5.1.2 Eye Movement and Smooth Pursuit 179
5.1.3 Disorders and Diseases Influencing Eye Tracking . . 181
5.1.4 Corrected-to-Normal Vision 183
5.2 Eye Tracking History 185
5.2.1 The Early Days 186
5.2.2 Progress in the Field 188
5.2.3 Eye Tracking Today 190
5.2.4 Companies, Technologies, and Devices 192
5.2.5 Application Fields 192
5.3 Eye Tracking Data Properties 197
5.3.1 Visual Stimuli 199

5.3.2 Gaze Points, Fixations, Saccades, and Scanpaths . . 202
5.3.3 Areas of Interest (AOIs) and Transitions 204
5.3.4 Physiological and Additional Measures 206
5.3.5 Derived Metrics 208
5.4 Examples of Eye Tracking Studies 209
5.4.1 Eye Tracking for Static Visualizations 210
5.4.2 Eye Tracking for Interaction Techniques 215
5.4.3 Eye Tracking for Text/Label/Code Reading 218
5.4.4 Eye Tracking for User Interfaces 221
5.4.5 Eye Tracking for Visual Analytics 223

6 Eye Tracking Data Analytics **229**
6.1 Data Preparation 230
6.1.1 Data Collection and Acquisition 231
6.1.2 Organization and Relevance 232
6.1.3 Data Annotation and Anonymization 234
6.1.4 Data Interpretation 235
6.1.5 Data Linking 236
6.2 Data Storage, Adaptation, and Transformation 237
6.2.1 Data Storage 238
6.2.2 Validation, Verification, and Cleaning 240
6.2.3 Data Enhancement and Enrichment 241
6.2.4 Data Transformation 242
6.3 Algorithmic Analyses 243
6.3.1 Ordering and Sorting 244
6.3.2 Data Clustering 245
6.3.3 Summarization, Classing, and Classification 247
6.3.4 Normalization and Aggregation 248
6.3.5 Projection and Dimensionality Reduction 249
6.3.6 Correlation and Trend Analysis 250
6.3.7 Pairwise or Multiple Sequence Alignment 252
6.3.8 Artificial Intelligence-Related Approaches 253
6.4 Visualization Techniques and Visual Analytics 254
6.4.1 Statistical Plots 256
6.4.2 Point-based Visualization Techniques 257
6.4.3 AOI-based Visualization Techniques 261
6.4.4 Eye Tracking Visual Analytics 263

7 Open Challenges, Problems, and Difficulties **267**

7.1 Eye Tracking Challenges 267
7.2 Eye Tracking Visual Analytics Challenges 269

References **273**

Index **335**

About the Author **347**

Preface

Visual analytics [277, 495] is a powerful concept that combines visualization techniques, algorithmic approaches, interaction aspects, as well as people's perceptual and cognitive abilities. It follows the goal of integrating the human observer into the data exploration process combined with automatic analyses to derive meaning, knowledge, and insights from large datasets. The building, refining, confirming, and rejecting of hypotheses plays a central role in all of these knowledge generation processes. However, understanding human behavior in this concept is a difficult task, since the human brain is a crucial parameter in efficiently and effectively finding solutions to the data analysis tasks at hand while the cognitive processes in the brain are still hard to extract. To gain insights into the strengths of a visual analytics system, user evaluation has to be considered, in the best case by recording more dependent variables than standard error rates and completion times. Eye tracking is a relatively novel technique for exploring the viewing behavior of spectators in information visualization and visual analytics; however, the vast amount of spatio-temporal data generated makes an analysis very challenging and complicated. Hence, visual analytics can again be a powerful concept to derive meaning from eye movement data, in particular if the data is complemented by additional data measurements like pupil dilations, galvanic skin responses, EEG, further physiological data, or qualitative feedback [44]. Following this principle, we can generate an iterative multiple step model starting with general visual analytics, user evaluation including eye tracking, and again visual analytics including algorithms, interactive visualizations, and human users to improve visual analytics, an idea that we refer to as dynamic visual analytics. This whole repeating process results in a cycle of visual stimuli, evaluations with users-in-the-loop, and again visualizations of the recorded evaluation data that might serve as visual stimuli for the next iteration, finally leading to valuable insights and hence improvements that would not have been possible without the application of eye tracking.

In this book, we describe the challenges and perspectives of dynamic visual analytics, i.e. we showcase the value of eye tracking for visual analytics and, in addition, the value of visual analytics for eye tracking. We first introduce visualization and visual analytics as methodologies to explore and analyze data with the user-in-the-loop, with and without automatic analyses and analytical reasoning. This process generates snapshots of visualizations that support humans due to rapid pattern detection, guiding further exploration processes like the choice of algorithmic approaches and applied interactions, and hence helping to build, refine, accept, or reject hypotheses.

Such visual snapshots – static or dynamic ones – serve as independent variables in controlled and uncontrolled user evaluations. Typically, those stimuli are varied, and dependent variables like error rate and completion times are recorded that are statistically evaluated as a post-process. The same could be done with eye movement data, although the evaluation is much more challenging due to the spatio-temporal aspect of the recorded data and the different stimuli properties. Moreover, additional data sources, and qualitative feedback, come into play, making such an analysis even more complicated. However, using visual analytics such heterogeneous data can be made explorable, in the case where right visualizations, interactions, and algorithmic approaches are chosen, also allowing human users to collaboratively and remotely identify insights, sharing them with others, and combining them into even stronger and more valuable insights.

Visual analytics combined with more advanced data science concepts like machine learning can be used to analyze recorded eye tracking data, either offline, after the recording of the data, or online, during the recording, making it a real-time evaluation process. The insights gained from these rapid analyses can be applied to the shown stimuli in order to improve them or adapt to the observers' requirements and needs. Consequently, visual analytics plays a crucial role, since it contains many useful methods for tackling upcoming challenges, although some are very hard and belong to future work. We conclude this book by several open problems in the field of eye tracking in general, but also in visual analytics applied to eye tracking data in particular.

List of Figures

Figure 1.1	Building, rejecting, confirming, and refining of hypotheses plays a key role in visual analytics. . . .	3
Figure 1.2	Inspired by "The Unexpected Visitor": a different kind of visual attention is reflected in the scanpaths depending on the tasks the spectators have to solve as given in scenarios (a), (b), (c), and (d) [539]. . .	5
Figure 1.3	Human faces can serve as visual stimuli in an eye tracking study, for example to identify dental imperfections [275]. Image provided by Pawel Kasprowski.	7
Figure 1.4	Software functions calling each other can change over time and can create a large and time-varying relational dataset worth investigating by a software engineer to identify bugs or performance issues in a software system [67].	8
Figure 1.5	An illustration of the synergy effect. Standard/ traditional user studies or eye tracking studies are conducted while the recorded data is statistically evaluated or explored by visual analytics concepts.	9
Figure 1.6	Eye tracking data in the visualization pipeline. . . .	10
Figure 1.7	Cognitive science and psychology are also important research fields to improve the design of eye tracking studies and the interpretation of the recorded data [305].	11
Figure 1.8	Dynamic visual analytics describes the process of evaluating a visual analytics stimulus by either a post process analysis or a real-time analysis. In a post process analysis, a second visual analytics system can be used to analyze the eye tracking data; in a real-time analysis, efficient algorithms must be used.	12

Figure 1.9 Two types of interactive stimuli: (a) a user interface of a ticket machine and (b) a more complex user interface of a visual analytics system. Image provided by Bram Cappers [106]. 13
Figure 1.10 Coordinating multiple views provides several perspectives on the data under exploration. In this case we see a visual analytics tool for analyzing eye movement data from video stimuli [308]. Image provided by Kuno Kurzhals. 15
Figure 2.1 (a) Part of a visualization for dynamic call graphs [75]. (b) A hierarchy visualization based on the space-reclaiming icicle plots [509]. 18
Figure 2.2 Connecting the list of 2D coordinates from Table 2.1 by straight lines in order transforms the raw data into a visual form, in this case a pentagon shape. . . 21
Figure 2.3 The shape of a pentagon as in Figure 2.2 can be drawn with just five points and five connecting lines, but the shape of a Christmas tree requires many more points. There are many more complex pattern examples in visualization. 21
Figure 2.4 Connect-the-dots is a popular game in newspapers and magazines. The human brain needs lines connecting the dots to successfully interpret the shape. 22
Figure 2.5 Visual variables are fundamental ways to distinguish visualizations. Jacques Bertin [37] described seven such variables and denoted them as retinal variables. 23
Figure 2.6 The task of judging and comparing sizes appears to be easier and more reliable with fewer errors in bar charts (b) than in pie charts (a). Visual variables are the cause of this effect, which are positions in a common scale in bar charts and angles in pie charts. 24
Figure 2.7 Visually depicting the four sets of 2D values reflects the real differences in the four example datasets. Statistics is a powerful concept but, due to aggregation effects, not all insights in the data variations can be found. 25

Figure 2.8 The outline of an animal found in the Cave of Altamira near Santander in Spain, also known as the Sistine Chapel of the cave painting. 29
Figure 2.9 Maps have been used in a variety of forms, including various visual variables: (a) a geographic map annotated with a grid-based overlay to faster detect the label information and the location of a place [371]. (b) Data from other application domains with a more abstract character have been visually encoded into maps, like trade relations [243]. Figure provided by Stephen Kobourov. 31
Figure 2.10 Today's pie charts are based on the ideas originally developed, for example, by Florence Nightingale. . 33
Figure 2.11 An example of a graphical user interface for visually exploring eye movement data [82]. Figure provided by Neil Timmermans. 35
Figure 2.12 A hierarchy visualization depicted on a powerwall display. The system allows collaborative interactions for several users equipped with tablets, or it serves as an overview of a large dataset [93, 441]. Pictures taken and provided by Christoph Müller. 37
Figure 2.13 Prominent visualization techniques for primitive data types already exist in several variants. The performance and visual attention strategies of human users while solving tasks with any of the visualization techniques can be analyzed by eye tracking studies: (a) a bar chart for quantitative data using the visual variable length for the quantities; (b) a dot plot for the ordinal data using the visual variable position to encode the order; (c) a scatter plot with varying colors and different circle sizes as visual variables to indicate the categorical nature of the data. 40
Figure 2.14 Jock Mackinlay gave an ordered list of the visual variables for each of the three primitive data types. He described the effectiveness of such a perceptual task in decreasing order [344]. 41

Figure 2.15 Three prominent visual metaphors for encoding the same graph dataset [29]: (a) a node-link diagram; (b) an adjacency matrix; (c) an adjacency list. . . . 42

Figure 2.16 Four major visual metaphors for hierarchical data exist: (a) explicit links; (b) nesting; (c) stacking; (d) indentation. 44

Figure 2.17 Multivariate data can be visualized in at least three major ways: (a) a glyph-based visualization, here in the form of Chernoff faces; (b) a scatterplot matrix (SPLOM); (c) a parallel coordinates plot. 45

Figure 2.18 Different kinds of movement data can be measured and visualized: (a) trajectories from bird movement [369]; (b) scanpaths from an eye movement study investigating the readability of public transport maps [372]. 45

Figure 2.19 There are various scenarios in which textual information is important: (a) label information on a public transport map [372] (Figure provided by Robin Woods, Communicarta Ltd); (b) an aggregated view on the occurrence frequencies of words in the DBLP, summarized as a prefix tag cloud [86]. 46

Figure 2.20 A set visualization based on the "bubble sets" approach [127]. Image provided by Christopher Collins. 47

Figure 2.21 Four examples that are typical data types in the domain of scientific visualization depicted by standard approaches: (a) a scalar field; (b) a vector field; (c) a tensor field; (d) a volumetric data visualization. 48

Figure 2.22 A part of the Eclipse software system and its hierarchical organization depicted as a node-link diagram with aligned orthogonal links to visually represent a list of quantitative values for certain derived attributes [62]. 49

Figure 2.23 (a) A time-varying graph dataset consisting of flight connections in the US from the year 2001 shown as a heat triangle [242]. (b) A Themeriver [218] representation for showing the evolving number of developers during software development [89]. . . . 50

Figure 2.24 An electrocardiogram consists of several time-dependent quantities that are shown as a line-based diagram, annotated with P, Q, R, S, T waves. This kind of diagram has also been investigated by an eye tracking study [144]. 51

Figure 2.25 A scatterplot with extra descriptions for the axes and the color coding, serving as metadata. 53

Figure 2.26 A visualization pipeline illustrates how raw data is transformed in a stepwise manner into a graphical output while the users can adapt and modify the steps and states [108]. 55

Figure 2.27 An interaction history can be modeled as a network of states; in the case of a visualization tool it consists of snapshots, each illustrating a certain parameter setting [68]. Node-link diagrams can depict the weighted state transitions and are interactive themselves. Here we see a network with additional thumbnails indicating the current view in the visualization tool. It may be noted that in this scenario some of the interaction steps cannot be undone, indicated by the directions of the links while some others are undirected. The thickness of the links might encode a transition probability for example. 55

Figure 2.28 Gaze-controlled buttons are used in a game environment to interact with a game character. After a user evaluation this was replaced by a simpler scenario due to unintended rotation issues [378]. Image provided by Veronica Sundstedt and Jonathan O'Donovan. 60

Figure 2.29 The user has the option to see one, two, or four views for hierarchy visualization techniques. Moreover, the views are exchangeable and support their own parameter settings while they are interactively linked [99]. 65

Figure 2.30 When designing a diagram we should take into account several general issues that can lead to problems when interpreting the diagram: (a) visual clutter; (b) chart junk; (c) lie factor. 67

Figure 2.31 (a) Emergence: a visual pattern (my son Colin) can pop out from a noisy background pattern. (b) Multistability: a visual pattern can carry several meanings and might be interpreted in several ways. (c) Invariance: a visual pattern can be deformed in various ways but it is still recognizable as a similar pattern as the original one. (d) Reification: a visual pattern can be completed although it is not shown completely on screen. 69

Figure 2.32 There are several ways of grouping visual elements described in the Gestalt principles. (a) Proximity. (b) Similarity. (c) Closure. (d) Symmetry. Further ones are given by the law of common fate, continuity, or good form. 70

Figure 2.33 Visual illusions can happen in a variety of forms including visual variables and the environments in which they are used: (a) Ebbinghaus illusion related to size effect, caused by the environment and surroundings. (b) Cafe wall illusion related to distance, caused by shifted black square patterns. (c) Herman grid illusion related to cognitive issues, i.e. visual elements are generated where no elements are. (d) Müller-Lyer illusion related to length, caused by extra visuals like arrow heads pointing in opposite directions [274]. Further well-known effects are the spinning dancer illusion related to movement, caused by missing reference points which seem to change the direction of movement, the Ponzo illusion related to depth, caused by the environment and additional effect with denser becoming parallel lines in the background like a railway track [274], or the checker shadow illusion related to color and the surrounding colors [228]. . 72

Figure 3.1 A quote by Albert Einstein or Leo Cherne describes the general ingredients of visual analytics: “Computers are incredibly fast, accurate, and stupid. Human beings are incredibly slow, inaccurate, and brilliant. Together they are powerful beyond imagination”. 76

Figure 3.2 Data-related concepts may happen at three stages in a visual analytics system in the form of preparing, checking and deriving, and advanced operations. Humans and computers play different roles in these stages and are involved to varying extents. 80

Figure 3.3 A node-link diagram in the field of graph visualization. (a) The relational data without clustering, just randomly placed nodes. (b) Computing a clustering of the same data as in (a) and, based on that, using a graph layout that takes into account the node clusters, encoded by spatial distances of the nodes. 84

Figure 3.4 Visual analytics is an interdisciplinary field that makes use of research disciplines involving the *computer, the humans, and also human–computer* interaction (HCI). 89

Figure 3.5 The visual analytics pipeline illustrates how data is transformed into patterns, correlations, or rules that can be regarded as tables filled with values or visual depictions. The users can adapt visual variables and refine parameters guided by their hypotheses and tasks, hopefully generating new insights that can be used to modify the data under exploration. 92

Figure 3.6 (a) Comparing the participants' scanpaths from an eye tracking study can generate pairwise similarity values shown in an adjacency matrix. (b) Applying a matrix reordering technique immediately shows a pattern in the matrix which is difficult to find by the pure textual values given in a table or 2D array [300]. 95

Figure 3.7 A parallel coordinates plot (PCP) for showing positive and negative correlations between pairs of metric attributes derived from eye movement data for selected study participants [299]. Axis filters are indicated to reduce the number of polylines. Image provided by Ayush Kumar. 96

Figure 3.8 (a) n-ary association rules (b) and n-ary sequence rules generated from eye movement data can express which general relations exist in eye movement data [63]. 97

Figure 3.9 Confirming or rejecting a given hypothesis in a simple bar chart can generate a lot of simple tasks which might be recognized if the eye movements of observers are recorded and overplotted on the bar chart stimulus in the form of a gaze plot [203]. . . . 98

Figure 3.10 Visual scenes illustrating several tasks. (a) Searching a red triangle in a sea of distracting visual objects. (b) Counting the number of red circles. (c) Reading labels attached to visual objects. (d) Judging the smallest bar in a bar chart. (e) Estimating the number of green squares. (f) Comparing the red and blue circle clusters with the blue triangle cluster. (g) Identifying patterns from groups of visual objects. (h) Identifying correlations between visual curves. (i) Finding a route in a network. (j) Detecting communities by similarity and proximity. 100

Figure 3.11 Changing the requirements for an algorithm can modify its output, which is seen in a visual result. In this case the layout algorithm for a generalized Pythagoras tree [31] for hierarchy visualization (top row) is changed to a force-directed one (bottom row) that creates a representation which is free of overlaps [363]. 103

Figure 3.12 Several dimensionality reduction algorithms applied to the same dataset consisting of eye movement scanpaths [79]. (a) t-distributed stochastic neighbor embedding (t-SNE). (b) Uniform manifold approximation and projection (UMAP). (c) Multidimensional scaling (MDS). (d) Principal component analysis (PCA). 108

Figure 3.13 Runtime performance chart for two algorithms generating visualizations. (a) A word cloud is generated for differently large dataset sizes, resulting in a linear-like runtime. (b) A pedigree tree is generated based on more and more people involved, standing in a family relationship, resulting in some exponential-like runtime. 113

Figure 3.14 During an algorithm execution, the runtimes (a) as well as the memory consumption (b) might differ from iteration to iteration. 114
Figure 3.15 A time-to-space mapping of the vertices and edges processed by a Dijkstra algorithm trying to find the shortest path in a network is visually represented in a bipartite layout [75]. The time axis runs from left to right. 115
Figure 3.16 A flight traffic dataset taking into account temporal clusters while a bipartite splatted vertically ordered layout is chosen to reflect static and dynamic patterns in the time-varying graphs [2]. Image provided by Moataz Abdelaal. 117
Figure 3.17 The graphical user interface of the pathology visual analytics tool with the image viewer, the image overview, a gallery with thumbnail images, a textual input to make reports, a scatterplot for showing correlations of bivariate data, and a view on sequential diagnostic data [134]. Image provided by Alberto Corvo. 118
Figure 3.18 EventPad [106] is based on a graphical user interface with several interactively linked views to support data analysts at specific tasks to explore malware activities. Image provided by Bram Cappers. 120
Figure 3.19 Visual analytics supports several views on the video data [232]: time navigation, video watching, snapshot sequence view, audio augmentation, statistical plots, graph views, schematic summaries, and filter graphs. Image provided by Benjamin Höferlin. 121
Figure 3.20 GazeStripes: visual analytics of visual attention behavior after several people have watched videos [309]. Images provided by Kuno Kurzhals. 122
Figure 4.1 The most important ingredients in a user study are the participants, the study type with the independent, confounding, and dependent variables, and the results in the form of statistics and visual depictions. 126

Figure 4.2 A comparison between Cartesian and radial diagrams in an uncontrolled user study recruiting several hundred participants in an online experiment [150]. Image provided by Stephan Diehl. 131

Figure 4.3 Examples from a visualization course at the Technical University of Eindhoven educating students in eye tracking and visual analytics [70]. (a) A visual attention map with contour lines. (b) An eye movement direction plot. 140

Figure 4.4 (a) A Snellen chart [470] can help to identify visual acuity issues. (b) An example plate of an Ishihara color perception test consisting of several pseudo-isochromatic plates [258]. 144

Figure 4.5 The way a stimulus is presented and the degree of freedom of the participant's position has an impact on the study design and the instrumentation. (a) A static stimulus, like a public transport map [372], inspected from a static position like sitting in front of a monitor. Image provided by Robin Woods. (b) A dynamic stimulus, like the game playing behavior of people recorded in a video [71], inspected from a static position. Image provided by Kuno Kurzhals. (c) A static stimulus, like a powerwall display [441], inspected from a dynamic position, allowing movement to change the perspective on the static stimulus. Image provided by Christoph Müller. (d) A dynamic stimulus, like driving a car with many other cars and pedestrians crossing our way while dynamically changing our positions [44]. 150

Figure 4.6 "Why is the road wet?" is a task that can be solved by watching a given abstract visual depiction of a scene (a). The visual scanning strategy to solve this task has to follow a certain visit order to grasp the information subsequently to solve the task (b). Eye tracking can give some insights into such viewing behavior [416]. 153

Figure 4.7 Varying the independent variable "link length" can have an impact on the dependent variables error rate and response time for the task of finding a route from a start to a destination node in a node-link diagram with a tapered edge representation style [97]. 154
Figure 4.8 Since eye movement data is composed of at least three data dimensions like space, time, and study participants, the visual representations also get more complex with many aligned and linked visual components supporting pattern identification in the data [298]. Here we see the x–y positions in the top row, the saccade lengths and orientations in the center row, and the filtered pairwise fixation distances in the bottom row while time is pointing from left to right. 159
Figure 4.9 Easy-to-understand diagrams are often preferred for depicting the results of a statistical dependent variable in a visual form. Such a variable could, for example, be a performance measure like response times or error rates or participants' individual feedback in the form of values indicated on a Likert scale: (a) a bar chart. (b) a pie chart. (c) a histogram. 163
Figure 4.10 A histogram can contain various patterns indicating a property of the distribution of the dependent variable under investigation. (a) Bell-shaped or normal. (b) Uniform. (c) Left-skewed. (d) Right-skewed. (e) Bimodal. (f) J-shaped. 164
Figure 4.11 A line chart is useful to depict several time-varying performances to identify trends as well as countertrends and to compare them for differences over time. 165
Figure 4.12 A box plot can show the distribution of a univariate dataset, for example the performance measure of the response time or the error rates. 165

Figure 4.13 Visual depictions of bivariate and multivariate data. (a) A scatter plot for showing the correlations between two variables. (b) A scatter plot matrix for depicting more than two variables. (c) A parallel coordinates plot (PCP) as an alternative to the scatter plot matrix for representing more than two variables. 166
Figure 4.14 A scatter plot enriched by error bars indicating the standard error of the means (SEM). The average saccade length is plotted on the y-axis while the average fixation duration is shown on the x-axis. (a) The complexity levels. (b) The task difficulty. . . . 166
Figure 4.15 The Euclidian distance to the start is plotted over time to show the progress of visual attention with respect to such a relevant point of interest in a visual stimulus. 167
Figure 4.16 Three hierarchy visualizations illustrating examples from a huge design space for depicting hierarchical data. (a) A bubble hierarchy. (b) A treemap. (c) A sunburst visualization. Images provided by the students from a design-based learning course in 2018 at Eindhoven University of Technology. . . . 168
Figure 4.17 Seven edge representation styles: (a) standard with arrow head; (b) tapered; (c) orthogonal; (d) color gradient; (e) dashed; (f) curved; (g) partial. 170
Figure 4.18 Using interaction techniques to adapt parameters in a contour line-based visual attention map. The public transport maps of (a) Zurich, Switzerland and (b) Tokyo, Japan. 171
Figure 5.1 The human eye is a complex organ that is important for the visual system [196]. Moreover, it builds the major ingredient for all eye tracking studies. 178
Figure 5.2 Cataracts [392] affect the lens of the eye in some kind of degeneration process causing clouded and unclear vision: (a) clear vision; (b) an eye with cataract issues. 182
Figure 5.3 No corrective lenses are needed for normal vision. . 183

Figure 5.4 Refractive errors: (a) nearsightedness (myopia) and (c) farsightedness (hyperopia) can be corrected by special lenses (b), (d). 184
Figure 5.5 Eye movements during a reading task consist of short stops (fixations) and rapid eye movements (saccades). This insight was found by Hering, Lamare, and Javal around 1879 [263]. 187
Figure 5.6 An example of an eye tracking device as we know it today, known as the Tobii Pro Glasses 3. Image provided by Lina Perdius (Tobii AB). 190
Figure 5.7 Eye tracking technologies can be useful in the field of aviation, in particular, when training pilots to land a plane [430, 432]. Image provided by David Rudi (Copyright ETH Zurich). 197
Figure 5.8 Eye movement data can be described as consisting of gaze points, which is the lowest level of granularity that is interesting for eye tracking in visual analytics. Those gaze points are spatially and temporally aggregated into fixations by modifiable value thresholds. The fixations with duration (encoded in the circle radius) contain saccades in-between, i.e. rapid eye movements. A scanpath is made from a sequence of fixations and saccades. Regions in a stimulus that are of particular interest are called areas of interest (AOIs). If we are only interested in fixations in a certain AOI we denote those by gazes. Between AOIs there can be a number of transitions indicated by the number of saccades between those AOIs [44]. 198
Figure 5.9 A car driving task generates a dynamic stimulus, in this case we see four snapshots at different time points (T1) to (T4) of a longitudinal eye tracking experiment with indicated points of visual attention [44]. 200

Figure 5.10 Selecting areas of interest in a static stimulus can reduce the amount of eye movement data and can impact the eye movement data analysis since each AOI is some kind of spatial aggregation: (a) AOI selection based on hot spots of the visual attention behavior; (b) AOI selection based on the semantics in the stimulus [100]. 205

Figure 5.11 Visualization techniques have been explored a lot by applying eye tracking. (a) Node-link tree visualizations [78]. (b) Trajectory visualizations for bird movements [369]. (c) Visual search support in geographic maps [371]. 211

Figure 5.12 Public transport maps for different cities in the world (in this case Venice in Italy) [372]. 213

Figure 5.13 Graph layouts with different kinds of link crossings, crossing angles, and the effects of geodesic-path tendency can have varying impacts on eye movements [245]. 214

Figure 5.14 Finding a bug in a source code typically requires to scan the whole piece of code before one concentrates on specific parts of it [454]. 220

Figure 5.15 A recommender system for scatter plot matrices equipped with eye tracking technologies to support the data analysts [451]. Image provided by Lin Shao. 225

Figure 6.1 Applying visual analytics as a combination of algorithmic analyses and interactive visualization to eye tracking data can provide useful insights into visual scanning behavior over space, time, and participants [309]. Image provided by Kuno Kurzhals. 230

Figure 6.2 A manual fixation annotation tool has been developed to step-by-step add extra information to the fixations, for example based on the semantics of a stimulus [370]. 234

Figure 6.3 The Antwerp public transport map was visually explored in an eye tracking study. The visual attention hot spots were used to split the static stimulus into sub-images which are then grouped by a force-directed layout taking into account the transition frequencies between the individual sub-images [98]. Different parameters can be modified such as cropping sizes, cluster radius, or number of sub-images displayed, for example. 246
Figure 6.4 Alignment of a set of scanpaths from an eye tracking study. First, the scanpaths are transformed into character sequences based on user input, before they are aligned [84]. 253
Figure 6.5 Statistical plots can be useful to get an overview of the quantitative values in an eye tracking dataset: (a) a bar chart; (b) a line graph; (c) a scatter plot. . . . 256
Figure 6.6 Splitting the fixations from a scanpath into their x- and y-coordinates: (a) the original scanpath; (b) a timeline for the y-coordinates; (c) a timeline for the x-coordinates. 258
Figure 6.7 Two different visual attention maps from a public transport map eye tracking study. In this case the hot spots of visual attention are indicated by contour lines [100]. Route finding tasks in the maps of: (a) Tokyo, Japan; (b) Hamburg, Germany. 259
Figure 6.8 Scanpath visualizations for (a) one participant and (b) 40 participants [100]. The scanpath visualization in (b) can hardly by used for data exploration. . . . 260
Figure 6.9 A space-time cube showing clustered gaze data for a given stimulus [311]. Image provided by Kuno Kurzhals. 261
Figure 6.10 AOI visits over time, either for one participant and three AOIs [414] (a) or three participants and four AOIs in parallel [417] (b). Extending these visualizations to many participants, many AOIs, and long scanpaths can lead to visual clutter effects. . . 262

Figure 6.11 Annotating a visual stimulus, overplotted with a contour visual attention map, with color coded AOIs (a); the AOI visits over time can be seen in a corresponding scarf plot (b) that uses the same color coding as in the annotation view. 263
Figure 6.12 The dynamic AOI transitions can be shown in an AOI river visualization [80] with an enhancement by Voronoi cells. 263
Figure 6.13 A graphical user interface showing several linked views for visually exploring eye movement data: a clustered fixation-based visual attention map, a timeline view on the visually attended AOIs, a scanpath visualization, a visual attention map with color coded hot spots, and a scarf plot for an overview about the inspected AOIs [22]. 264

List of Tables

Table 2.1 A table of raw data consisting of five subsequent 2D coordinates. 20
Table 2.2 Anscombe's quartet [15] is based on four sets of 2D coordinates. 25
Table 4.1 Examples of user studies focusing on aspects in visualization, interaction, and visual analytics: comparative (CP), laboratory (LB), controlled (CT), qualitative (QL), quantitative (QN) 174
Table 5.1 Eye tracking companies with respect to hardware and software developments as well as focused applications, described by major buzz words 193
Table 5.2 Examples of eye tracking studies focusing on aspects in visualization, interaction, text reading, user interface design, as well as visual analytics 226

List of Abbreviations

2D	Two-dimensional
3D	Three-dimensional
AI	Artificial Intelligence
ANOVA	Analysis of variance
AOI	Area of interest
AR	Augmented reality
BCE	Before the common era
CP	Comparative
CT	Controlled
DBLP	The Digital Bibliography & Library Project
DNA	Deoxyribonucleic acid
ECG	Electrocardiogram
EEG	Electroencephalography
EMG	Electromyography
ETRA	Symposium on Eye Tracking Research & Applications
GGobi	Free statistical software
GPS	Global positioning system
GUI	Graphical user interface
HCI	Human-computer interaction
HD	High definition
Hz	Hertz
IEEE VIS	The premier forum for advances in visualization and visual analytics
JASP	Open source statistics program
KDD	Knowledge discovery in databases
LB	Laboratory
MATLAB	Programming and numeric computing platform
MDS	Multidimensional scaling
MRI	Magnetic resonance imaging
MTurk	Mechanical Turk
NASA	National Aeronautics and Space Administration

NCBI	National Center for Biotechnology Information
NP	Nondeterministic polynomial time
NSF	National science foundation
OLAP	Optimal linear arrangement problem
PCA	Principal component analysis
PCP	Parallel coordinates plot
PNNL	Pacific Northwest National Laboratory
PRISM	Commercial statistics software
PSPP	Program for statistical analysis
QL	Qualitative
QN	Quantitative
R	Free software environment for statistical computing and graphics
RapidMiner	Environment for machine learning and data mining
RNA	Ribonucleic acid
ROI	Region of interest
SAS	Statistical analysis software
SD	Standard deviation
SDK	Software development kit
SEM	Standard error of the means
SMI	SensoMotoric Instruments
SPLOM	Scatterplot matrix
SPSS	Software platform for statistics
Stata	Software for statistics and data science
STATISTICA	Software for statistical methods
t-SNE	t-distributed stochastic neighbor embedding
UI	User interface
UMAP	Uniform manifold approximation and projection
US	United States
USB	Universal serial bus
UX	User experience
Var	Variance
VAST	Visual analytics science and technology
VR	Virtual reality

1

Introduction

Data can be considered as one of the major ingredients in computer and data science. Devices measure and algorithms simulate and produce data at ever increasing rates, eventually bringing the term big data [42] into play. Exploring such data with the goal of finding insights and deriving meaning has become a challenging task due to the fact that the currently available data science concepts cannot be efficiently applied to any kind of data due to their heterogeneity, complexity, and size. Moreover, in many cases human observers are not able to precisely describe what they are looking for. Consequently, any kind of algorithm cannot be applied because of a lack of parameter descriptions and missing details on what we are precisely looking for.

Visualization, on the other hand, is a valuable concept although it cannot solve the problems; however, due to the strengths of human perceptual abilities [219, 521, 522], visualization builds a great means to guide a data analyst and to uncover patterns or anomalies in data, even partially if the wrong visual metaphor is initially chosen. Visualization can give hints about something we are not aware of. Those visual insights can then be used to more or less guide the exploration process and, hence, visualization can build a starting point for further data exploration. This starting point helps to build hypotheses, to confirm or reject them, and also to refine them, if interactions are supported, allowing the visual observer to change parameters or apply different kinds of algorithms. However, visualization gives no guarantee that the patterns found in the data lead to the correct conclusions or confirm what was to be expected.

Visual analytics is actually equipped with several powerful concepts like information visualization, human–computer interaction (HCI), algorithmic and statistical approaches, and most importantly, human users who are able to start any kinds of supported processes with the final goal of finding insights in a dataset. Moreover, it is not limited to an individual user but can be applied in a collaborative manner [441], exploiting the perceptual strengths and the manpower of several people in stepwise consecutive processes or even simultaneously, being in the same room or sitting at different places in the world connected via a well-designed online visual analytics system.

Although visual analytics concepts have been and are still developed for nearly any research field involving data, it is still an open challenge to figure out what the best or at least a suitable configuration of such a system is. For example, the displayed visualization techniques, the interactions to be applied, or the algorithms to transform or project data are very vague. This brings evaluation into play which allows us to measure and record dependent variables from human observers while solving given tasks based on the application of a visual analytics system. However, standard evaluation only tells half of the truth due to the limitation to typical standard variables like completion times and error rates, or in some cases, qualitative verbal feedback. This evaluation data is useful, but it does not explain much about the intervening visual and interaction processes, i.e. the fine granular visual attention behavior that people show while solving a given task.

Eye tracking, on the other hand, is an advanced technique that supports the recording of spatio-temporal eye movement behavior during small instances of time, hence providing insights into the where and when questions [306]. Moreover, extra variables can be recorded, meaning eye tracking provides information in addition to the standard evaluation measures like completion times and error rates. On the negative side, the setup of eye tracking studies is typically much more complicated since it demands a proper calibration phase to avoid errors in the recorded data. Moreover, the recorded data is much larger and more complex than standard evaluation measures, not only because of the finer time granularity during the measurements, but also because the scanpaths can have different lengths consisting of more or less fixations and the fact that the viewing behavior can have significant differences over space and time.

A challenging issue with eye tracking data is the analysis of the recorded data, which is important to find insights, to eventually enhance the shown stimulus. For traditional evaluations in the form of user studies, the recorded dependent variables are statistically evaluated, leading to confirming or rejecting formerly stated hypotheses or research questions. This straightforward analysis is difficult with eye tracking data since it has a spatio-temporal nature. But, fortunately, visual analytics can be of great help to analyze the more complex eye tracking data, again involving visualizations, interactions, algorithms, and the human users.

Applying eye tracking evaluation in a visual analytics system in an iterative manner can be useful to collect eye movement behavior at different development phases. Exploring this data step by step can lead to insights in the form of design flaws that can be mitigated one after the other. However, for this to work properly, vast amounts of eye movements have to be recorded and analyzed while several tasks have to be tested. Dynamic visual analytics describes the process of applying eye tracking to visual analytics and vice versa, i.e. applying visual analytics to eye tracking, generating some kind of synergy effect for both research fields.

1.1 Tasks, Hypotheses, and Human Observers

Although visual analytics can provide many insights into a dataset, this is only possible with the support of the human users with their tasks in mind and the strength of their perceptual abilities [219] to rapidly identify patterns. Based on those patterns, further hypotheses can be built or already existing ones can be refined that guide the data exploration process (see for example Figure 1.1). This can be reflected in the chosen visualization techniques, the interactions, but also in the algorithms that can be adapted to certain parameter settings. Hence, the human observers play a crucial role in this whole exploration process. Understanding the visual scanning strategies and behavior can give suitable hints to the usefulness of a visualization technique but also in more complex visual analytics systems, for example, in which visualization and interaction techniques are powerful for solving certain tasks, but also in which algorithms can be useful under certain parameter

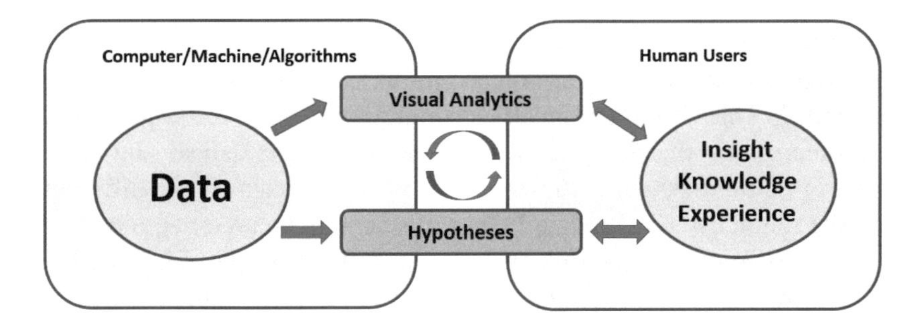

Figure 1.1 Building, rejecting, confirming, and refining of hypotheses plays a key role in visual analytics.

settings. Without evaluation, in particular eye tracking, such insights would not be possible. Moreover, the human users themselves have large differences like experts vs. non-experts, different genders, different age groups, or even different viewing abilities affected by color deficiency, visual acuity, or other perceptual issues. All of these aspects must be taken into account when evaluating recorded eye movement data, so that wrong or incomplete conclusions are not made.

Using a visual analytics system typically involves a certain kind of dataset that is investigated for patterns or anomalies, i.e. knowledge and insights into the data that support the decision making. When starting a visual analytics system, normally the users have certain tasks in mind that they plan to solve or for which they wish to see some indication to more efficiently search in the right direction. Such tasks could be, for example, comparison, counting, estimation, or just general pattern finding tasks [304], in case it is not absolutely clear what to search for. Based on each task there are hypotheses, for example a comparison task might cause a hypothesis stating that one stimulus region is more visually attended than another one, derivable from the denser point cloud on a visual attention map [50]. Each hypothesis typically falls into a certain task category, sometimes in several, for example a comparison task might include counting tasks. In the best case, the comparison task is supported by visual differences in one or several views provided by the visual analytics system. Hence, the users get some support to faster find a solution, for example, by applying interaction techniques that filter parts of the dataset to better spot the differences. Unfortunately, the application of the right interaction technique is not clear right from the beginning. If a visual output of the data is seen, we might consider further operations or views to come closer to what we are looking for, sometimes creating novel insights that we have not expected before or that we would have never been aware of.

The tasks in mind have a crucial impact on the visual attention behavior [539]. This difference in the scanning strategies (see Figure 1.2) causes problems for the analysis of the eye movement dataset, but also brings new opportunities to investigate the recorded study data under different aspects. One might even argue that the visual scanning behavior applied to a certain stimulus might be used to identify the task behind it, i.e. what the study participant, the eye tracked person, was looking for which could be important to adapt a (graphical) user interface (UI) automatically based on the eye movement behavior. Algorithmically identifying such a task, based on various people's eye movement behavior, might be a good strategy to guide

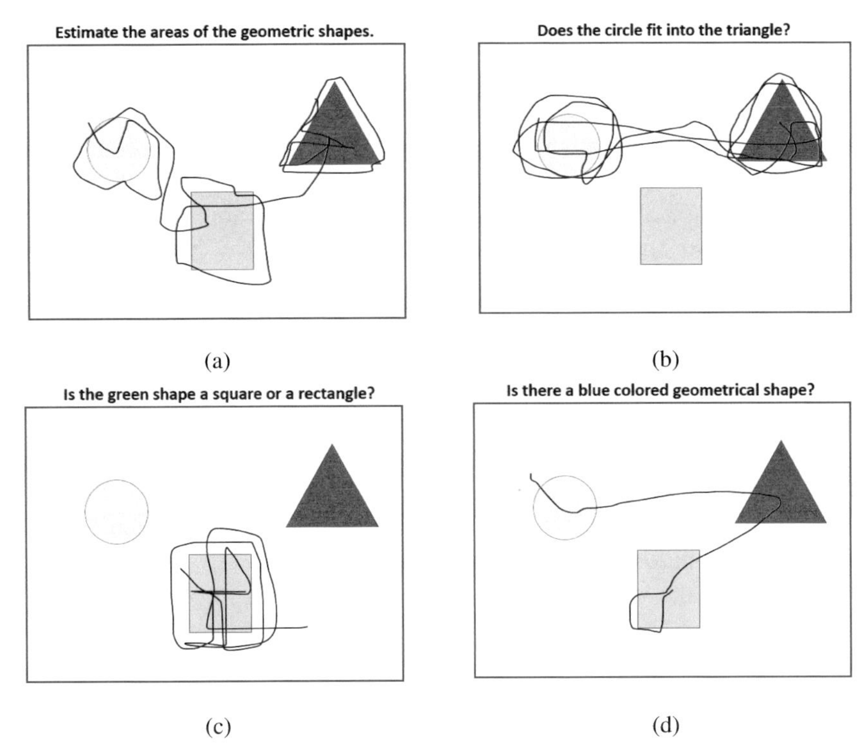

Figure 1.2 Inspired by "The Unexpected Visitor": a different kind of visual attention is reflected in the scanpaths depending on the tasks the spectators have to solve as given in scenarios (a), (b), (c), and (d) [539].

the appearance of a user interface for example, but for a rapid adaptation, a real-time analysis of the eye movement data is required. However, the scanning differences might not only be caused by the tasks at hand but also by the experience level of a person who was eye tracked. Consequently, by just algorithmically deciding or predicting, such a task could become error-prone, hence a combination of visual analytics and a human observer can be a great way to achieve more reliable results. This again shows the value of eye tracking for visual analytics [306] but also the value of visual analytics for eye tracking [14].

Considering the reliability of the results brings the number of eye tracked people into play. There is no clear definition for that in the corresponding literature, but since the recorded eye movement data is quite complex, consisting of several data types, the number of study participants cannot be large enough. One rule might be the more the better; however, we have to take into account the applied algorithms for the data analysis and the fact that the generated visualizations can suffer from scalability issues, i.e. algorithmic as well as visual scalability issues in this special case. Scalability means that the algorithms and visualizations in use can handle larger growing datasets, for example, growing in logarithmic or linear time with the increase of the data. Typical scenarios for algorithms are much worse, for example, when they fall into the class of NP-hard problems [195], or when they have quadratic or cubic runtime complexity. Heuristics are required in this case, although the generated results are no longer optimal. But, as a negative consequence of a poorly performing algorithm, the interactive responsiveness of the visual analytics system will also suffer. Who wants to wait for a few seconds, or even minutes, for a clustering algorithm to produce an optimal visual grouping? For visualization techniques we are typically confronted by the visual clutter problem [426] if the data to be visualized grows too large. This effect is regarded as "the state in which excess items or their disorganization leads to a degradation of performance at some task".

The number of the involved eye tracked people is definitely a challenge, either for recruiting them or for analyzing their scanpaths over space and time. In addition, it is questionable whether the eye movements can be really used or if privacy issues and ethical reasons make an algorithmic and visual analysis problematic or even impossible, in particular when sharing the results with other people. This can even be a major problem if the data is anonymized due to the fact that at least some of the unknown data might be recovered by clever algorithms, hence also eye movement data has to be taken with care. Eye movement data recorded in future user interfaces might be used to manipulate people's decisions, i.e. when we know where people are paying visual attention to. Privacy can even be an issue for the provided stimuli, for example if other people serve as stimuli to be watched by human observers to explore where we pay visual attention to, with or without tasks to be solved. For example, which regions are visually observed for certain facial expressions might be an interesting study, including infants. Moreover, dental imperfections might be worth investigating from a visual attention perspective [275] (see Figure 1.3).

As a negative consequence, privacy issues lead to complications when a large number of eye tracked people are needed with various additional measures, for example, to get reliable, statistically significant results to successfully adapt a scenario or improve a design flaw, even more if this has to happen in real-time. A large number of scanpaths would be required for

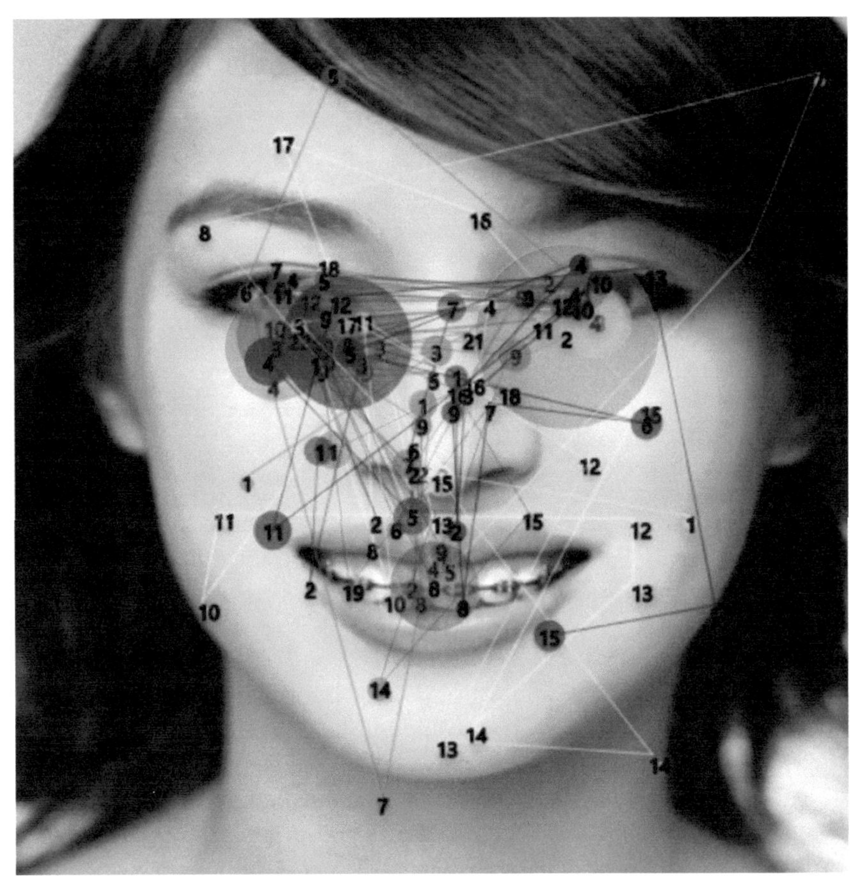

Figure 1.3 Human faces can serve as visual stimuli in an eye tracking study, for example to identify dental imperfections [275]. Image provided by Pawel Kasprowski.

tasks like car driving or shopping, however, smart phones might offer a way to provide various datasets, just like in a crowdsourcing experiment.

1.2 Synergy Effects

Typically, many research fields do not exist alone, but are linked to other well-known disciplines and take benefits from all of them to form some kind of new research area: a synergy effect. A popular example can be seen in the fields of biology and computer science that are partially merged into a field known as bioinformatics [226] which has as one of its major goals

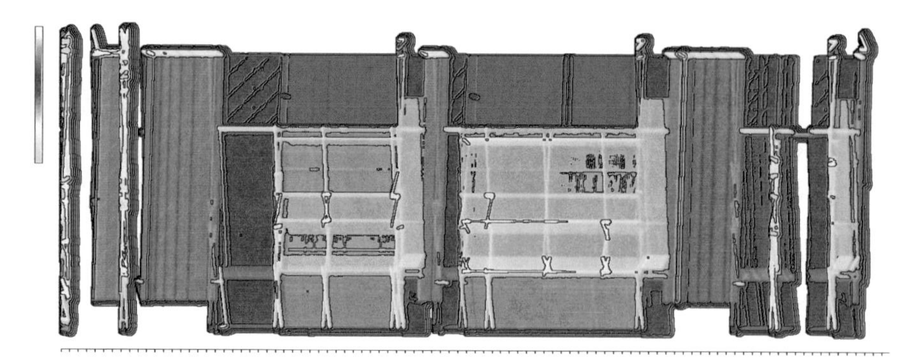

Figure 1.4 Software functions calling each other can change over time and can create a large and time-varying relational dataset worth investigating by a software engineer to identify bugs or performance issues in a software system [67].

the sequencing and comparing of DNA [383]. Multiple sequence alignment algorithms [493] have to be applied to make this happen efficiently while expert knowledge from biology is important to interpret or classify the gained results.

Another famous example is software visualization [149] which originated from the fields of software engineering and information visualization. Software developers typically create a wealth of data, in general source code which is text-based, but also relations between software entities, named call graphs (see Figure 1.4 for a visualization of dynamic call graphs). The data is mostly time-varying and can become large and complex, in particular, if data mining is applied to generate rules based on developer activities in the code [74]. Hence, information visualization is important to gain insights from the software data or the transformed data that appears as association or sequence rules. Visualization experts and software engineers work side-by-side to develop useful techniques that support the software development process which saves time and money.

There are plenty of such examples. Visual analytics can also be regarded as a combination of several fields like visualization, human–computer interaction, data management, perception, or algorithmics just to mention the most important ones. But, positively, visual analytics can be further combined with usability testing, a field that involves human users with their strengths and weaknesses to perform certain tasks. These tasks can be done more or less efficiently depending on a variety of factors, with the visual analytics system in first place. However, the system is too complex to gain insights into visual attention behavior, the differences between several system states, and which

Figure 1.5 An illustration of the synergy effect. Standard/traditional user studies or eye tracking studies are conducted while the recorded data is statistically evaluated or explored by visual analytics concepts.

states cause problems or have design flaws, algorithmically, interactively, visually, as well as perceptually. Consequently, eye tracking is a powerful concept that builds another synergy effect with the goal of precisely looking into the viewing behavior over space and time while responding to a given task at hand. In our case the synergy effect goes in a bidirectional way: visual analytics profits from eye tracking [306] while eye tracking profits from visual analytics [14]. Figure 1.5 illustrates the data being recorded in traditional user studies in contrast to eye tracking studies as well as the evaluation, also by means of visual analytics concepts.

The general idea behind the extension of standard user studies to eye tracking comes from the fact that traditional studies just record an aggregated value for the task completion times. With these it is possible to compare two scenarios in a comparative user study, but it is unclear what happened "in between". For example, the users might have the same completion time on average for both scenarios, but it is unclear if the visual attention was a different one. One group of users might have started to solve the given task at a different region in the stimulus than the other group. In general, the viewing behavior might differ between participant groups, over space and time which can provide many more insights into finding design flaws in the stimulus, but on the other hand creates a massive challenge for the data analysis, and at exactly this point we can make use of visual analytics.

A user study, controlled or uncontrolled, with a few users or a crowdsourcing experiment, with laymen or domain experts, typically involves stimuli and tasks. The stimuli can be static like a poster or diagram, or they can be dynamic like an interactive user interface, a video, or a real-life scene. Visual analytics is some kind of interactive dynamic stimulus while the users try to solve given tasks and while doing that, they change the views, i.e. the dynamic stimulus to some extent. These scenarios can be regarded as the independent variables in the study for which we explore

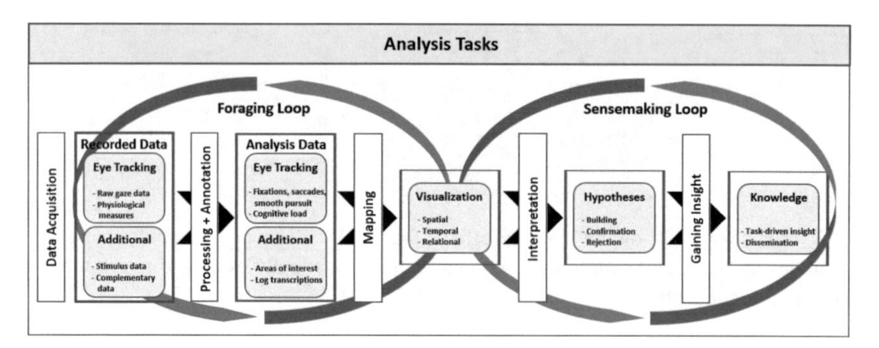

Figure 1.6 Eye tracking data in the visualization pipeline.

variations and their impact on measurements, i.e. the dependent variables in the study. The dependent variables can be error rates, completion times, emotions, qualitative feedback, and many more. If eye tracking is applied in the user study we get additional measurements in the form of the spatio-temporal eye movements, but also physiological data, facial expressions, and an endless list of extra measurements depending on what we are investigating in the study. Finally, the recorded data, no matter what type it is, has to be analyzed, otherwise the recorded data would uselessly sleep somewhere in a database or on a text file (see Figure 1.6 for eye tracking data incorporated in a visualization pipeline). The analysis tasks play a crucial role in choosing which algorithms, parameters, visualizations, and interactions are used, and in which order.

This whole problem gets really challenging if a real-time analysis of the eye movement data has to be done, for example, by comparing the currently recorded eye movement data with a huge database of existing data. Then the new data could be classified based on the existing one and the stimulus could be adapted dynamically. In such a scenario a database might store data for different scenarios, i.e. the independent variables, and a clever algorithm searches for the best option in which a new eye movement recording might fall in, i.e. the dependent variables. In such a case we are more interested in a fast algorithmic solution; the visual analytics approach is more suited for a post-processing of the data, i.e. a situation in which the data analyst has enough time to explore the data. However, visual analytics could be used for the already recorded data, to prepare the data to allow better predictions in the real-time scenario.

In addition, research fields like cognitive science and psychology build synergy effects with the field of eye tracking [305]. Figure 1.7 illustrates some

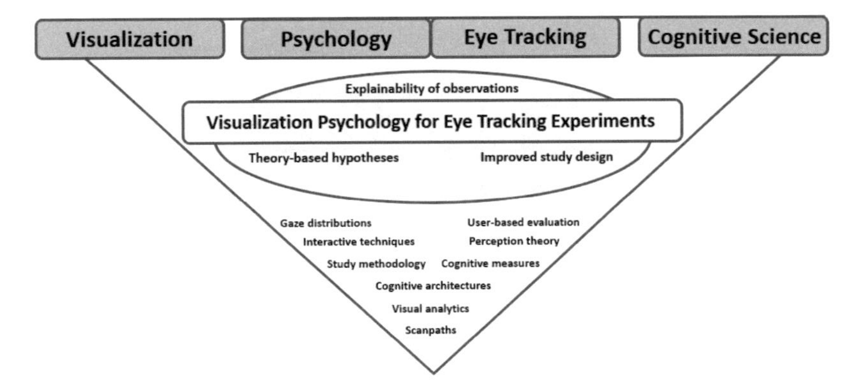

Figure 1.7 Cognitive science and psychology are also important research fields to improve the design of eye tracking studies and the interpretation of the recorded data [305].

of the major concepts that are most powerful for detecting valuable claims about human behavior applied to provided stimuli, static or dynamic ones, in case they are combined in a meaningful way. This way, tasks solved in an eye tracking study are not only dependent on the shown stimulus but also on important aspects like cognitive load [512] in the cognitive processes [268] or the working memory [386] which is typically not seen in the recorded eye movement data. The so-called eye–mind hypothesis [269] plays a role in this context, stating that people visually fixate what they process, having its origin in reading research [268]. Also machine learning, statistics, and data science play a crucial role in finding insights, patterns, correlations, and knowledge in the eye tracking data efficiently, also for classifying scanpaths or predicting visual behavior based on existing data.

1.3 Dynamic Visual Analytics

Eye tracking could be used to enhance or adapt the shown visual stimulus, either by changing the visual content after the eye tracking data has been recorded, as a post process [14], or in real-time [395]. Both approaches have in common that they make the stimulus, with static or dynamic content, a dynamically changing visual representation in a way that it is modified after each iteration, either by inspecting the data in a post process or very quickly, i.e. in real-time (see Figure 1.8). The real-time approach can be applied to a visual analytics system equipped with an integrated eye tracker, with algorithms running in the background analyzing the recorded eye tracking

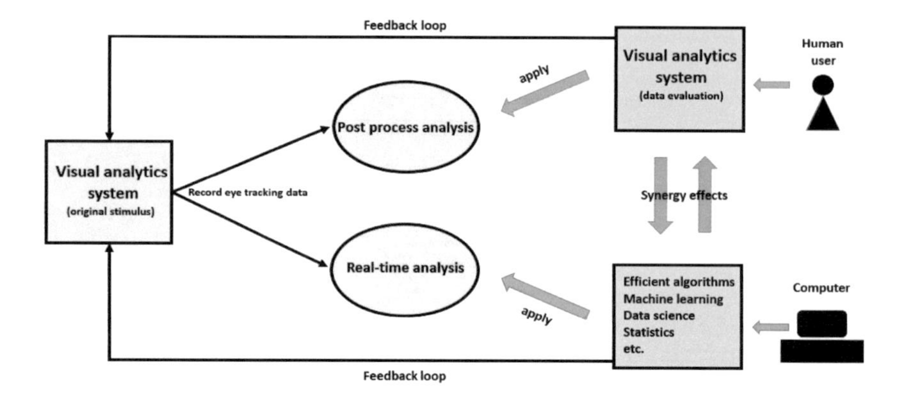

Figure 1.8 Dynamic visual analytics describes the process of evaluating a visual analytics stimulus by either a post process analysis or a real-time analysis. In a post process analysis, a second visual analytics system can be used to analyze the eye tracking data; in a real-time analysis, efficient algorithms must be used.

data for patterns and changing the visual content when required based on the output of the efficient algorithms. In a second concept, visual analytics can be used to analyze the data recorded in an eye tracking experiment with another visual analytics system as a visual stimulus [46]. This concept typically does not work in real-time due to the fact that the human user is involved in visual analytics and the eye tracking data has to be analyzed by a combination of interactive visualizations, algorithms, and the human observers to build hypotheses about the recognized visual patterns. The visual content of the static stimulus or a temporal snapshot of the dynamic stimulus is important to guide the observer, also leading to different applications of interaction techniques or varying parameter settings for the applied algorithms or to using completely different classes and variations of supported algorithms.

Exploring the recorded eye tracking data using a second visual analytics system allows insights into the data, but it also gives us information about how well the second visual analytics system works as some kind of user evaluation. This information, on the other hand, can be used to find design flaws in the second system while the actual task was to adapt or improve the first visual analytics system, i.e. the original dynamic stimulus in the eye tracking study. Creating such a dynamic visual analytics system sounds like an easy task to tackle, however, it involves several challenges to be solved. For example, a visual analytics system can also be equipped with

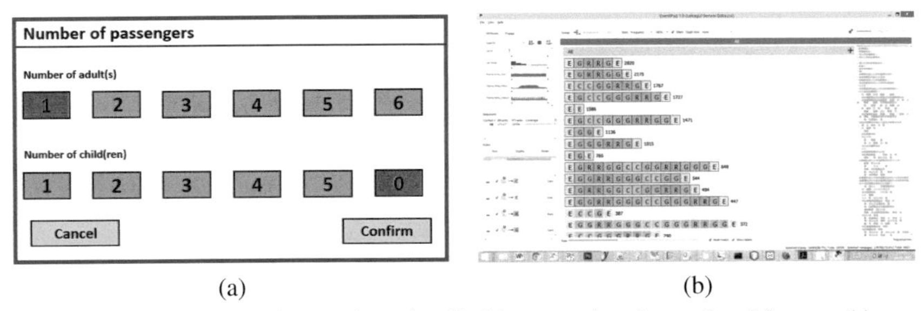

(a) (b)

Figure 1.9 Two types of interactive stimuli: (a) a user interface of a ticket machine and (b) a more complex user interface of a visual analytics system. Image provided by Bram Cappers [106].

gaze-based interaction, combined with several more input media like voice [510], mouse [532], keyboard [114], touch [284], gestures [110], and so on. Such extra interaction media build a source for additionally recorded data [45] worth investigating and incorporating into the whole evaluation process. Facial expressions [373] might also be recorded to extract information about the human user in front of the visual stimulus. The user might be young or old, male or female, wear glasses or contact lenses, be laughing, smiling, or angry, and the like. All these variations of aspects allow for a more reliable analysis of the data and build a means to hopefully adapt the shown system or at least identify design flaws, either as a post process or in real-time. But negatively, they also increase the burden of building an efficient and effective eye tracking data analysis tool.

A concrete example for the post process model, but with a simpler visual stimulus as a traditional visual analytics system would be the user interface of a ticket machine (see Figure 1.9(a)). If we imagine that hundreds or thousands of people use this interface to buy a ticket for their favorite train connection every day we might have a large database of scanpaths if the ticket machine had an integrated eye tracking device. If it was equipped with gaze-based interaction we would have even more additional data. The ticket machine designers might be interested in the eye movements of the people to investigate whether their user interface is designed in a way to provide understandable and easy-to-apply functions to get the requested service, or if several design flaws have occurred, leading to dissatisfied passengers due to the bad service. For this scenario the recorded eye movement data can be analyzed in a post process by using a visual analytics system that shows typical user behavior patterns. A user interface for a ticket machine is much simpler than for a visual analytics system (see Figure 1.9(b)) since it is equipped with less functionality, fewer algorithmic approaches, less visual output, and fewer complex interactions; however, it requires a substantial number of analyses, algorithmic and visual, to understand the data patterns and correlations.

In a corresponding real-time scenario of a ticket machine exploiting eye movement behavior, people's scanpaths must be permanently recorded. These scanpaths have to be compared to existing and previously recorded eye movement data as well as additional data sources with extra information. Efficient algorithms run in the background and analyze the data of the new eye tracked persons. One consequence might be that the user behavior can be classified and mapped to a group of people that show similar behavior. This class of people's visual attention behavior is used to adapt the user interface to a certain appearance with only the required information to solve the task, for example buying a ticket for a certain person group (adult or infant). The adapted appearance of the user interface depends on several factors, which is a challenging task. We argue that such a real-time approach might be useful but successfully building such a system is a long way off, primarily based on eye movement behavior.

A visual analytics system is typically much more complex than a user interface for a ticket machine. Although the ticket machine already provides some kind of user interaction, typically based on touch, the user interface is designed in a more task-driven way, focusing on providing a specific service for efficiently buying one or more tickets. In visual analytics we are offered a complex graphical user interface (GUI) consisting of several interactively linked views, so-called multiple coordinated views [374] (see Figure 1.10). Algorithms run in the background and the user adapts the parameters of those and decides which visualizations in which layouts and settings to show. Visual analytics systems have a variety of functionality, supporting hypothesis building with the goal of allowing analytical reasoning. A visual analytics system can be used by several people simultaneously, in the same room, but also remotely, located all over the world in a collaborative interaction manner. Even more, the display medium can be small-, medium-, or large-sized and the interaction styles can be manifold, ranging from touch on mobile phones, gestures in front of a high-resolution powerwall display, or just mouse and keyboard interactions with a standard computer monitor. Recording eye tracking data of several people is a suitable means of understanding visual attention behavior to improve the visual analytics system; however, typically only a few people's scanpaths can be recorded

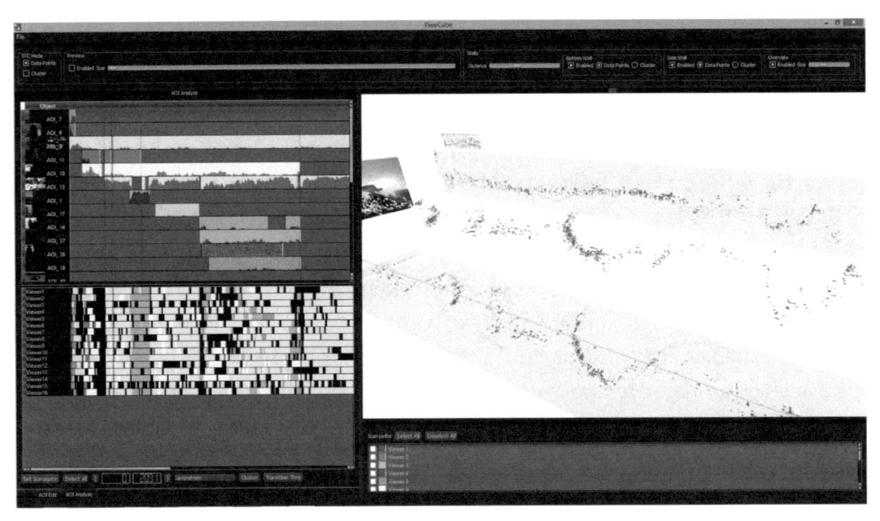

Figure 1.10 Coordinating multiple views provides several perspectives on the data under exploration. In this case we see a visual analytics tool for analyzing eye movement data from video stimuli [308]. Image provided by Kuno Kurzhals.

compared to the ticket machine scenario in which thousands of scanpaths can be recorded easily in a short period of time.

Analyzing the recorded eye movement data in the visual analytics scenario turns out to be a complicated task due to the fact that it has a spatio-temporal nature, i.e. changing over space and time. Moreover, the displayed stimulus can also be time-varying, for example if videos have to be watched in a video surveillance system [234, 308], in an animation [505], or in an interactive stimulus like in a visual analytics system. Videos and animations are not that complicated as a visual analytics system because they only provide a sequence of frames, i.e. a linear sequence of static stimuli. In a visual analytics system, the users can decide which static stimulus to watch next, i.e. if we interpret the static stimuli as snapshots of the system we reach a situation in which the users decide from which snapshot to move to another one. This is similar to a graph structure. The many possible snapshots create a huge graph with eye movement data for every graph node [68]. However, a post process of such eye movement data is still simpler than the real-time scenario for which we also require a lot of users of a visual analytics system which we just do not have these days, but it may be possible in the future. The analysis of the eye tracking data for a post process can be done by another visual analytics system that has to be adapted to the situation at hand in an iterative way.

To reach the real-time scenario we also need a visual analytics system that contains an integrated eye tracking device, transmitting the recorded data to a server on which the efficient analyses run. Efficiency is very important in this scenario to keep up with the flood of incoming data that has to be processed and the adaptations to the user interface that have to be made. Important approaches from data science include machine learning concepts as well as data mining, in particular to learn from the data already generated and to use that to classify or predict new results. Data mining could particularly be applied to generate rules between people, user interface components, and events, attached with a certain occurrence probability. In the case where the data is stored on a remote server, aspects like a reliable and stable internet connection also have to be considered. Moreover, privacy issues might play a crucial role, depending on the country in which the data is recorded.

The remainder of this book is as follows: in Chapter 1 we introduce the general research questions and which stages are required to come closer to solutions. Chapter 2 discusses typical visualization techniques in information visualization as well as visual variables to provide a spectrum for independent variables for user studies and eye tracking experiments. Chapter 3 extends this idea by additional interaction techniques and algorithms with the goal of creating a visual analytics system, including the human observer, to gain insight and to find knowledge in the data. In Chapter 4 we explain traditional evaluation methods in information visualization and visual analytics that are extended by eye tracking in Chapter 5. The analysis and visualization of the recorded eye tracking data is described in Chapter 6 and we conclude the book with an outlook on open gaps, challenges, and further difficulties in Chapter 7.

2

Visualization

Before describing visual analytics it is important to mention the most important concepts in the field of visualization [182, 513]. The stimuli that we explore in an eye tracking study are initially of a visual form; the interactions and algorithms are second, but are equally important. The representation of data in a graphical form has been a focus for many years now, producing a variety of different visual encodings based on a large repertoire of visual metaphors. Information visualization is the scientific field that deals with studies and researches the visual representation of abstract data [108]. This data can come in many forms like numerical or non-numerical data. On the other hand, scientific visualization deals more with the visual depiction of spatial data. This difference in the definitions of both subfields of visualization sometimes causes confusion [364]. However, both fields have in common that a suitable visual representation for data is required to support an observer to detect patterns in data, no matter if the data is abstract or spatial. The common goals in visualization are the presentation of results of analyses, or already existing non-transformed data, to a larger audience or readership, support for confirming formerly built hypotheses, or the means to explore a dataset visually and to interactively search for data patterns by first transforming the whole pattern-including dataset into a suitable visual form and by re-mapping those identified visual patterns to corresponding data patterns with the goal of rapidly detecting them in a larger dataset.

In particular, the research field of visualization combines other fields like computer science, computer graphics, visual design, human–computer interaction, as well as psychology, to mention the most important ones. Typical areas in which visualization is useful and for which it can bring a lot of benefits are software engineering, biology, bioinformatics, social networking, sports, geography, eye tracking, and many more (see Figure 2.1 for examples). No matter which area it focuses on, the human eye plays

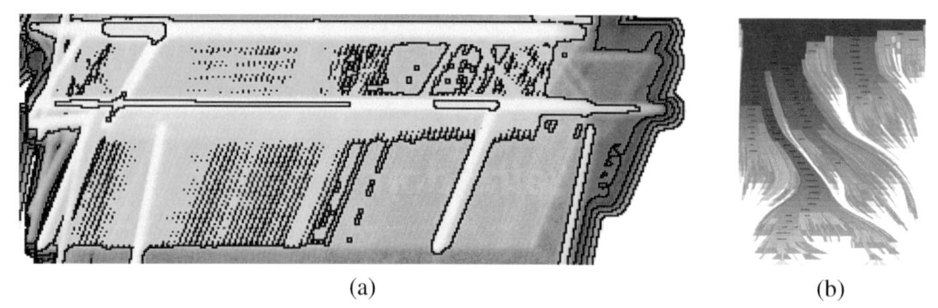

(a) (b)

Figure 2.1 (a) Part of a visualization for dynamic call graphs [75]. (b) A hierarchy visualization based on the space-reclaiming icicle plots [509].

a crucial role in the rapid pattern finding process which typically makes visualization a suitable concept, in many cases performing better than if pure algorithmic approaches are applied that require an exact definition of the input parameters. This means that visualization can be helpful in situations in which the data analysis problem cannot be defined in enough detail. The human user decides which parts of the data are interesting and require further attention, given that the chosen visualization is an adequate one.

Typically, the human observer is not left alone with the static diagram, but interaction techniques [476, 544] awaken the otherwise static visual representation, providing us with several other perspectives on the data, even in multiple coordinated views [374], linking various views and giving us the full potential a visualization can have. All this can only be achieved by good design of the whole graphical user interface [461] in which the linked views are laid out as well as the design, layout, and arrangement of the visual elements in each of the visualization techniques provided in each individual view. Hence, it is important to have a good understanding of and some experience in visualization before starting to develop advanced visualization tools. But it is also crucial to have knowledge about the concepts of computer and data science, like powerful algorithms that transform or project data into a pattern-preserving and usable form. The efficiency and effectiveness of the incorporated algorithms are important to achieve responsive interactions that allow the views to be adapted and to rapidly inspect the data from different visual perspectives.

Performance evaluation explores how fast the individual algorithms are and whether certain algorithmic scalability issues might occur that make real-time interactions hard to be applied [195]. In this case a better data preprocessing and data handling stage might be required to first bring the data into a form that allows efficient operations, making the interactive visualization tool acceptably fast. User evaluation [397], in particular, if many users are involved like in crowdsourcing experiments [52], with or without eye tracking, can provide further insights into the usefulness of a visualization tool, i.e. if there are design flaws that hinder spectators from quickly getting insights into the data. For example, a chosen visualization technique might be more difficult to understand than another equivalent one. This has already been evaluated for simple graphical representations [124]. Moreover, the arrangement or layout of visual components or views in a graphical user interface might not be perfectly chosen for the tasks at hand. Even visual variables like color coding or font sizes might cause problems in the data exploration process, which can be detected by user studies; however, the parameter space is so huge that not all aspects can be studied in such experiments, even the simplest ones. Positively, this makes user evaluation a research topic on its own with all its facets and variations. Eye tracking brings an even bigger challenge into play since it produces spatio-temporal data that is hard to analyze for patterns focusing on improving visual designs. But, on the other hand, it provides a great opportunity to dig deeper into visual attention behavior and hopefully combining eye tracking successfully with cognitive psychology [305] one day, to tap the full potential of user evaluation.

2.1 Motivating Examples

Data is the core ingredient of information visualization as we know it as a means to find patterns and correlations, and to derive meaning and knowledge, for example to understand a list of statistical values or the relationships between a group of people. The human users with their tasks in mind are responsible for identifying patterns and correlations, but also the way the visualization is presented, i.e. the visual metaphor or the visual encoding with all of its visual variables [39] plays a crucial role in the usefulness of a visualization technique. The list of visual variables has been extended over the years and also simple evaluations for ordering them based on a certain task have been conducted and the results presented [344].

To start with the major ingredient of visualization, there is, however, a certain difference between data and information. Data needs some kind of process to turn the raw, unorganized, and unstructured entities into something that is understandable and organized according to specific (mostly user-defined) requirements, maybe equipped with additional meaning in a certain context, finally transforming data into information. Clearly understanding the difference between data and information is important for the field of visualization, in particular information visualization since the major principle in information visualization is to allow information to be derived from data [476]. An important aspect of information visualization is also the communication of information, i.e. after having seen a dataset in a visual form we should be able to discuss the visual appearance with other people. This means we try to share the insights in a different form than the raw data, in a language that the conversation partners understand. The original raw data is first interpreted and a new kind of pattern is derived that is understandable by all people who speak the same language of visualization.

As an example for raw data we might be confronted with a table of values as in Table 2.1. Reading this raw data does not help us identify any pattern. But if we know the general context from which the data stems and interpret it as 2D coordinates, we might try to plot the 2D points and connect subsequent ones from the table by a line while we also draw a line between the last one and the first one. This transforms the raw data into a visual form that can be seen in Figure 2.2. What we did here is make use of a visual variable that encodes the data property of consecutive coordinates ordered into a line-based diagram based on the visual variable of connectedness by a direct linking with lines. Moreover, it visually transforms the coordinate values into positions in space, in this case x- and y-positions in a given coordinate system. In summary, we used some kind of visual transformation, i.e. visual variables like connectedness and position. Everybody who is familiar with geometric objects would interpret this pattern as a pentagon shape. This pattern could not be detected from the raw coordinates, even if we tried very hard. It may be argued that the lines are not needed to interpret the coordinate sequence as a pentagon. However, if the coordinate list gets larger and larger, the lines are definitely needed because our brain cannot derive a shape immediately, it needs some help by explicitly drawn lines (see Figure 2.3).

Table 2.1 A table of raw data consisting of five subsequent 2D coordinates.

Number	x	y
1	15	47
2	38	46
3	47	25
4	27	6
5	6	25

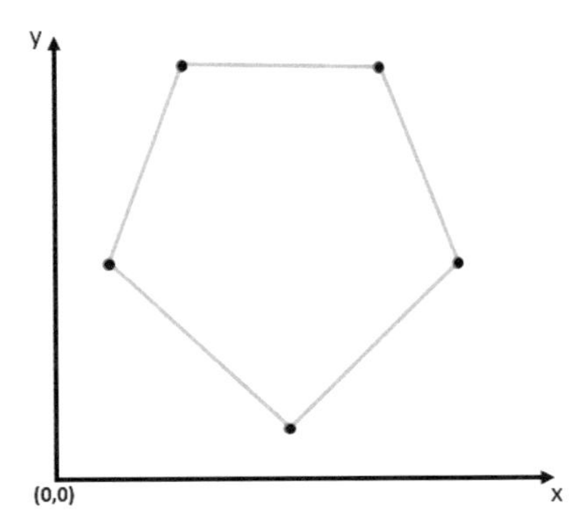

Figure 2.2 Connecting the list of 2D coordinates from Table 2.1 by straight lines in order transforms the raw data into a visual form, in this case a pentagon shape.

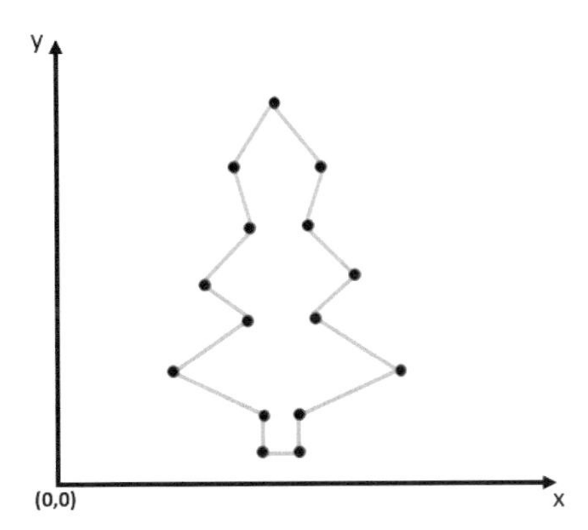

Figure 2.3 The shape of a pentagon as in Figure 2.2 can be drawn with just five points and five connecting lines, but the shape of a Christmas tree requires many more points. There are many more complex pattern examples in visualization.

This more complex example, consisting of many more 2D coordinates, would create some kind of Christmas tree when connected by straight lines in the right order, given the fact that the observer knows this visual pattern, being able to map it to a pattern already seen before based on experience. More complex examples are sometimes provided in newspapers and magazines where the goal is to connect a point cloud by straight lines in the right order (indicated by small numbers, see Figure 2.4) while the task is to interpret the resulting shape.

Figure 2.4 Connect-the-dots is a popular game in newspapers and magazines. The human brain needs lines connecting the dots to successfully interpret the shape.

The created shapes we saw here were just simple data examples consisting of a handful of points, easily manageable by hand with pencil and paper. Visualization these days, supported by a computer, has the benefit that it can visually encode hundreds, thousands, or even millions of data points very exactly in a short space of time, in the best case reflecting hidden patterns in the data. In addition, if the visualization of all the data points at once is not possible due to space limitations or a lot of overdraw and visual clutter, aggregation or projection methods can be applied to reduce the amount of data while preserving most of the hidden patterns, but at the cost of a loss of information.

If we have never seen the pattern of a pentagon or a Christmas tree we are not able to interpret it, or communicate it to someone else because we have no common term for it. Visualization is hence based on a common language of known patterns, just like letters that form words which form sentences and, finally, an entire story, but the repertoire of visual patterns can easily be extended (the number of letters is fixed). However, the visual pattern repertoire must be extended in everybody's pattern repertoire, otherwise we cannot easily discuss the findings and communicate the insights. Nearly any kind of shape can be generated from a list of 2D coordinates and it depends on the repertoire of known patterns as to which ones can be successfully interpreted and are useful for a suitable communicative visualization.

There is a larger list of visual variables that can be used to build any kind of complex visual representations of data. A visual variable is a container for a certain visual value that can change, but that still has the same type as all the values fitting in the same visual variable. A visual variable describes to which visual elements of the same type a data value is mapped. An example of a visual variable is the color hue. The hue can be changed but it is still a

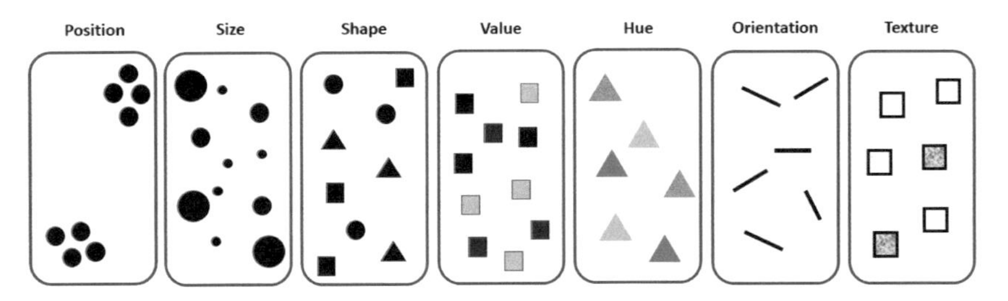

Figure 2.5 Visual variables are fundamental ways to distinguish visualizations. Jacques Bertin [37] described seven such variables and denoted them as retinal variables.

hue. Examples for visual variables are illustrated in Figure 2.5 and are color, size, position, shape, value, orientation, arrangement, texture, and several more [39, 344]. For color we can have a finer categorization into hue, value, or saturation for example. There are also some more modern ones like crispness, resolution, or transparency [342].

Before starting to design and implement complex visualizations, it is important to understand the fundamentals or the basic rules in visualization first. Two pioneers in this domain were William S. Cleveland and Robert McGill who tried to figure out which kinds of visuals lead to good performances in terms of error rates. In their experiments [124], they showed that some visual variables are better than others for a given task in terms of performance, like error rates. However, they were not applying eye tracking to record the visual attention paid to their stimuli. The spatio-temporal eye movement data would have given them many more insights, in which case they would have been able to analyze it. Eye tracking devices and also visual analytics were either not that well researched or even did not exist as they are known today in their advanced forms. The steady progress in hardware technologies and software engineering has led to better, faster, and more efficient solutions. However, what they found out with traditional user experiments was the fact that the visual variable position in a common scale leads to better user performance than using the angle for the task of judging the size of a data point for example.

This effect appears in the very common and simple visualization techniques like bar charts and pie charts, which were used as polar area diagrams by Florence Nightingale during the Crimean War (1853–1856) to show deaths in British military hospitals [130], although it is said that they were invented much earlier by William Playfair [188] (see Figure 2.6). Pie charts are very common representations in newspapers or magazines for the

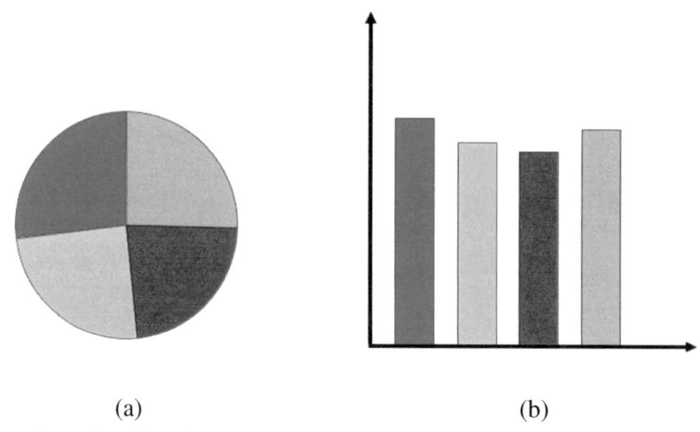

(a) (b)

Figure 2.6 The task of judging and comparing sizes appears to be easier and more reliable with fewer errors in bar charts (b) than in pie charts (a). Visual variables are the cause of this effect, which are positions in a common scale in bar charts and angles in pie charts.

everyday visualization consumer, i.e. the non-expert in visualization, but bar charts of the same data lead to much better performance for the task of judging the different sizes of the represented values [102]. For example, the result of an election and the number of votes is typically shown as a pie chart, with color coding indicating the different parties and the visual variable angle encoding the percentage of votes based on the total number of votes in the election. In a bar chart the color is used for the same encoding for the parties, but the data variable number of votes is encoded in the visual variable position in common scale instead, which makes the tasks of judging the values and comparing them much easier (fewer error rates and faster responses). This slight change in the choice of visual variables shows that we can obtain a large effect in user performance, even for the simplest diagram types.

The real power of a visualization technique can be recognized if we rely on the visual variable position in a common scale for representing two quantitative values, i.e. two variables in the data also called bivariate data. The resulting visualization technique is referred to as a scatter plot [189] if each of the variables is placed on one axis in a Cartesian coordinate system. Those plots are beneficial since they are easy to create, without complex mathematical background and programming experience, and reflect a lot of visual patterns that can hint at positive and/or negative correlations in bivariate data (data consisting of two variables for each object).

If we consider an example of four sets of 2D values (see Table 2.2) and do some statistics to get derived values, one might get the impression that

Table 2.2 Anscombe's quartet [15] is based on four sets of 2D coordinates.

x	y	x	y	x	y	x	y
4.0	4.26	4.0	3.10	4.0	5.39	8.0	5.25
5.0	5.68	5.0	4.74	5.0	5.73	8.0	5.56
6.0	7.24	6.0	6.13	6.0	6.08	8.0	5.76
7.0	4.82	7.0	7.26	7.0	6.42	8.0	6.58
8.0	6.95	8.0	8.14	8.0	6.77	8.0	6.89
9.0	8.81	9.0	8.77	9.0	7.11	8.0	7.04
10.0	8.04	10.0	9.14	10.0	7.46	8.0	7.71
11.0	8.33	11.0	9.26	11.0	7.81	8.0	7.91
12.0	10.84	12.0	9.13	12.0	8.15	8.0	8.47
13.0	7.58	13.0	8.74	13.0	12.74	8.0	8.84
14.0	9.96	14.0	8.10	14.0	8.84	19.0	12.50

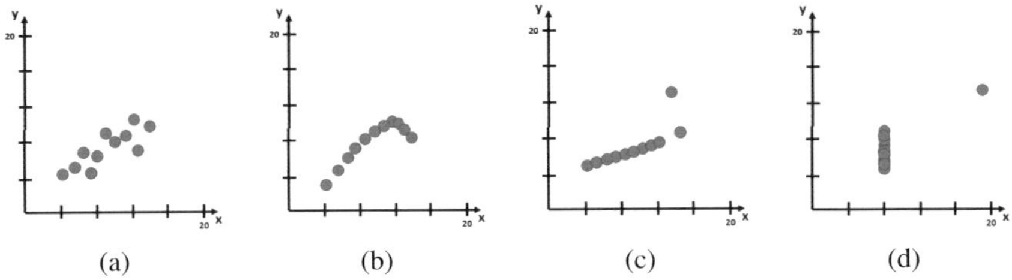

(a) (b) (c) (d)

Figure 2.7 Visually depicting the four sets of 2D values reflects the real differences in the four example datasets. Statistics is a powerful concept but, due to aggregation effects, not all insights in the data variations can be found.

all of the value lists are the same, or at least similar. The benefit of the derived values is that we get rid of the large number of values by aggregating them into one number; however, aggregation is always prone to data loss. Statistics is a powerful concept, but only applying statistical approaches to detect insights into a dataset is a process only showing half of the truth and can lead to various misinterpretations of the underlying data.

Representing the values from Table 2.2 in 2D scatterplots and inspecting their distributions, i.e. visual patterns, reveals something totally different than reflected by the pure statistics. The visual depiction of this simple bivariate dataset provides more insights than the statistical summaries into individual aggregated values (see Figure 2.7). This does not mean that statistics is useless or error-prone, but a look at a corresponding visualization can help to see the data under a different light, maybe leading to more detail and more fine-granular hypotheses. This effect was studied by Anscombe [15], and is known as the Anscombe quartet.

Most of the data situations do not allow such a simple depiction as a scatterplot, but rely on more complex visual scenarios, inspired by well-known concepts (maybe visual patterns in nature). This mapping from a dataset scenario to a visual encoding is oftentimes referred to as a visual metaphor which can be understood as a mapping of data to a visual concept where the data points are represented as graphical primitives. A visual metaphor is based on familiar symbols in order to make it understandable. It represents some kind of analogy to something well known from another field.

Examples for visual metaphors are trees, i.e. the depiction of hierarchical data into naturally growing tree shapes following a parent–child relationship. Moreover, a word cloud puts words with their occurrence frequencies shown as varying font sizes in groups of varying proximity. Also computer science makes use of various visual metaphors, for example when introducing the concept of Turing machines to students, which is composed of bands for manipulating symbols based on a set of rules. Everybody who has studied computer science has this picture in mind of a machine painted on a blackboard or projected onto a wall, although the mathematical computations have nothing to do with a machine.

To conclude this section, there are various examples of visualizations, diagrams, plots, charts, or whatever we would like to call them. If the reader needs inspiration for static visualizations, a Google image search is recommended since a picture is worth a thousand words. Typing in the term "visual" provides many results which, in case one is already familiar with a small repertoire of visualization candidates, build a basis for further ideas. Scrolling down the endless list of visual depictions while trying to understand which visual variables are chosen for a data representation might lead to the final goal of either finding the candidate one is looking for or getting completely new ideas for the dataset under investigation and the tasks at hand.

Dynamic, i.e. interactive visualizations, can also be found using the image search if we search for animated gifs for example. To get a longer description and explanations of the variety of features in an interactive visualization tool, YouTube is recommended while reading the endless list of academic research papers in the field of visualization, which can be supported by the DBLP, the Digital Bibliography & Library project, hosted at the University of Trier [327]. The DBLP contains, while writing this book, more than 5 million papers and articles from the fields of computer and data science, while filtering for the word "visual" still results in more than 85,000 hits [90, 92], each being its own small subtopic in the visualization domain with its various conferences and journals bridging the communities in those subdomains and trying to create synergy effects. Visualization has the great benefit that it can be applied to nearly any kind of discipline that produces, transforms, or deals with data, which at some point wants a visual depiction of it [513].

The research field of information visualization is understood as the use of computer-supported, interactive, visual representations of abstract data to amplify cognition [108]. Information visualizations attempt to efficiently map data variables onto visual dimensions in order to create graphic representations [198]. Information visualization is the communication of abstract data through the use of interactive visual interfaces [279]. The purpose of information visualization is to amplify cognitive performance, not just create interesting pictures. Information visualizations should do for the mind what automobiles do for the feet [107].

The previous definitions are just a short list of a larger repertoire, all expressing similar aspects, mostly amplifying cognition in the best case to support the pattern recognition process in large data, in the case that this data is efficiently mapped to visual variables to allow interpretations to reliably solve the tasks at hand or build, refine, confirm, or reject hypotheses.

Over the years the field of visualization has split into several subfields, such as software visualization, graph visualization, biovisualization and so on, but also more general overlapping areas such as information visualization and scientific visualization. No matter which subfield we are in, there is always overlap and synergy effects with other, even less related subfields and research disciplines. Mostly, the difference depends on the data to be visualized, but also on the domain from which it stems. It is said that scientific visualization deals more with continuous and spatial data whereas information visualization researches abstract, non-spatial, and discrete data, but this definition always leads to misunderstandings and causes trouble.

2.2 Historical Background

Visualization goes back in history to long before the invention of the computer [188]. For many years, humans tried to visually mark, remember, or disseminate information from scenes, daily environments and circumstances, or rituals for the next generations. This visual language was understandable in a quick manner, in some cases warning people about something, a visual language whose visual intention is still incorporated today in traffic signs for example. Drawing by hand was a time-consuming process and errors could not be corrected in an easy and fast way as would be possible in today's computer programs or in a graphical user interface supporting drawing and painting, but once the visual scenes were created they could be instances of the famous saying "a picture is worth a thousand words" [399], which first appeared in a 1911 newspaper article [57].

However, although the creation of visualizations is definitely possible without devices such as the computer [37, 39], it would probably take much more time to generate them, in particular if the underlying dataset consists of a variety of data entities. Moreover, the visual result will be static, meaning it cannot be modified in a fraction of a second, keeping pace with the perceptual abilities of the human brain to rapidly detect patterns. In particular, many of those visual patterns, watched from various perspectives make a visualization tool a powerful concept to keep up with the vast amounts of data generated in these days. Without the use of the computer combined with the experience and knowledge of human users, much of the data would hide most of the patterns and hence the informational insights that are contained in it. Consequently, the visualizations designed and sketched before the invention of the computer could not tap the full potential for data visualization of the big data era that we find today, consisting of various data sources, being static or changing over time.

In this section we will focus on the history of visualization, giving a brief overview of how the field has developed over time to get a rough impression of why the application of eye tracking to visual concepts is very important but also challenging due to the fact that many more visual variables are combined, while interactions allow us to quickly switch between them based on users' demands.

2.2.1 Early Forms of Visualizations

We will start the journey of visualization history with famous cave paintings. Those visual depictions can typically be found on cave walls and ceilings, mostly being of a prehistoric origin. One of the earliest of such paintings dates back to approximately 40,000 years ago, the Aurignacian period in Europe, found in the El Castillo cave in Cantabria in Spain [125]. Figure 2.8 gives an impression of what typical animal drawings in such cave paintings might look like, here inspired by a painting from a cave close to Santander in Spain. The spectators can detect by themselves which scene it illustrates visually since the scene reflects some kind of natural circumstances. There are even older ones, typically illustrating hunting scenes, depicting large and wild animals like horses, deer, or bison, while abstract patterns showing human hand-like shapes or time-relevant weapons give a visual impression that those animals might get caught by humans.

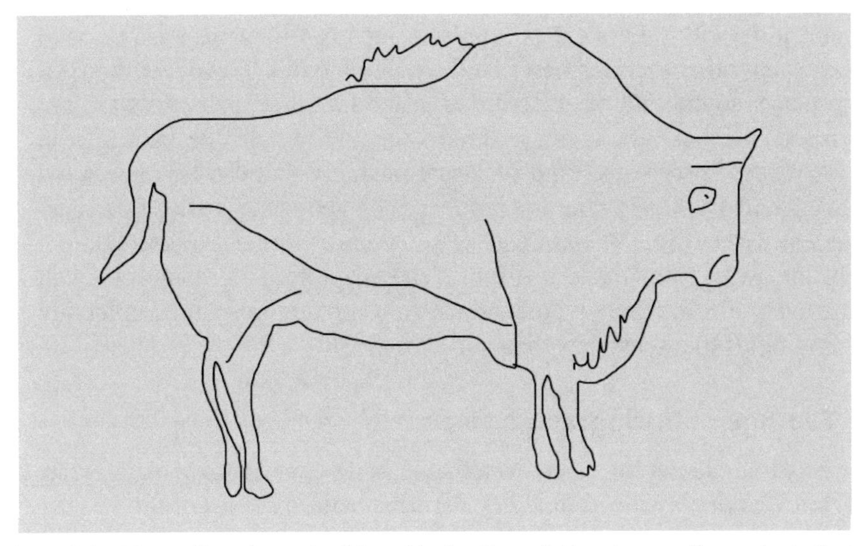

Figure 2.8 The outline of an animal found in the Cave of Altamira near Santander in Spain, also known as the Sistine Chapel of the cave painting.

In these prehistoric times, the human eye also played a crucial role, for example, when rapidly detecting a predator hunting for food. The strength of human sight was one key to survival in the wild, for example during hunting it was important to rapidly distinguish the berries in a bush from the eyes of a wild animal like a tiger with food in mind. Touch, hearing, smell, and taste were also important in these times but no sense has a higher bandwidth than sight transmitting information faster than any other sense to the brain to react on situations that might possibly have had a bad impact on a human life.

Apart from inspecting natural scenes, the human eye was also important for interpreting behavior among humans, by seeing where a gaze was directed one could guess the mood of another human being. In this context, the so-called cooperative eye hypothesis [290] was researched as a means to describe visible characteristics to facilitate humans when following another human's gazes while they communicate simultaneously, or while they collaboratively solve certain tasks. The research in this field was guided by scientists from the Max Planck Institute for Evolutionary Anthropology in Germany investigating effects of head and eye movements on gaze direction changes, comparing humans and great apes.

In particular, for the cave paintings, some theories have been developed over the years. One of them regards the paintings as a means of communicating with others, or as religious, mostly following a ceremonial purpose. Some other theories describe the paintings rather as decorations of a living place, some kind of art, but due to the fact that such caves are not easily reachable those theories are quite vague. However, the paintings in these early days carry some kind of visual meaning that can be interpreted and understood by those who are familiar with the same visual language. Consequently, they can be classified as some early forms of visualization, but still far away from those diagrams, charts, plots, or graphics we can generate today in a fraction of a second with a computer program, graphically supporting fields like science or art.

2.2.2 The Age of Cartographic Maps

The prehistoric depictions can even be classified as data visualizations although it was first assumed that they describe natural scenes or art, instead of more data-related aspects as we know them today. However, it is said that the earliest known data visualizations stem from the field of cartography [188] and those have also been found in cave paintings. Hence, some of those map-, route-, or geography-related prehistoric depictions could also be considered to be real data visualizations, i.e. the first ones of their kind. Some recent work [388] has shown that several of the maps found describe hunting areas or maps of the stars.

To successfully create visual depictions, the designers in those days had to transform the geographic positions, their connections, as well as extra attributes attached to the positions and the routes, into a graphical representation by using visual variables and by focusing on some kind of visual metaphor, in this case a map metaphor as we would call it today, mirroring the spatial information in the data as naturally as possible. Whether the designed maps were useful and interpretable was dependent on the success the hunters had after having observed them, hence some kind of user evaluation has also been conducted, without explicitly taking that into account, to enhance the maps.

For example, it is known that at around 6300–6100 BCE people sketched map-like plans, like the one containing buildings and a volcano that was found in Anatolia [442]. Moreover, the Egyptians created maps on papyrus to better navigate and travel in their empire, and to battle for winning new territories [170]. Those practical ideas supported military and administrative challenges as a suitable way to have an overview of the economy and finance in certain regions under their command. The major ideas in cartography and

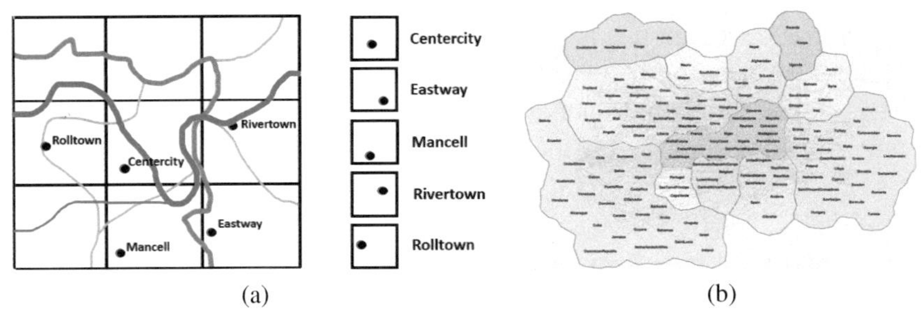

Figure 2.9 Maps have been used in a variety of forms, including various visual variables: (a) a geographic map annotated with a grid-based overlay to faster detect the label information and the location of a place [371]. (b) Data from other application domains with a more abstract character have been visually encoded into maps, like trade relations [243]. Figure provided by Stephen Kobourov.

geography were developed by the Greeks, including scientists like Ptolemy, Herodotus, or Eratosthenes. This progress in geographic mapping, including visual variables like area, position, or shape to visually express data aspects like geographic extents, locations, or forms, for example, produced the ingredients that are required to create an understandable geographic map as we know it today [342, 343] (see Figure 2.9(a)).

The early maps were still far away from the visualization concepts used today but many of the visual variables applied in these ancient times are still being used, hence they have been easily interpreted due to the experience of users creating or reading them over many centuries. However, these developments led to the maps we use in the 21st century consisting of regions with boundaries as well as road networks connecting the major places in these regions. A major difference to the ancient maps is the fact that we can generate a map on demand, in a fraction of a second based on a huge dataset, with a certain spatial granularity with dynamic non-overlapping labels supporting interactions like zooming and panning [544]. Prominent examples are Google Maps [546] or Google Earth [537] that make the data quickly accessible and support various tasks in this context while providing the means to interact with geographic data from nearly the whole world based on satellite imagery, meaning we can forget the existence of devices like compasses and sextants.

The map metaphor is, by the way, also used today not only for geographic data, but also for abstract data like social networks or trade relations as some kind of overview representation [243] (see Figure 2.9(b)). If a different dataset than geographical data is incorporated in the map-generating process, some kind of user evaluation is needed to investigate whether people can still interpret the data visually to solve the tasks at hand. Eye tracking could be useful here, since it brings into play spatial visual attention behavior that can be visually encoded directly on the map to see where the design flaws or problems might occur during paying visual attention [371].

2.2.3 Visualization During Industrialization

There was a time when people started to create visual depictions for various kinds of relevant data, trying to graphically depict what was going on. This trend could be seen in many application fields like healthcare, public transport, epidemiology, astronomy, economy, or military, to mention a few. The most common visualizations were based on bar, area, and pie charts, line graphs, geographic maps with visual annotations, or even Sankey diagrams [422], obviously counting for one of the more complex visualizations. Most of the visual encoding used just a small number of visual variables to transform data into a visual output, providing an overview or serving as an illustration, like that later used in data journalism, i.e. printed for the public as some kind of dissemination process. This new aspect also meant that the graphical depictions in the form of diagrams had to be simple charts to make them quickly understandable for the everyday consumer and not primarily for the expert in visualization.

Famous diagram examples from these times have been developed by people like Harry Beck, John Snow, Florence Nightingale, Charles Joseph Minard, William Playfair, and Francis Galton, to mention the prominent ones. These people had different educations, jobs, and obligations in their daily lives when they took on the challenging burden of designing expressive diagram types that have persisted until today, and that are well-known inspirational bases for many of the developed visualization and visual analytics tools today. The largest burden, however, was the fact that those diagrams had to be created without the use of a computer, i.e. they were typically hand made and their creation took quite a long time, although the underlying dataset was quite small compared to the megabytes of data we have to visually encode these days.

Sometimes the diagrams from these times were even referred to as infographics [338], which are some kind of visual illustrations that are easily absorbed by the user [503]. Infographics are typically based on a simple visual metaphor, visually enhanced and decorated, for example, by textual descriptions as well as extra surroundings, or inscribed graphics that indicate

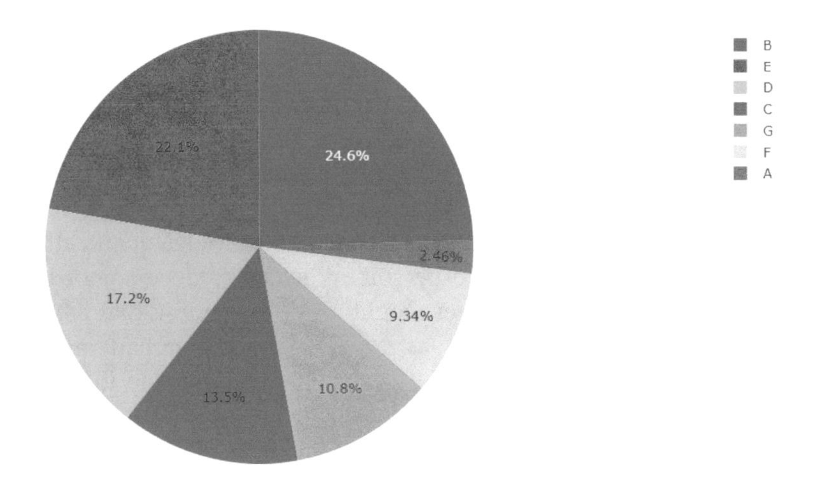

Figure 2.10 Today's pie charts are based on the ideas originally developed, for example, by Florence Nightingale.

the application field or topic that the infographic is about, sometimes denoted by the term reference graphic. Although it is said that they originated many centuries ago, they had their high tide in the years of industrialization due to the fact that the data was quite small and no computer was available to quickly produce or print them. However, they are still created in the 21st century because they typically attract the attention of the human observers and are quickly, easily, and effortlessly understood.

There was also no kind of advanced user evaluation giving feedback about the usefulness of the diagrams, about their interpretability as well as understandability and effectiveness, for example taking into account the encoded visual variables or perceptual issues. However, it seems that user evaluations, even for the simplest infographics, are conducted today as progress in technology and a growing visualization community, in particular eye tracking studies [346], become more and more of interest, since they can uncover design flaws over space and time, not only aggregated to measures like response times and error rates as in traditional user experiments.

There are various hand-made examples from the early days of infographics. For reasons of completeness and for illustrative purposes the most important ones will be described here. "The best statistical graphic ever drawn" was a famous quote by Edward Tufte that illustrated the expressive power in the so-called Minard map [503]. The graphic designed by Charles Joseph Minard (1781–1870) shows Napoleon's army marching to Moscow and illustrates some Sankey-based visualization of the army size in terms of soldiers on their way to Moscow and the retreat. Various visual variables are encoded in this diagram as well as merging and splitting behavior, temperatures, and geographic aspects such as rivers. Another famous example from the field of healthcare and medicine was developed by Florence Nightingale (1820–1910). She visually depicted the deaths among British soldiers using a diagram type referred to as a coxcomb or wedge chart called a polar-area diagram, which seems to be some kind of predecessor of the pie chart that we know today. However, Nightingale used the radial extent as well, instead of just the angular one as it is typical for pie charts as we know them today (see Figure 2.10). Another example falling into the field of healthcare or even epidemiology was designed by John Snow (1813–1858). Based on his spot map, people were able to close a water pump in Soho close to Broad Street that seemed to be the cause of various cholera infections. Public transport also turned out to be an important topic for creating visual representations to facilitate traveling in a city or to support route finding tasks. Harry Beck (1902–1974) designed a variant of the London Underground Tube map that helps to identify stations and routes by distorting the geographic locations in a way that the crowded information in the city center is reduced. He is popular for his famous quote: “It does not matter where you are when you are underground.” The design from these early days is still in use today, although mostly in an adapted style.

2.2.4 After the Invention of the Computer

The computer changed our lives in many respects. The field of visualization has also gained even more popularity than before due to the power and incredible exactness of programs that can create diagrams of more and more data at even faster rates. Moreover, diagrams are no longer static but awake to life, letting the users play around by applying more and more advanced interaction techniques [544]. Modern technology allows us to produce a chart with just a few mouse clicks while the implementation just required several lines of program code to turn a huge dataset consisting of various values measured for certain attributes into a simple understandable graphic in a fraction of a second.

Not only can the visualization perspective on the data be changed quickly, the data itself can also be recorded, stored, manipulated, processed, and structured at an increasingly fast rate. The data-related issues can also have an impact on the visualization itself by observing the data with it and then

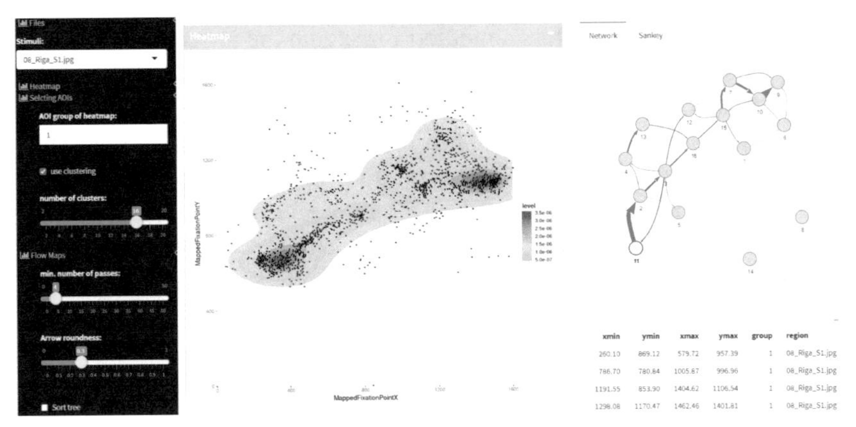

Figure 2.11 An example of a graphical user interface for visually exploring eye movement data [82]. Figure provided by Neil Timmermans.

slightly adapt it to the needs given by the newly computed structure of the data and the users' tasks at hand. The amazing power of the computer brings a new topic into play but also even more new challenges: real-time data visualization [533]. Not only is real-time data a challenge for visualizers but also data that is changing over time, allowing to play back or replay several times, i.e. after the dynamic data has been recorded and preprocessed.

We discussed time-dependent data in infographics in Section 2.2.3, for example the march to Moscow and the retreat, which has an implicitly inscribed timeline, but shows the same underlying data as an animated diagram with the aforementioned features. This was impossible in the years before the invention of the computer, at least not with an equal outcome. Moreover, several visualizations could be shown next to each other and linked, a concept that is known as multiple coordinated views [424]. The graphical user interfaces (GUIs) in which typical visualization tools are integrated consist of a variety of additional functionalities like buttons, sliders, menus, and many more, all targeting the common goal of supporting users to find insights into a dataset (see for example Figure 2.11 for an eye movement data visualization user interface). In general, the advent of the computer meant a huge step in the field of visualization, and the field is still progressing and outputting lots of research ideas, also based on new technologies.

The very first definition for the term visualization was given in an NSF panel in 1987 by McCormick, DeFanti, and Brown [146, 349] stating: "Visualization is a method of computing. It transforms the symbolic into the geometric, enabling researchers to observe their simulations and computations. Visualization offers a method for seeing the unseen...It studies those mechanisms in humans and computers which allow them in concert to perceive, use and communicate visual information." This definition after the invention of the computer shows that the focus was no longer on creating infographics, but it was focused on observation tasks for simulations and computations, which reflects that the datasets in use have increased a lot in size compared to those before the invention of the computer.

Soon after that the whole field brought more and more research ideas, presented in famous journals, conferences, and workshops. The variety of all of these topics from different application areas led to the effect that the whole visualization community split off into several subcommunities having overlapping interests. For example, areas like scientific visualization and information visualization showed this effect, but there is some kind of controversial debate on how those fields could be merged since many researched topics share common ideas and one might benefit from the other, building some kind of synergy effect. The merge could partially be recognized in the field of visual analytics [495] which makes use of concepts from both but goes even a step further by including human–computer interaction (HCI), statistics, data science, algorithmics, perception, and many more relevant fields. However, from a user experience point of view it gets more and more challenging to identify design flaws, compare visualizations with each other, and enhance them by the feedback and recorded measures of real users, either laymen or domain experts, with or without eye tracking.

2.2.5 Visualization Today

The visualization community has grown a lot in the past few years, consisting of several thousand researchers from various application fields. Popular events like the IEEE VIS, EuroVis, the Symposium on Pacific Visualization, and many more have built a platform for interested scientists to communicate, share, discuss, and publish their innovative ideas. Research topics range from abstract data visualization for tree, network, text/document, high-dimensional/multivariate, non-numeric, stream/time-varying, as well as geo-spatial data to more spatial data like scalar/vector/tensor fields, unstructured, or volumetric data, to mention a few from a large repertoire of topics. Interaction techniques also play a crucial role these days, including coordinated multiple views [424] as well as the display types, such as small, medium, or large ones with a high resolution, even in combination, also having an impact on the way interaction techniques are incorporated [385].

Figure 2.12 A hierarchy visualization depicted on a powerwall display. The system allows collaborative interactions for several users equipped with tablets, or it serves as an overview of a large dataset [93, 441]. Pictures taken and provided by Christoph Müller.

Figure 2.12 illustrates a hierarchy visualization in the form of a generalized Pythagoras tree [31, 363] that supports collaborative interactions between several users [93, 441]. Data visualization was also tried in virtual and augmented reality environments [276], trying to bring the users even closer to the data, a field that comes into play here is immersive analytics [168].

More and more algorithmic approaches including concepts from artificial intelligence are covered by the field, but they also bring new challenges and opportunities for research such as ethical issues and data privacy and security [478]. Finally, user evaluation is a steadily growing subfield that shows its strengths in nearly any niche of visualization [111], not only because visualization relies on the perceptual abilities of an individual human user but also because it has to uncover the problems that several users might have in a multi-user system in a collaborative interaction setting for example. Moreover, the various parameters in a visualization system demand extensive scientific research in the domain of user evaluation, also taking more and more into account visual attention behavior [306, 307] that is recorded by eye tracking devices whose progress has to keep pace with the enhancing visual and interaction technologies.

These days we see an increase in substantive and impressively challenging problems and data structures, more and more including the industry with a lot of money involved. There is a growing interest in combining real-world problems that companies face with the basic knowledge from researchers working and teaching at universities all over the world. The education of young people is, at the same time, important [87] as is the involvement of the vast amount of money offered by industry to further strengthen the field and to build a platform for further inspirations and powerful developments supporting real-life problems in an effective way, always keeping the real user in the development process in some kind of feedback loop. All of these stages and involved persons with varying knowledge and experience levels make visualization the powerful field that it has become today, grown from a small number of interested researchers and practitioners who met at renowned symposia, conferences, and workshops to share their ideas. Also technological infrastructures like the world wide web accelerated this process and the exchange of ideas among researchers while they also provided a new means for sharing interactive visualization systems, for example in a web-based environment based on existing visualization libraries and frameworks.

2.3 Data Types and Visual Encodings

We generate data at faster rates than we can analyze, visualize, or make sense of in any way [486]. Data can be measured or simulated while we are able to store massive amounts of it, creating an information overload. The field of big data brings new challenges, but on the positive side also opportunities for researchers working at a university or in industry. The field of big data [42] tries to tackle the daily problems that arise in the data domain, for example data storage, data transformation, data analysis, but also data linkage for heterogeneous data sources, furthermore concepts like data ethics, privacy, or provenance [478]. In the context of big data we often read the three terms volume, variety, and velocity, expressing the sheer size data can have these days. However, it may be noted that classification if a dataset is big data also depends on the application domain, not only on the number of bytes a dataset requires on a hard disc. It does not matter how big the data is when we have to decide which visual encoding to choose from the existing repertoire, i.e. which visual variables to combine. It is more important which ingredients a dataset has and what tasks should be solved if an interactive visualization tool is designed and implemented.

In this book we describe the most frequently occurring data types that have been visualized and that have been investigated in user evaluations, in particular with eye tracking technologies. We start with primitive data types, proceed with complex data types, and finally explain combinations of data types as well as time-dependent data, also taking a look at metadata because it is important for a self-explanatory diagram. In the following we provide visual examples for representing data categorized by the type to which they belong. We describe the included visual variables in visual depictions and discuss the pros and cons whenever we assume that it is required. In some scenarios we also explain for which tasks a visualization of data of a certain type is particularly beneficial. The visual encoding and the tasks play a crucial role in user studies since they play the role of the independent variables for which dependent variables are recorded, like error rates, response times, or spatio-temporal eye movement data, in case visual attention during the task solving process is of interest.

2.3.1 Primitive Data

There are three different primitive data types: quantitative, ordinal, and categorical/nominal. Categorical data is in some cases also referred to as nominal data because it can have some kind of textual description like a label. But numbers can also serve as a textual description, although it seems as if we could do arithmetic operations with them but not in a meaningful way, which again makes them belong to the class of categorical and not to the class of quantitative data.

In the following a clearer impression of primitive data types will be given together with some real-world examples as well as visualization candidates from a large repertoire. Typically, depending on which kind of data is involved in a visualization technique it has an impact on the complexity of a user study since the number of variable parameters allows for a more flexible study design and for many more recruited participants to obtain reliable or statistically significant results.

- **Quantitative data:** a set S or list L of elements $\{x_1, \ldots, x_n\}$ belongs to the class of quantitative data if we are able to do arithmetic operations with them in a meaningful way, i.e. multiplying, dividing, adding, or subtracting the elements in S or L. A real-world scenario can be file sizes in a file system. Each individual file has its own file size in bytes. Computing the size of the directory with all the files inside means summing up all the file sizes of the files contained in it. In this

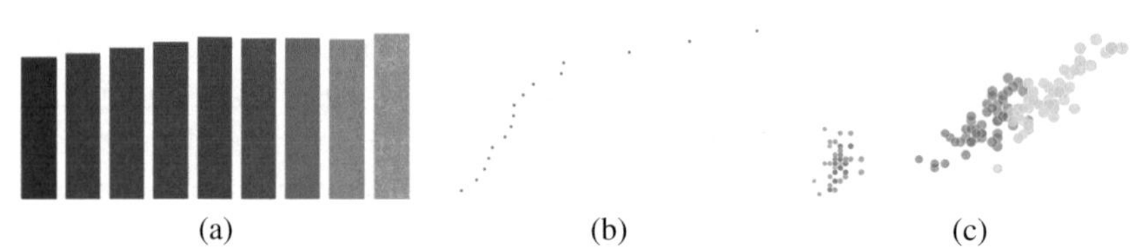

(a) (b) (c)

Figure 2.13 Prominent visualization techniques for primitive data types already exist in several variants. The performance and visual attention strategies of human users while solving tasks with any of the visualization techniques can be analyzed by eye tracking studies: (a) a bar chart for quantitative data using the visual variable length for the quantities; (b) a dot plot for the ordinal data using the visual variable position to encode the order; (c) a scatter plot with varying colors and different circle sizes as visual variables to indicate the categorical nature of the data.

scenario, arithmetic operations make sense, like additions or summing up; consequently file sizes belong to the class of quantitative data.
A prominent visualization for quantities are bar charts making use of the visual variable length (see Figure 2.13(a)), but pie charts or line plots can be found in the literature; however not all of them are equally effective when it comes to the task of comparing or ordering the values by size, or in the case of a line chart, they connect discrete values by lines although this does not make sense.

- **Ordinal data:** a set S or list L of elements $\{x_1, \ldots, x_n\}$ belongs to the class of ordinal data if we are able to bring all the elements in some kind of meaningful order, i.e. to apply some kind of sorting algorithm that takes the elements as the input parameter and terminates after finitely many steps with the sorted elements. A real-world scenario can be found for shoe sizes. Here, arithmetic operations make no sense, i.e. adding the shoe sizes is a meaningless operation although the numbers would allow it. But ordering the shoes by their sizes makes sense, for example in a shoe store.
A prominent visualization for this kind of data would be a dot plot that uses the visual variable position in a common scale (see Figure 2.13(b)). The order of the elements can be read from the vertical axis by checking the corresponding value of each point and visually comparing it with the others. A sorting algorithm could also improve the plotting order of the elements before they are visualized. It may be noted that also time has an implicit temporal order, i.e. time steps are of an ordinal nature.
- **Categorical/nominal data:** a set S or a list L of elements $\{x_1, \ldots, x_n\}$ belongs to the class of categorical data if arithmetic and also the

generation of an order do not make sense. Although it seems that we can apply arithmetic operations to them, like adding bus line 11 to bus line 7 summing up to bus line 18, however, this is not meaningful. Furthermore, ordering the bus lines increasingly is also not a meaningful solution. It is not clear which order criterion would make sense for bus lines. Bus lines are just categories and placeholders or representatives for routes that a bus takes.

Prominent visualizations for categorical data exist in various forms. Textures, shapes, or colors are typically used as visual variables to indicate a category, and iconic graphical depictions are sometimes used, for example for car brands, football clubs, or national flags. In a scatter plot we could indicate with color and circle size or different shapes that a value pair belongs to a certain category (see Figure 2.13(c)). In a public transport map, the train lines are typically color coded which also generates the effect that the lines can be traced easier, faster, and with fewer errors as in grayscale maps [371].

Jacques Bertin described a list of visual variables [37] which was later extended by Jock Mackinlay [344] who also ordered them for the three primitive data types to be visualized (see Figure 2.14). Mackinlay conjectured the effectiveness of the visual encoding for each of the primitive data types; however, nowadays an eye tracking experiment might give insights in the visual attention behavior when solving such perceptual tasks as he called them.

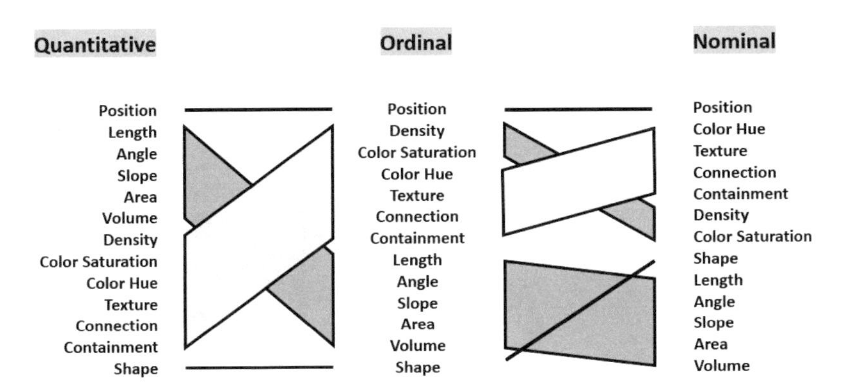

Figure 2.14 Jock Mackinlay gave an ordered list of the visual variables for each of the three primitive data types. He described the effectiveness of such a perceptual task in decreasing order [344].

2.3.2 Complex Data

Visualization for just primitive data is already difficult and can lead to misinterpretation problems or performance issues in case the wrong visual metaphor for solving a task is chosen. This can be seen in the example comparing bar charts and pie charts (Section 2.1) for the task of visually ordering quantities. An eye tracking study would show the design flaw in an increased response time for the pie charts, a higher error rate, and more chaotic eye movement behavior.

If we leave the scenario of primitive data types and move to more complex ones, it seems as if we have a larger repertoire of visualization candidates, but those are also much more complex and finding the best candidate for a certain task or task group is a challenging procedure.

- For example, graphs or networks consist of two types of entities in its simplest form. Vertices model the objects or persons that are related while edges indicate that a pair of them is related or not, hence a graph is modeled as $G = (V, E)$ with a vertex set V and an edge set $E \subseteq V \times V$. There can be an endless list of additional attributes attached to the vertices or edges, in its simplest form the edges might have a weight or the vertices might carry a textual description, i.e. a label. But as said before, there is no limitation to the extras that can be attached to the graph entities.
  In its raw form a graph can be represented as a node-link diagram (see Figure 2.15(a)), an invention dating back to Leonard Euler [179] when trying to find a mathematical abstraction to the problem of the "Seven Bridges of Königsberg". However, in these modern days, the graph data

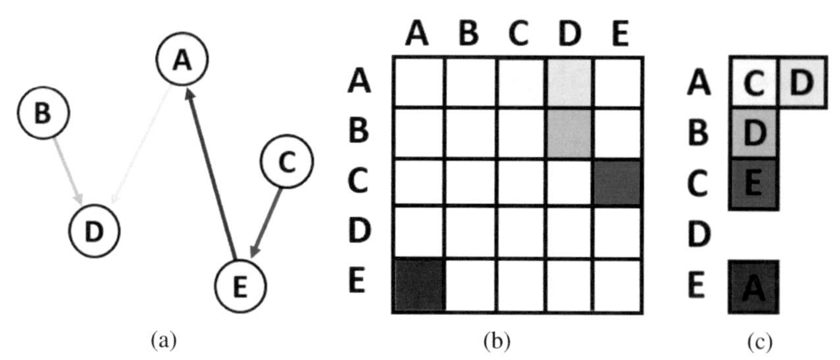

Figure 2.15 Three prominent visual metaphors for encoding the same graph dataset [29]: (a) a node-link diagram; (b) an adjacency matrix; (c) an adjacency list.

gets so large that node-link diagrams produce visual clutter [426] and are no longer readable, even after applying an advanced layout algorithm focusing on aesthetic graph drawing criteria [406, 409] or changing the edge representation style [237], and even bundling them together [236, 239]. Hence, adjacency matrices [29] were invented (see Figure 2.15(b)) to get a more visually scalable, clutter-free representation of relational data, evaluated for typical graph interpretation tasks [200]; however, the rearrangement of the matrix rows and columns [34] has a large impact on the identification of patterns, typically vertex groups and clusters. A third rarely used visualization technique is an adjacency list [229] that has some benefits but it is quite hard to read paths and identify clusters (see Figure 2.15(c)). If a graph is locally dense and globally sparse, hybrid methods are used such as the NodeTrix representation [227].

- Hierarchies are another relational data type with the differences that they can be drawn as a node-link diagram in 2D in a planar way without link crossings, that the corresponding graph has no cycles, and the fact that they have a designated root node. Hierarchical data is built on parent–child relationships defining the level a vertex is located and also aspects like its depth and branching factor. Visualizations for this type of data exist in a variety of forms [445, 446]. For this type of relations there are also various attributes to be attached to vertices and edges, like the sizes of files in a hierarchically organized file system or the evolutionary distances of related species living on earth in a so-called NCBI taxonomy [509]. For visualizing hierarchical data there exist at least four major types of visual metaphors (see Figure 2.16) that come in the form of explicit links [78, 419], nesting [266, 458], stacking [295, 509], and indentation [91, 95]. Each of these types contains various variants exploiting different combinations of visual variables, even hybrids of these types and variants exist. User studies, in particular with eye tracking, can find design flaws and perceptual problems for typical hierarchy-related tasks like finding the least common ancestor for example [72, 78].
- Data coming in the form of a table consisting of rows and columns is said to be multivariate or multi-dimensional data. The rows are called the observations or cases while the columns are the variables or attributes. One major task for this kind of data is the identification of correlations, positive or negative ones, or more complex ones as well as outliers and anomalies.

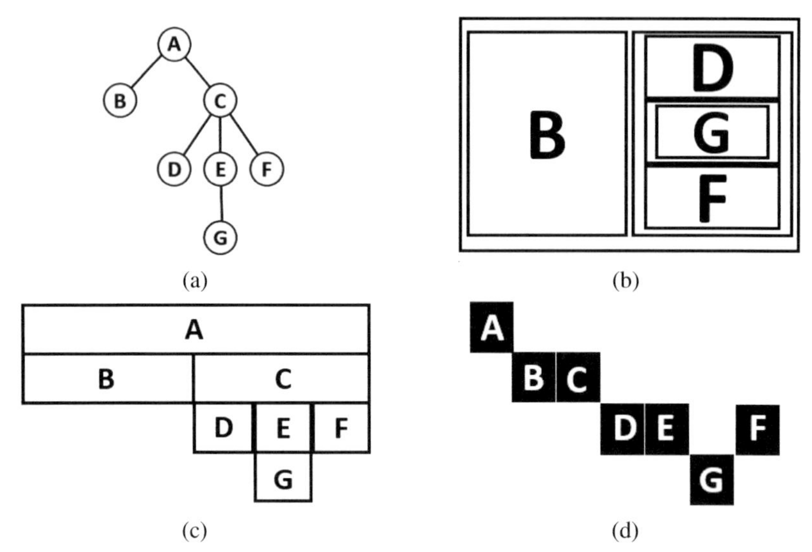

Figure 2.16 Four major visual metaphors for hierarchical data exist: (a) explicit links; (b) nesting; (c) stacking; (d) indentation.

There exist three major visual metaphors for this kind of data which come in the form of glyph-based representations [325, 428], scatterplot matrices [172], and parallel coordinates plots [224, 255] that have been evaluated in several ways in the past [265]. Figure 2.17 illustrates an example for a multivariate dataset with the same observations and attributes. We see that the glyph-based representations rely on a combination of visual variables, one for each attribute, combined in a single visual entity. The scatterplot matrices consist of a quadratic scheme of individual scatterplots, each making use of the visual variable position and maybe color or shape if the points carry categorical information as well. The parallel coordinates are based on the visual variables position in common scales and connectedness by straight links, forming polylines.

- Trajectories are generated by moving objects, animals, or people, leading to some kind of spatio-temporal data since they change their locations over space and time. Eye movement data [161, 235] also falls into this category. If we only take into account a pure scanpath, i.e. the movements the eye makes, we receive the simplest scenario of such an eye movement trajectory. In general, eye movement data is much more complex with extra attributes attached, for example fixation duration,

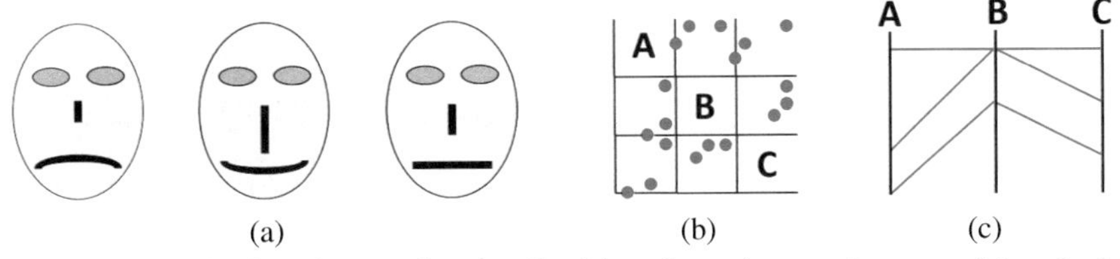

Figure 2.17 Multivariate data can be visualized in at least three major ways: (a) a glyph-based visualization, here in the form of Chernoff faces; (b) a scatterplot matrix (SPLOM); (c) a parallel coordinates plot.

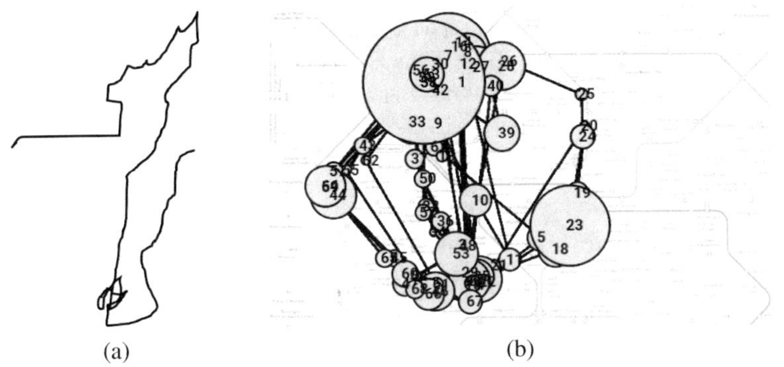

Figure 2.18 Different kinds of movement data can be measured and visualized: (a) trajectories from bird movement [369]; (b) scanpaths from an eye movement study investigating the readability of public transport maps [372].

calibration details, physiological measures, face expressions, or verbal feedback [44].

Typical visualization techniques for trajectories show the spatial information as a geographic map or, in case of eye movement data, the visual (static) stimulus while overplotting it with a line-based visualization, denoted by a gaze plot. This results, like nearly all line-based diagrams, in visual clutter effects [426]. Figure 2.18 illustrates two scenarios for a trajectory visualization of bird movement [369] (a) and a gaze plot [203] that shows the scanpath of several eye tracked people overplotted on a static public transport map stimulus [372] (b).

- Text or document data consists of a sequence of words that typically carry some meaning and that, in most cases, already present some kind of structure. DNA sequences are typically modeled as one string

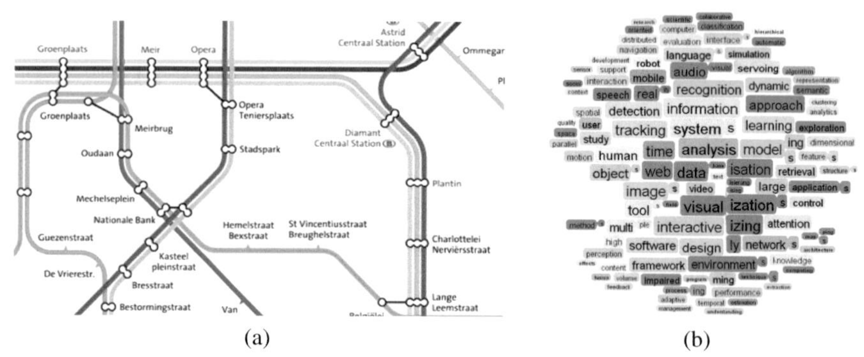

Figure 2.19 There are various scenarios in which textual information is important: (a) label information on a public transport map [372] (Figure provided by Robin Woods, Communicarta Ltd); (b) an aggregated view on the occurrence frequencies of words in the DBLP, summarized as a prefix tag cloud [86].

consisting of a finite set of characters. However, in some scenarios only a single word or text fragment is enough information to provide the detail needed to better understand a visual depiction of a dataset or a part of it, for example a label information in a map [371] indicating a street, a village, or a public transport station name (see Figure 2.19(a)). Moreover, a text or document might be summarized by counting the number of words contained in it. Those word summaries can be illustrated as quantitative information in a tag or word cloud [220], sometimes even aggregated in a prefix tree-like fashion [86] while the word occurrence frequencies are mostly mapped to the visual variable font size, sometimes color. Figure 2.19(b) gives an impression of word frequencies closely related to the word "visualization" based on data from the digital bibliography and library project (DBLP) [327, 328] that collects most of the publications in the field of computer and data science, hence itself being a source for several texts and documents, worth visualizing [92].

- Data elements can be in a set relation, meaning they form some kind of entity following a similar well-defined property. If several of those sets exist they can be combined by union, intersection, or difference operations forming new sets or subsets. Visualizing sets [8] has become a topic with various research ideas, not only because sets appear in many application fields but also because they are powerful concepts in mathematics. Moreover, they can be used in interaction techniques to

Figure 2.20 A set visualization based on the "bubble sets" approach [127]. Image provided by Christopher Collins.

filter data, navigate in it, or collapse and expand the data to allow more space for the remaining visible elements.

Figure 2.20 shows an example of a complex visualization based on the so-called "bubble sets" approach [127]. Numerous visual variables are used to create fancy but also complex diagrams that require some time to learn and to interpret the data that is visualized.

- In the field of scientific visualization in particular, we have to deal with complex data such as 2D/3D scalar, vector, or tensor fields as well as volumetric data, just to mention a few data type examples that can easily get more complex by adding more data attributes, and also time dependency. Moreover, compared to information visualization the datasets under examination are typically much larger and generate continuous data values instead of discrete ones, either for the static spatial data or in the dynamic time-varying case [281].

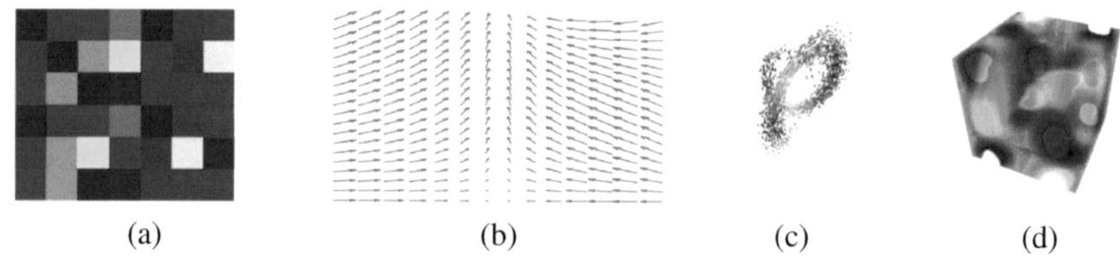

(a) (b) (c) (d)

Figure 2.21 Four examples that are typical data types in the domain of scientific visualization depicted by standard approaches: (a) a scalar field; (b) a vector field; (c) a tensor field; (d) a volumetric data visualization.

Figure 2.21 illustrates four examples from the field of scientific visualization. Typical data types are 2D scalar fields that are represented by visual variable color (a) while 3D versions can also be found. In addition, a visual attention map [50] with which the hot spots of eye movements are depicted falls into the category represented by scalar field visualizations. The vector fields are represented by glyphs; in the case of Figure 2.21(b) those glyphs are arrow-based, indicating at least two data variables such as direction or extent by the visual variables line orientation and line length. Sometimes the arrow heads have different sizes as well to indicate another data variable visually. In (c) we see a tensor field representation that visualizes some kind of relationship between algebraic objects. A volumetric data visualization is shown in (d), typically suffering from occlusion effects, hence mathematical/algorithmic, visual, and interaction concepts have been developed to find insights in this kind of data, even if most of the information is occluded.

2.3.3 Mixture of Data

The data-in-data concept makes the visualization much more complicated than if just one basic data type exists to be visualized. The major challenge with this scenario is the fact that we have to figure out which data of the combined data stands in focus, i.e. plays the primary role. This has a major impact on the appearance of the final visualization, the secondary, tertiary and further data types have to be incorporated in another way by extra visual variables. Hence, before starting with a visual design of such a mixture-of-data example we definitely should be aware of the task or tasks at hand to be solved. This task order will typically generate another order among the data types that guides and influences the appearance and combinations of

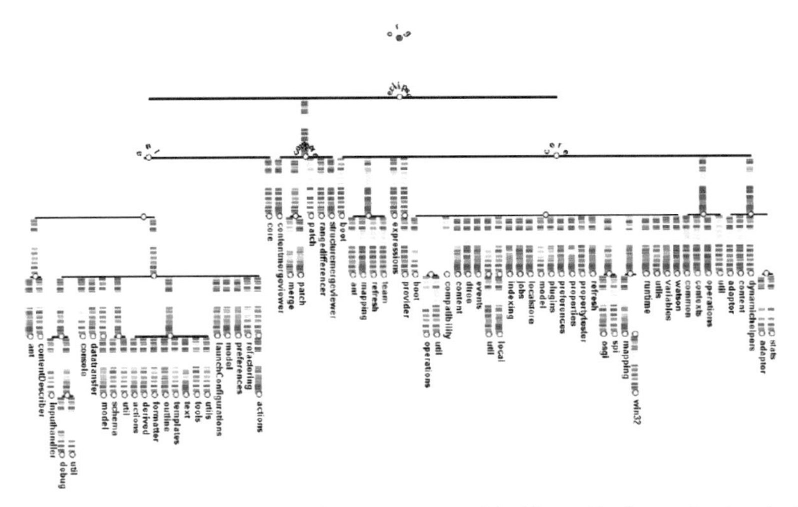

Figure 2.22 A part of the Eclipse software system and its hierarchical organization depicted as a node-link diagram with aligned orthogonal links to visually represent a list of quantitative values for certain derived attributes [62].

the visual variables for the final visual design, i.e. that decides about the most effective data-to-visualization mapping. However, in some scenarios this order is not that clear right from the beginning, offering the users the choice to reconfigure the visualization in a way that adapts the roles and orders of the variables in use.

A popular example is the field of software engineering that produces a wealth of data based on the source code, the involved developers, comments, bug reports, check-in information, call dependencies resulting in a call graph, code–developer relations, the hierarchical organization of the software system, and even more, for example the time-varying aspect bringing the field of software evolution into play as well as software visualization [149]. In Figure 2.22 we see an example from a software system for which the hierarchical organization stands in focus, i.e. plays the primary role. The secondary role is given to a list of quantitative attributes which generate some kind of multivariate dataset, with the files being the observations. For the visual depiction we chose a node-link diagram in a top-to-bottom layout indicating the parent–child relationships while the attributes are reflected in the color coding represented in an aligned fashion to allow comparison tasks on different levels of hierarchical granularity, looking similar to bar codes [62].

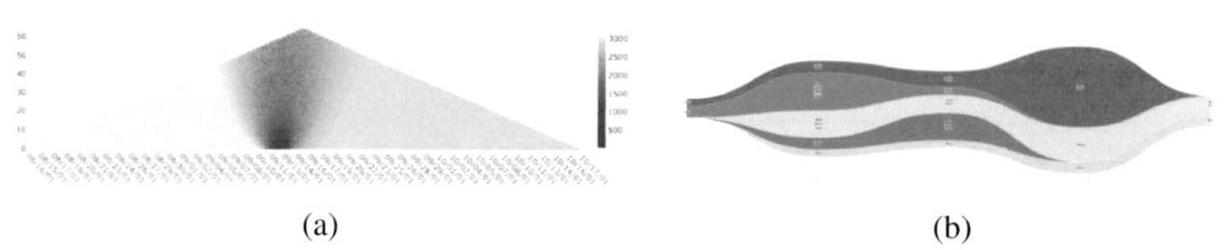

(a) (b)

Figure 2.23 (a) A time-varying graph dataset consisting of flight connections in the US from the year 2001 shown as a heat triangle [242]. (b) A Themeriver [218] representation for showing the evolving number of developers during software development [89].

2.3.4 Dynamic Data

With temporal evolution we can add an extra data dimension to any kind of dataset. This means that a certain variable or several of those exist in a time-varying behavior, either measured discretely, i.e. time step by time step, or continuously, i.e. for infinitely small instances of time there exists a data value, typically modeled by a mathematical function and not measured as in the discrete case. Moreover, not only time can be involved to make a dataset consisting of several subsequent instances but also just a time-independent sequence, or even a set of instances without an explicit order among the data instances. No matter which kind of scenario we are confronted with, the visual metaphor to be chosen is typically a different one than for the static counterpart of the same data type scenario.

With dynamic data we normally have to take into account an additional task which is related to the comparisons of data values over time. This makes the chosen visual metaphor a different one from the one chosen for static diagrams, but one might say we can just copy the metaphor for the static visualization and put one side-by-side for each instance of the dynamic case. This is a suitable and oftentimes applied solution but we have to keep in mind that we only have a limited display space that causes visual scalability problems [4]. This time-to-space mapping can also be replaced by a time-to-time mapping in which the temporal or sequential information is mapped to physical time, typically illustrated in an animation. Although this is a valuable and well-researched concept it comes with its drawbacks [505] which are cognitive issues when comparing subsequent values to make a claim or hypothesis about a time-varying behavior in a dataset like a trend, a countertrend, oscillating or alternating behavior, or even combinations thereof, as well as outliers and anomalies [101].

A mixture between time-to-space and time-to-time mappings was introduced by Spence et al. [475] who investigated a concept denoted by rapid

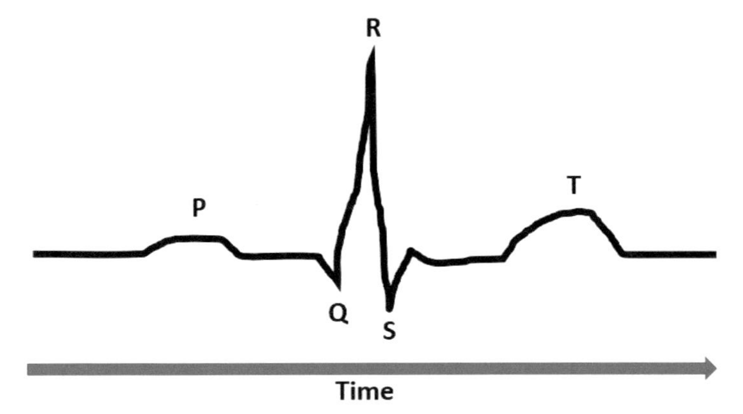

Figure 2.24 An electrocardiogram consists of several time-dependent quantities that are shown as a line-based diagram, annotated with P, Q, R, S, T waves. This kind of diagram has also been investigated by an eye tracking study [144].

serial visual presentation (RSVP) often used for text reading tasks [113] or inspecting large image collections [477] in a rapid way. The concept supports tasks like browsing a large static dataset or a time-varying dataset consisting of several instances of data values visualized in some way. Moreover, weighted browsing tasks describe the way in which we search for a certain element for what we have an approximate visual pattern in mind. No matter which concept we chose for displaying dynamic data, it is definitely more challenging to effectively visualize, and further, an eye tracking study has to take an additional parameter into account which comes in the form of the sequential behavior [309]. A video might even fall into this category since it consists of a sequence of static image frames that are watched one after the other in a rapid serial visual presentation; however, eye tracking experiments have already been conducted for this kind of scenario [310].

Examples for dynamic data visualizations exist a lot due to the prevalence of the temporal aspect in nearly any kind of application domain that stores, measures, or simulates data at different instances or time steps. Figure 2.23(a) illustrates an example from dynamic graph visualization showing the flight behavior in the US in the year 2001 depicted as a heat triangle [242]. One can clearly detect a visual anomaly pattern which was caused by the terror attacks on 9/11. Another popular example (b) stems from the field of topic exploration which results in a diagram denoted by Themeriver [218] showing a list of time-dependent quantities (here a developer river [89]), but negatively also bringing issues for value comparison tasks into play [124]. Moreover, a simpler diagram showing quantities as a time-series plot is known as the electrocardiogram (ECG) [144] with which every medical doctor is familiar (see Figure 2.24). All of the presented diagrams use the time-to-space mapping concept, i.e. displaying the individual time steps next to each other to support comparison tasks.

2.3.5 Metadata

Every dataset requires some kind of additional descriptions that explain and give more details about the data, extra information that is needed to understand the context of the data. The data about data is important for a visualization to pick the right visual metaphor, for example the correct scale given by the units of the measurements. For the user the metadata is crucial information to interpret the visual patterns in the right context and scale [317]. This data defines how and when data is measured, where it is stored, who is the owner, in which environment it was measured, which device was used, and so on. For example, when taking a picture with a camera, the picture is the data of interest while the date and time when it was taken, the label of the picture, the resolution, or further extra information, describe the metadata about it. Moreover, in the field of eye tracking, metadata might describe the quality of the recordings, giving a hint about how trustworthy and reliable the data is. This quality can even be measured for every data entity, for example at different time instances or for every individual participant in an eye tracking study [444].

For a visualization it is crucial to add the metadata to give the users the opportunity to interpret what is actually depicted. Without the metadata it might get misinterpreted or the metadata information must be derived from another context, making the interpretation process quite time-consuming. Figure 2.25 illustrates an example of a scatterplot with additional information about the attributes shown at the axes as well as the color coding of the points. Without them the visual patterns might be observable but they cannot be put into the context of an already known scenario with which it should be compared.

The intention of Section 2.3 is not to introduce and discuss all data types; that would generate an endless list of examples. It is more given in a way to provide an overview about the most common data types and possible visual depictions of them as well as some additional remarks about drawbacks and benefits of one technique compared to another one. Later, these techniques

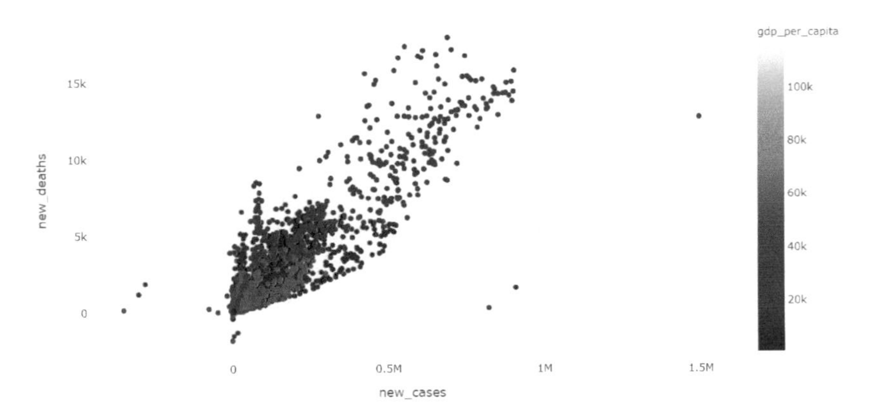

Figure 2.25 A scatterplot with extra descriptions for the axes and the color coding, serving as metadata.

will be examined for user evaluation, in particular if eye tracking was used to investigate their usefulness, readability, and interpretability by human users based on visual attention paid over space and time, hence explaining the value of eye tracking for these techniques.

2.4 Interaction Techniques

To not leave users with a static representation, as in the years before the invention of the computer, interaction techniques typically add life and flexibility to a diagram, plot, chart, or any kind of visual output that needs to be modified [152]. These varying perspectives on a dataset may shed some light on aspects that might not have been detected if just the static version of the visualization was presented [476]. Interaction is particularly important if views on real-time data under different parameter settings are required. This allows fast and adaptive user requests within the functionality of a visualization tool, making the whole process some kind of bidirectional dialogue in which the user inputs some information and the visualization tool outputs (hopefully) the desired information. Powerful input and output devices are required, and human users with their perceptual abilities to rapidly detect patterns [219, 521, 522] and who react on those, demanding some cognitive skills and technology experience. Interaction in visualization is hence a real-time action with an expected quick response while the visualization itself is not.

Interaction has its origins in the field of human–computer interaction (HCI) [107, 109] and is somehow regarded as the little brother of visualization, which is due to the fact that more information flows from the visualization to the user and not vice versa [476, 496]. Moreover, in many scenarios, interaction is so fundamental that it is no longer recognized but is rather a standard visualization ingredient. Interaction is a goal-oriented activity, required to modify a diagram until it meets the needs of the human users to solve a given task or at least give some hints about its solution. A visualization system reacts with a response, change, textual, or visual output, in a reasonable and acceptable time, and some studies' results have revealed a time limitation of 20 ms [337] to achieve an interactive feeling. Bertin [38] describes interaction as a diagram being constructed and reconstructed, leaving the interaction choice to the decision makers themselves. A good diagram must be flexible in a way that interactive modification allows refinements of hypotheses and insights. This is also supported in many of the designed visualization pipelines [359] which, in the best case, allow interactive feedback in any of the intermediate stages.

In this section an overview of interaction categories, physical input and output devices, as well as human users-in-the-loop is given, without explicitly stating that a complete list of all these aspects is provided. From the perspective of eye tracking it makes a difference if gaze-assisted interaction is evaluated in which the visual attention over space and time is recorded at the same time as someone is interacting with a visualization tool or if another kind of interaction without gaze support is applied.

2.4.1 Interaction Categories

Nowadays, a visualization tool is equipped with various options to interact. The intention is in most cases to create a new perspective on the represented data by refining parameters in a way that the resulting new view on the data supports the identification of what one is looking for. This typically happens in several, sometimes repeating, stages while applying interactions from a given certain tool-supported repertoire (see Figure 2.26 for a visualization pipeline [108] that incorporates human interaction). Apart from just the individual interactions, undo/redo options would be desirable, in case an interaction was not well chosen and caused the opposite effect to the one intended or an individual interaction technique has to be repeated several times, which would be a daunting task if it had to be applied one after the other by always first setting all the parameters again. To reliably step back to

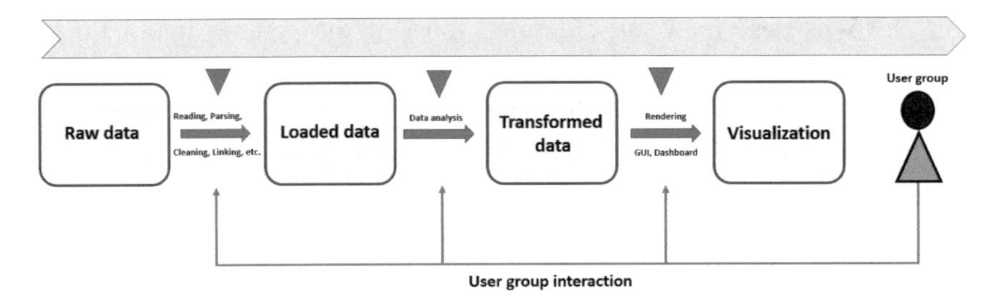

Figure 2.26 A visualization pipeline illustrates how raw data is transformed in a stepwise manner into a graphical output while the users can adapt and modify the steps and states [108].

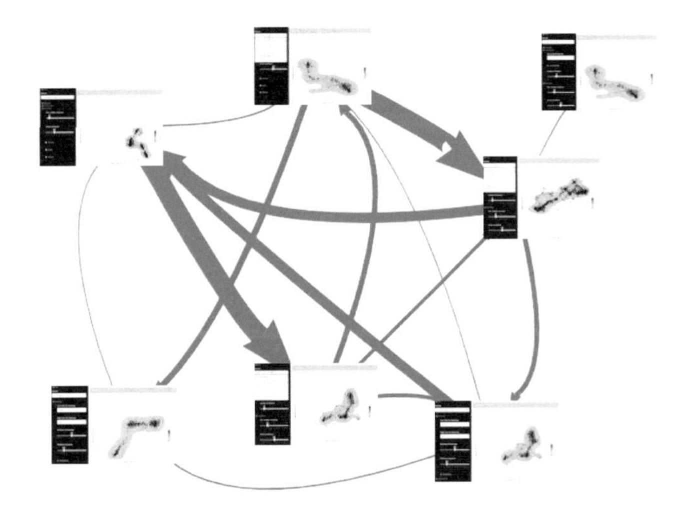

Figure 2.27 An interaction history can be modeled as a network of states; in the case of a visualization tool it consists of snapshots, each illustrating a certain parameter setting [68]. Node-link diagrams can depict the weighted state transitions and are interactive themselves. Here we see a network with additional thumbnails indicating the current view in the visualization tool. It may be noted that in this scenario some of the interaction steps cannot be undone, indicated by the directions of the links while some others are undirected. The thickness of the links might encode a transition probability for example.

previous states of a visualization system, to keep track, and to get an overview of the exploration stages, some kind of interactive interaction history [68] would be beneficial that supports the mental map of the intermediate modifications (see Figure 2.27).

Yi et al. [544] described and classified most of the general interaction concepts supported in visualization into seven categories. They argued that the most prominent techniques fall into categories like select, explore, reconfigure, encode, abstract/elaborate, filter, and connect.

- **Select.** Certain visual elements are selected, marked, and highlighted to keep track of them in case a view is modified, i.e. we either see a new arrangement of the visual elements or a completely new visual encoding of the same visual elements. A selection also has the benefit that the elements of interest can be treated separately from the non-selected ones, for example, applying further interactions to them while the rest remains unchanged.
- **Explore.** If only a small number of elements can be presented visually but the users wish to see the invisible part they start to explore. This means that they move to a different part of the dataset under investigation by, for example, panning to a different location or area. This introduces the concept of smooth changes, i.e. the view is not changed abruptly, but to preserve the viewers' mental map this is done smoothly. Moreover, the smooth transitions support the keeping track of what is present on the way from the old to the new position.
- **Reconfigure.** Rearranging the placements or layouts of the visual elements are important principles to obtain new perspectives on the data that might show insights that would have been hidden in the case in which just one static representation was given. For quantitative values this could be achieved by a sorting function while for graph visualization a switch between certain layouts might fulfill the task. For rearrangements it is crucial to keep in mind that the users' mental map should be preserved.
- **Encode.** In some situations it is a powerful idea to even change the visual metaphor or visual encoding by applying a different set of visual variables. This could include just the change of the values of the visual variables like using a different kind of color scale, but it could also switch from one visual variable to a completely new one, like switching between size and color for representing quantities. In the best case a multiple coordinated view is supported allowing the users to compare the visual encodings side-by-side.
- **Abstract/elaborate.** Starting with a visual overview about a dataset is a good concept, however, the users wish to dig deeper in a dataset by changing the level of granularity or abstraction, until they reach

a detail level in which the tasks at hand can be finally solved or at least a hint for a possible solution is given. The overview-and-detail is a well-known problem in visualization due to the fact that in many situations the details alone are not helpful, but the overview is required as contextual information. This challenge is denoted by the focus-and-context problem.

- **Filter.** Certain conditions or properties, typically guided by the user tasks, can be applied to a dataset, having the impact that only a part of the dataset is visually represented. All of the data elements not falling into the filtered part either disappear completely or are shown as grayed-out contextual information. Typically, the visual metaphor or the arrangement is not changed after applying a filter interaction; in some cases the won free display space is filled with the remaining visual elements by smoothly enlarging them, fitting the whole area.
- **Connect.** If there exist relations, dependencies, or associations between certain data elements, those should be indicated visually in the best case, supporting a user to quickly recognize those. This scenario could occur for a single view in which elements could be linked visually like in a node-link graph visualization. But it is useful even more in a situation of a multiple coordinated view in which the users select a visual element in one view and the same visual element is highlighted in all other views, in case it is currently visible. This challenge typically occurs if brushing and linking is supported [256].

Although these are very general descriptions of such interactions, the "how" they are applied also depends on aspects like the input and output devices as well as the persons who are starting an interaction; for example, what their experience level is or if they suffer from visual deficiencies or color blindness, making them, for example, use a special color coding adapted for their needs or even some kind of braille technology [288] to interact, as well as audio feedback to get additional support and hints with which component they have interacted.

Apart from the seven categories discussed by Yi et al. [544] there are several more that could be counted as interaction techniques. In many cases the users wish to apply undo/redo options while seeing their flow of states in an interaction history as already described above. For visualization tools this would be a helpful visual extra but for visual analytics systems that allow many more complex exploration processes this would definitely support in many ways. Such an interaction history can generate another dataset, a huge graph consisting of states and transitions, meaning also for this kind of dataset we need an advanced interactive visualization technique, otherwise it would not be useful at all.

A further interaction technique can be the editing of a textual description, like changing a label or adding an annotation, although annotations could be classified as connect interactions since they link a textual description to a visual element. Such a modification, update, or even correction of text could even be applied directly to visual elements, for example by exchanging a visual variable or removing a visual pattern that is superfluous which would again fall into the category of encode. Moreover, more visualization-independent interactions might be imaginable like adding extra data sources or requesting a data update if the visualization tool is not using the latest version. However, such an update could demand a longer preprocessing time to bring the new or missing data into the same format as the already loaded one. Even more visualization-independent interactions might be opening and closing menus, menu items, or whole views containing a visualization. Also interactions related to the graphical user interface in which all the visualizations and their functionality live can build their own category; for example, maximizing or rearranging views, adding more views, highlighting or decorating a view with additional information, and many more.

The way in which we interact in a visualization also depends on further factors, for example, based on the visual variables directly. A 2D interaction is typically different from a 3D interaction in which occlusion and perspective distortion effects have to be taken into account. This could, in particular, be interesting and challenging for VR/AR environments; for example, in immersive analytics, a field that is more related to visual analytics since it supports complex exploration processes including many technological concepts. Furthermore, interacting with a static visualization requires different interaction techniques than when interacting with animated, dynamic visualizations. Those need to be stopped, replayed, slowed down, or sped up on user demand.

2.4.2 Physical Devices

The physical devices required for successful interactions could be split into input and output devices. This is in particular important if we inspect them under the light of eye tracking technologies. An input device is one that is controlled by the user while an output device is required by the visualization tool to reflect the feedback, i.e. the visual output based on the user interaction.

In these modern days there exist various input devices while the number of output devices is not that large in comparison. Evaluating the impact of input devices on the usability of a visualization tool by eye tracking is similarly challenging as evaluating the impact of the output devices. However, for the output devices we obtain the primary stimulus, i.e. most of our visual attention is paid on the output (the visual representation) rather than on the input, hence most of the recorded eye movement data contains output device information. Typically, the information from the input device has to be collected by additional sensors, apart from gaze-assisted interaction in which the input data comes from the eye.

Most of the input devices focus on hand control, for example mouse, keyboard, touch, pen, or joystick, to mention a few. For touch devices, the input and output typically happens on the same display, for example the computer monitor or a mobile phone to provide a faster and more natural feeling during interacting. Touch supports direct connection of the user's finger with a graphical element while a mouse does not and requires some synchronous action with the eye and the hand to efficiently work. Hence, mouse interaction needs some training compared to touch interaction. Touch is, these days, integrated in most of the public user interfaces, for example, in ticket machines, information monitors, interactive public transport maps [96], and so on. However, for a visualization technique it might lead to occlusion problems due to the size of a human finger compared to a mouse cursor, for example. This could, in particular, be a problem for eye movement studies because the visual attention behind the finger is hard to track reliably.

Apart from hand-driven interactions, further input modalities have been researched and developed including gestures, voice, or gaze [381]; even a combination thereof can be efficient if it is designed in a user-friendly way. For example, for gaze interaction, which is useful for disabled people who cannot just use standard interactions; the popular Midas touch problem [260, 516] typically makes a pure gaze-assisted interaction challenging, hence some approaches combine gaze and voice interaction (see Figure 2.28). For game playing such a setup was evaluated, comparing gaze and voice interaction with mouse and keyboard [378]. For gaze-assisted interaction, an eye movement data evaluation can be started right away because the eye movement data is already recorded which is typically not the case if just a standard non-gaze-based user interaction is supported. For even more complicated settings such as 3D interactions [191], for example on mobile devices [104] as well as walking in 3D immersive virtual reality environments [145] or tangible visualization [287], standard interaction

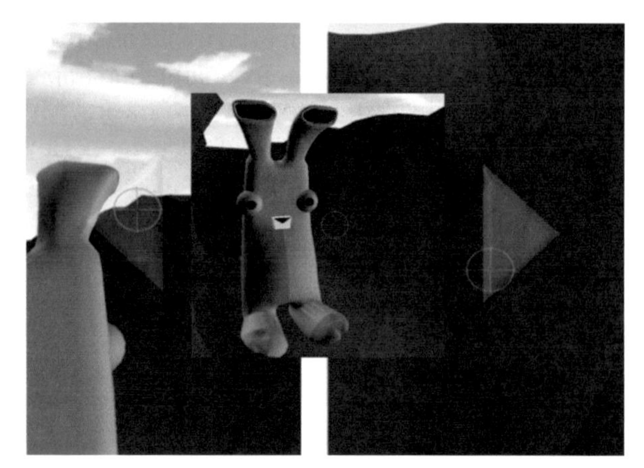

Figure 2.28 Gaze-controlled buttons are used in a game environment to interact with a game character. After a user evaluation this was replaced by a simpler scenario due to unintended rotation issues [378]. *Image provided by Veronica Sundstedt and Jonathan O'Donovan.*

techniques have their limitations and have to be exchanged with more advanced techniques; however, their evaluation, in particular with eye tracking, is difficult. More natural human-based interactions like body and head movements, gestures, or speech and verbal conversations have to be taken into account. We see that, depending on the provided visual stimulus, 2D or 3D, static or dynamic/animated, and also the way in which interaction techniques are incorporated as well as their complexity must be adapted.

Each input device for applying interaction techniques requires certain output devices to illustrate the modifications of the views or visualizations to the users. For visualization tools the standard output device is a computer monitor, but even smartphones [56] or powerwall high resolution displays [531] are more and more frequently used. Such small-, medium-, and large-scale interaction settings allow for a variety of alternatives, even in combination, focusing on the goal to providing views and perspectives on the data while supporting changes, modifications, and user requests. Visually impaired people might use braille technology [288] as the output device, while stereoscopic displays [352], VR/AR environments [524], as well as a hybrid output [294], even on several output devices, are imaginable. The type of output device in use makes a difference for eye tracking studies since the visual stimulus appears in a different environment. This has an impact on the eye tracking device, the user groups, as well as the eye tracking data analytics technique applied.

2.4.3 Users-in-the-Loop

Human users are responsible for initiating interaction techniques. They awake a visualization to life, guided by tasks at hand that require some kind of solution, in most cases raising new questions falling in a certain task category. The human users cannot only intervene in the final visual output but in typical visualization tools, they have the option to interact at any stage illustrated by the visualization pipeline (see Figure 2.26), making them users-in-the-loop. This pipeline can even be run through in an iterative manner, adjusting and adapting parameters, views, and perspectives at any stage, until the final desired configuration is hopefully reached, generating an "aha" effect, i.e. either providing a solution to a task or at least giving a hint where to search further. Interaction can be subdivided into subinteraction sequences in which all interactions should occur quickly, typically in a fraction of a second from starting the interaction to the final output, i.e. until the interaction has ended, before a new one from the sequence starts. The human user decides the order of subinteractions and how this sequence is created, changing the parameter setting in a visualization tool step-by-step. Some interactions from the sequence might be executed several times in a row, some might be reversed or be undone. There is some debate about an acceptable time for an interaction technique to be considered interactive. Some results indicate that 20 ms [337] might be an acceptable response time, but that surely depends on the users' knowledge and experience levels as well as the application. However, no matter how long it takes for a visualization tool to react on user input, it is important that a user feels comfortable, engaged, and entertained, aspects that ease the burden of using a visualization tool with the goal of deriving knowledge and insights for the tasks at hand. Such properties are hard to measure with an eye tracking device but additional data sources might give a hint about them as well as cognitive and psychological issues [305] worth integrating into a data analytics process to identify design flaws in a visualization tool.

There is even a difference if an individual user interacts with a visualization or if the tool supports collaborative interaction [60], i.e. letting several users work with the tool, either at the same place in front of the same output device or remotely if they are physically separated, even one after the other due to different time zones or maybe in temporally overlapping processes. However, such a scenario typically requires a web-based visualization tool [82] that offers easy possibilities to store and organize the results found by many users, all, just a few, or even an individual one.

Eye tracking can be a powerful concept for collaborative interaction to figure out how people communicate information and how they solve a common task in a cooperative way. The eye movement data could be recorded at physically different places, matched by the same or at least similar stimuli, i.e. tool settings. This eye movement data could flow into the data analysis since it provides hints about visual attention and visual scanning strategies while finding a solution to a certain task. The eye movements from successful viewers could be shared with other remote users to get some guidance in the exploration process or to recap the found solution as some kind of learning goal. The most important aspect for collaborative interaction is the fact that the found insights can be merged in some way to faster find insights or even find more insights that one individual user would never find. To reach these goals the found insights must be shared and communicated among users and a consensus must be reached to coordinate the further investigations, i.e. split the general task among the users, maybe as subtasks.

Visualization tool users could have various properties. They could belong to a group of visualization experts or non-experts, they could be application domain experts or not, or they could stem from different age groups and have different genders with varying judgments of visual or perceptual issues. Moreover, perceptual abilities, color deficiencies, or visual acuity problems could have an impact on the design of a visualization tool and its effectiveness for the users. Some of them might be handicapped, have senso-motoric issues, be blind or visually impaired to a certain degree, and so on. All of these challenges have to be taken into account when integrating interaction techniques into a visualization tool. Gaze-assisted interaction is a way to support handicapped users in case they suffer from certain disabilities that other users might not have.

2.5 Design Principles

Creating a readable, intuitive, and interpretable diagram is a challenging task, but following certain design principles can be of great support to generate a user-friendly, powerful, and valuable visualization [513]. Those principles describe facts that have been studied and evaluated in the past to more and more enhance a visualization by taking into account human observers and their perceptual abilities [521, 522]. If a visualization technique or tool is created we have to take into account the visual variables and their combination, the supported interaction techniques, the input and output devices, as well as the human users with their tasks in mind, properties, background knowledge, and experience levels. Designing a suitable visualization tool that is able to support all tasks is an impossible endeavor. On the other hand, creating a poorly designed visualization can lead to a time-consuming visual and cognitive process to understand the key ideas to derive meaning and furthermore, this drawback might lead to misinterpretations when exploring the data depicted in the graphics.

However, the design principles discussed in this section describe some general powerful concepts and even no-goes that should be followed as criteria before creating such a tool. It may be noted that the creation process of a visualization tool is some kind of never ending story with the designers and the end users in the loop, adapting and modifying the current version by steady feedback, but reaching a stable state with which everybody is confident in the end is a challenging and time-consuming problem, in particular, if a proper user evaluation, maybe even with eye tracking, is considered as a way to enhance the implemented visualization tool. Dix described this process in a quote stating: "you may not be able to design for the unexpected, but you can design to allow the unexpected" [154], also meaning that the tool might be equipped with various functionalities that cover nearly all aspects, but the human users decide which ones to use. Hence, the decisions are somehow left to the users, not to the designers and developers. However, such a strategy demands for a lot of implementation work to cover all possible directions to support nearly all user tasks.

2.5.1 Visual Enhancements and Decorations

Visually depicting data in the plain vanilla form might already show some data patterns, but to make a diagram readable and interpretable we have to take into account that additional visual enhancements or visual decorations have to be attached in a suitable way to get the ultimate depiction for an accurate and reliable visual analysis of the underlying dataset.

Such extra, but definitely required, decorations are, for example, legends that either visually or textually describe the visual variables or any extra signage used in the technique, for example, iconic representations, typically used for visually encoding categorical data. Legends are put next to a diagram where the user can keep an eye on them whenever needed. Some situations are self-explanatory, meaning legends are not needed for one user, but depending on the users' background knowledge and experience levels they might be an absolutely necessary information to read a diagram. Color legends, for example, depict the range of the color mapped to quantitative values indicating minimum and maximum or they show the individual colors used for a certain category like, for example, public transport lines.

Axes scales are important, in particular, if an axis is composed of several scales; for example, if there is a large range between the minimum and maximum values. In addition, axes descriptions and corresponding units are required as well as guiding lines that perceptually help to quickly read the represented value. In a scenario in which the vertical axis is used to indicate quantities, but those differ in size a lot, a logarithmic or a scale-stack is of special interest [230], depending on the user tasks. As a negative issue, these types of stacked scales have to be learned; however, the first attempt to make them interpretable is by attaching an intuitive scale description.

It is also important that textual information is added in enough detail, for example labels that give extra hints about scale values. Adding too many is as bad as adding too little; a good balance is required and texts should be readable in an acceptable font size and font style. Moreover, a left-to-right or a slightly varying reading direction should be used for users from Western civilized countries, for others the reading direction has to be adaptable in the design. The users' visual acuity plays a crucial role for text reading tasks [470], while text reading tasks have been evaluated by eye tracking a lot in the past [41]. Any additional information is useful, but the diagram should not be too crowded to avoid an information overflow. However, if extra textual or visual information is presented, overlaps and occlusion should be avoided as well as distortions. The choice of the right color scale depends on the data to be displayed as well as effects related to the user, for example color blindness [348, 521] or color deficiencies [380].

There could be two extreme situations for data visualization that have to be treated with care, and this depends on the user tasks. The first issue comes from the fact that data elements might be missing or erroneous right from the beginning. Those data elements might be ignored in the visualization or they might be indicated very clearly to alert the user to those issues. If they have to be shown in some way, they should be color coded in a gray scale, indicating that there are missing or wrong values, but they are depicted in a different way to contextual information. The same holds if those values are aggregated with other surrounding values or interpolated, and all of these hints attempt to avoid misinterpretation of the data. A second issue comes from the fact that some values are considered outliers or anomalies. These should play a special role in the visualization and should be highlighted in a way that makes them pre-attentively detectable [219, 500]. This means they pop out from the display, with just one glance at the monitor, without paying a lot of attention to them.

No matter which kind of visual enhancements are chosen, the readability, intuitiveness, and understandability play major roles when designing powerful visualizations, and aesthetics [24] plays a nearly equally important role, but should still be considered a minor second option to further improve a diagram after it fulfills the task solution and explorative functionalities that a visualization tool should have.

2.5.2 Visual Structuring and Organization

Apart from visual enhancements and decorations the design should also take into account the structure and order of visual variables contained in a visualization, but also the views and components provided by a graphical user interface. In particular, if a multiple coordinated view [424] is used to show several perspectives on a dataset, those views should have a suitable order that considers the user tasks. The most important view should be centered and in the best case should be mapped to the largest display region. The views should be interactively exchangeable (see Figure 2.29 for the GUI of the VizWick tool) and the users should be able to select a certain number of views from a given repertoire as for example shown in the VizWick tool for hierarchy visualization [99]. Whenever possible, a hierarchical layout of the

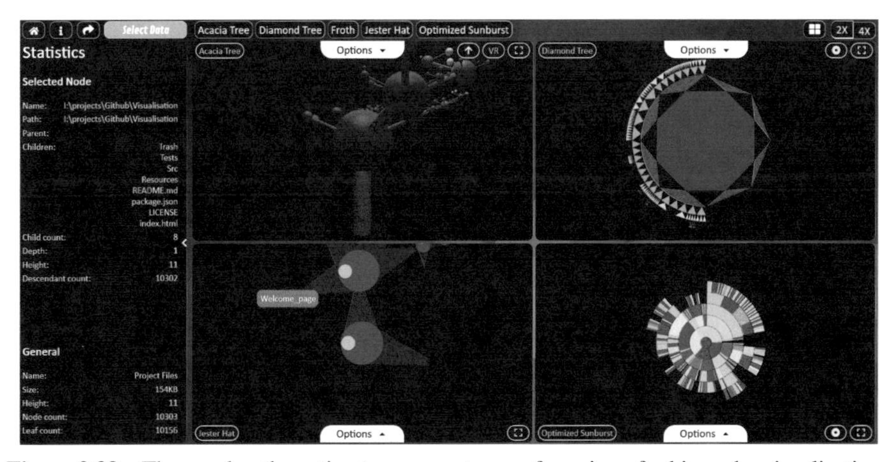

Figure 2.29 The user has the option to see one, two, or four views for hierarchy visualization techniques. Moreover, the views are exchangeable and support their own parameter settings while they are interactively linked [99].

views is adequate, starting with the most important view as the root node and following a hierarchical exploration task order. In any case the views should be clearly distinguishable, separable, and a good layering should be used.

The visual information seeking mantra [459] is a good concept to follow in a visual design. For a data visualization it is important to provide an overview as a starting point for further exploration processes. After a user sees the whole dataset or, due to scalability and display limitation issues, at least a large portion of it [503], further interactions should be supported like zooming and filtering and finally, details-on-demand. Another option for reducing the amount of data to be displayed is by using small graphical elements like pixel-based representations whenever that is possible. For multivariate data, it could be projected first to a lower dimension and then visualized, supported by various projection algorithms [176]. In such a projection process it must be guaranteed that formerly similar data points in high-dimensional space are still similar in the low-dimensional one to preserve the patterns and hence, the interpretation of the data in a reliable way.

No matter how the data is visually depicted, the visualization should at least try to tell a story [324]. A diagram should be as self-explanatory as possible [362]; if a lot of textual descriptions are given a good way to bring them to life by visuals is by using sparklines [429], which are tiny graphics that are embedded in a text, being so small that even visual enhancements or decorations are left out. In these modern days, a special focus could be on a web-based visualization, making it accessible by using a smartphone, reaching thousands of users; however, the display space is really small and hence the visualization must be organized in a way that it makes the most of the small display area. The opposite effect occurs if we need high resolution, for example showing tiny visual features that are crucial in the visual depiction to explore the underlying dataset, but for such a scenario the number of users is typically small, i.e. it is created for domain experts. In any case the design should be balanced, using symmetrical layouts, to make it aesthetically appealing, maybe by making use of radial forms instead of Cartesian ones [24].

2.5.3 General Design Flaws

Although a picture can say a thousand words, meaning a static representation can already show a lot of visual patterns, interactions should be taken into account to obtain a well-designed visualization [476]. Otherwise the

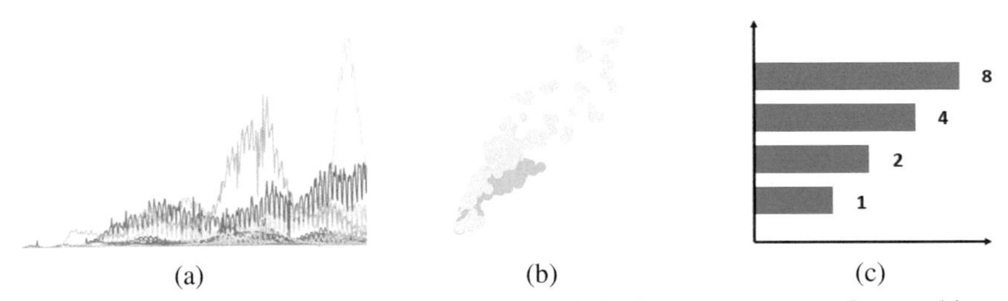

Figure 2.30 When designing a diagram we should take into account several general issues that can lead to problems when interpreting the diagram: (a) visual clutter; (b) chart junk; (c) lie factor.

usability would be limited in a way that the users cannot really adapt the presented diagram to their needs by, for example, changing parameters. No matter which kind of visualization is created, the experience levels and further properties of the users have to be taken into account to not design the visualization for the wrong user group, making it useless. Moreover, interaction is one way to avoid boring visualizations, but even further, visual variables and in particular color [440] should be chosen in a way to create enjoyable diagrams, a fact that is also related to aesthetics.

For a static visualization there are already issues that can decrease the value enormously. Those come in the form of visual clutter which is "a state in which excess items or their disorganization leads to a degradation of performance at some task" [426]. This could, for example, happen if line-based diagrams are used and if there are too many crossing lines like in a time-series visualization based on line plots (see Figure 2.30(a)). Node-link diagrams for graph data mostly suffer from this issue. For this reason, graph drawings have to consider aesthetic rules like reducing the number of link crossings, avoiding node–link and node–node overlaps, or reducing link lengths, to just mention a few from a longer list [406, 409]. A data variable should in the best case only be mapped to one visual variable, otherwise the additional decoration of a diagram will no longer be a visual enhancement, but could generate the opposite effect, leading to misinterpretations. Less is oftentimes better in visualization, leading to some kind of simplistic or minimalistic diagrams [91]. This is related to the data-to-ink ratio [254, 503] although this just describes that as little ink as possible should be used for drawing a diagram, which also includes that second, third, or more visual variables should not be used for encoding the same data variable, which is denoted by the term chart junk [26, 503] (see Figure 2.30(b)). Finally, the lie factor [503] describes the ratio of the extent of the effect shown in a visualization and the extent of the effect existing in the data. This proportional issue typically occurs for displaying comparable values. The chosen visual variable for the numeric variable should treat each value in the same way, i.e. a value that is twice as large in the data should also be shown with a twice as large effect in the visual variable (see Figure 2.30(c) for a strong lie factor).

For dynamic data [4] it is important that two representation choices for the time dimension can be considered, which are denoted by either time-to-time mapping or time-to-space mapping [75]. The time-to-time mapping describes the concept of displaying each time instance in the data to physical time, for example in an animation, typically called smooth animation in visualization. Time-to-space mappings try to encode as many time instances as possible to the display space. While time-to-time mappings generate cognitive and change blindness [376] problems when comparing time steps due to limitations of the short term memory [410], time-to-space mappings typically lead to visual scalability issues due to display space limitations [4]. Another negative issue is that for comparison tasks the individual visual elements have to be mapped to each other in each time instance to reliably do the comparison. Moreover, both mappings have to rely on dynamic stability [151] to allow the preservation of the users' mental maps [18] which is even more important in animated diagrams. In non-animated diagrams we do not speak of dynamic stability but more of a temporal alignment, meaning that the time axis should be the same for all varying variable values which leads to better performances for comparison tasks. Both concepts have their benefits and drawbacks [505] while rapid serial visual presentation (RSVP) [475] is some kind of hybrid concept that includes ideas from both concepts.

2.5.4 Gestalt Laws

There are some general rules that hold when watching a scene, picture, or a visualization that we apply without really paying attention to them. Those rules have been denoted as Gestalt laws [292] and they describe how we build visual objects from perceiving surrounding smaller objects or pieces thereof as well as their environment. In Gestalt psychology we often hear the quote "the whole is greater than the sum of its parts" [485] meaning that the humans' visual systems more or less automatically create visual patterns without explicitly paying attention to them, hence reducing the cognitive burden for our brains when identifying patterns. The illustrated principles and laws in Gestalt theory are important for visualization design as well since they explain which patterns are easily perceivable and which ones could cause confusions and misinterpretations. This is in particular useful if we take into account several visual variables of which a diagram is composed. This does not only hold for static diagrams but also for dynamic, animated ones that model change by movement, typically causing issues when keeping track of visual elements, either individual ones or whole groups of them, merging with each other and splitting off again after some time. The most important principles are summarized by emergence, reification, multistability, invariance, and grouping.

The principle of emergence is probably one of the most relevant ones in visualization since it states that visual patterns might emerge from a visual depiction, for example, separating them from noise or from chaotic patterns that do not carry any meaning (see Figure 2.31(a)). Without this principle the data-to-visualization mapping is useless because we might not be able to detect visual patterns that can be remapped to data patterns, hence the visual exploration process would be impossible in this case. Multistability is important to let the users see several visual patterns from the same arrangement of visual elements, meaning several perspectives on the visual depiction of data are possible (see Figure 2.31(b)). The principle of invariance allows the detection of deformed visual patterns, for example if they are rotated, stretched, or scaled in any direction (see Figure 2.31(c)). The perceptual abilities are good enough to recognize similarities which is one of the strengths of visualizations since a pure algorithmic approach cannot be used due to the fact that we do not know how to describe the similarity between two or more visual patterns. Reification is a principle in visualization that describes how patterns can be completed virtually although they are not completely visible on screen (see Figure 2.31(d)). For example, a link in a graph visualization might be dashed or incomplete, or even overplotted by a larger area.

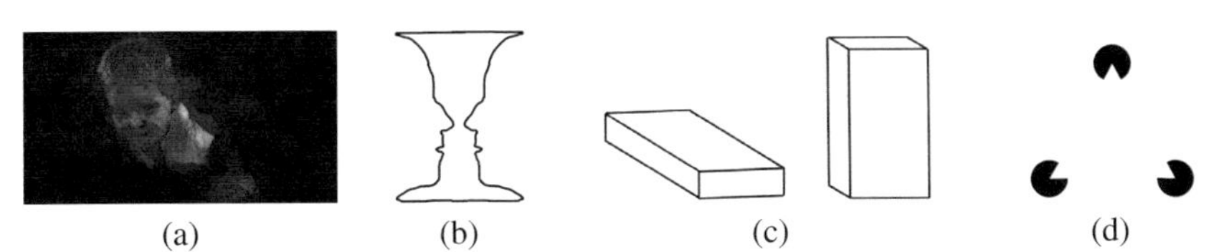

Figure 2.31 (a) Emergence: a visual pattern (my son Colin) can pop out from a noisy background pattern. (b) Multistability: a visual pattern can carry several meanings and might be interpreted in several ways. (c) Invariance: a visual pattern can be deformed in various ways but it is still recognizable as a similar pattern as the original one. (d) Reification: a visual pattern can be completed although it is not shown completely on screen.

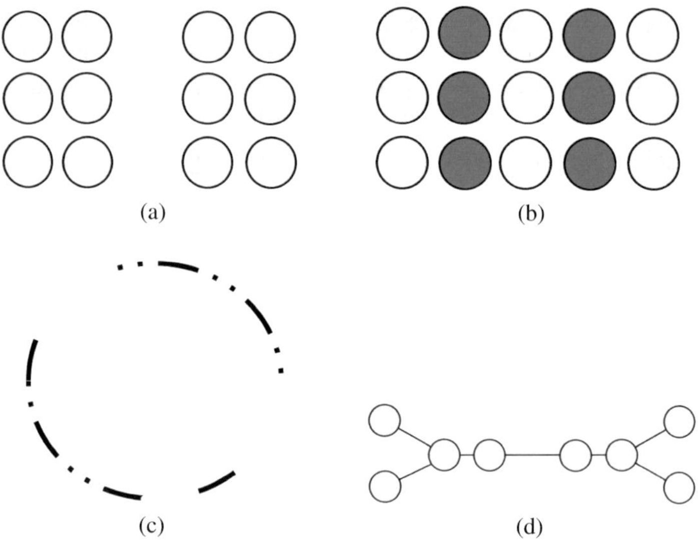

Figure 2.32 There are several ways of grouping visual elements described in the Gestalt principles. (a) Proximity. (b) Similarity. (c) Closure. (d) Symmetry. Further ones are given by the law of common fate, continuity, or good form.

Grouping is typically a visual process that supports the detection of visual elements belonging together like clusters, i.e. visual elements with a small distance between each other, obviously standing in some kind of relation behavior (see Figure 2.32). This is in particular a crucial property if dimensionality reduction methods project a high-dimensional dataset to a lower-dimensional space, typically 2D. Without detecting group patterns, such a projection method would not tap the full data analysis potential. Gestalt psychology describes major groupings as laws of proximity, similarity, closure, symmetry, common fate, continuity, and good form.

Proximity (Figure 2.32(a)) is important for detecting groups of nodes in a graph visualization forming a cluster. The human eye is able to separate several groups by identifying gaps between them, even if the groups have different shapes and partly overlap each other. Similarity (Figure 2.32(b)) uses a certain visual variable, for example color, to make a group of visual elements different from another one, which could be useful for the detection of several areas or regions to make them distinguishable quickly. Closure (Figure 2.32(c)) describes the effect of perceiving objects as a whole even if they are not completely shown. This effect could be important in a visualization if there are overlapping visual objects or dashed and partial links. Symmetry (Figure 2.32(d)) is some kind of aesthetic criterion that should be followed whenever possible to let users identify similar patterns in a visualization, in particular, in graph visualization this law could lead to faster perception and comparisons of node clusters. Further ones, like common fate is useful for animations, for example in a dynamic dataset in which groups of visual elements move around in the display space and have to be tracked over time like in a dynamic graph visualization [368]. This effect is also perceivable in swarm behavior, for example, groups of birds flying around. We perceive each of the groups as a whole not as individuals. Continuity leads to the benefit that crossing lines can be perceived correctly by following easily with a reduced probability to become distracted and misled at the crossing point. Finally, good form supports the identification and separation of several patterns by taking into account already learned and experienced patterns.

2.5.5 Optical Illusions

If the visual depiction uses the correct visual variables but the perceptually wrong effect as intended is created we speak of some kind of optical illusion, also visual illusion, to indicate that it is mostly caused by the visual system. In particular, for visualization such illusions can come in a variety of forms related to color, distortion, depth, size, distance, movement, or cognitive illusions to mention a few from a long list [201], of which all are crucial to design a powerful diagram. Normally, such illusions rarely happen but we should be aware of them in case certain ingredients, properties, or environments are given that might cause problems we do not like. The wrong mixture of visual variables can, hence, cause a representation that generates an opposite effect to that intended. This section neither focuses on giving a complete explanation of all categories of visual illusions nor on providing a complete list of examples in any of the categories. The research field of optical illusions is so huge that we can only literally describe less than the tip of the iceberg. However, to illustrate some of the major issues occurring when mixing up visual variables we come up with a somewhat condensed perspective on this fascinating field of research.

Figure 2.33 shows some general visual illusion examples with which we might be confronted when designing diagrams, static or even dynamic ones. The Ebbinghaus illusion (a) illustrates some kind of perceptual lie factor. In a scenario in which twice the same numeric value has to be represented by a circular area visual variable, the properties of the surrounding visual elements should be known, otherwise the area variable might get interpreted differently

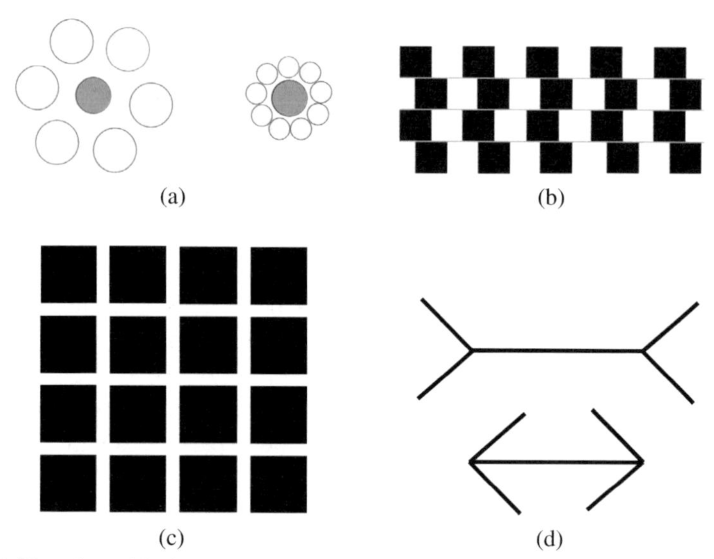

Figure 2.33 Visual illusions can happen in a variety of forms including visual variables and the environments in which they are used: (a) Ebbinghaus illusion related to size effect, caused by the environment and surroundings. (b) Cafe wall illusion related to distance, caused by shifted black square patterns. (c) Herman grid illusion related to cognitive issues, i.e. visual elements are generated where no elements are. (d) Müller-Lyer illusion related to length, caused by extra visuals like arrow heads pointing in opposite directions [274]. Further well-known effects are the spinning dancer illusion related to movement, caused by missing reference points which seem to change the direction of movement, the Ponzo illusion related to depth, caused by the environment and additional effect with denser becoming parallel lines in the background like a railway track [274], or the checker shadow illusion related to color and the surrounding colors [228].

for the same numeric value. This effect could happen in bubble treemaps or node-link diagrams using circular shapes for the nodes with node weights visually encoded by node size. Illustrating parallelism (b) could become a problem for process visualization in which two processes are depicted as running in parallel. Seeing patterns where no patterns exist (c) can lead to misinterpretations of data and a longer response time due to visual elements that have to be checked although they do not exist. Also the visual attention might be misled which could be investigated by an eye tracking study. If the length visual variable (d) is chosen to depict quantities, for example in a histogram, those lines should not be attached by visual extras like arrow heads to avoid confusion and wrong visual results.

Further visual illusions can be found that are worth mentioning. For example, for a dynamic dataset it would be a disaster if the movement effect was interpreted in several ways by different user groups or even the same user. If depth is used to indicate older time steps or visual elements not in focus we should be careful with comparison tasks that might lead to wrong conclusions. Color is frequently used in visualizations to encode a variety of data variables. If those are applied in certain environments, typically combined with other colors, this might lead to an effect that causes misinterpretations when judging and comparing the values of the colors, hence the underlying data could be misinterpreted.

3

Visual Analytics

Compared to visualization, the field of visual analytics [495] seems to be much more complex, including many more technical and technological aspects, disciplines, theories, models, and practical problems, making it an interdisciplinary field [279]. Moreover, it seems as if it supports many more real-world applications related to data, with respect to small, big, static, and time-varying issues, and also distributed in several sources in various formats with a mixture of data types with various additional attributes, like conflicting, erroneous, missing values, not explicitly given links, and correlations between the datasets. In some respect this is true, but visual analytics contains at least one major ingredient from the field of visualization which is the visual depiction of data enhanced by interaction techniques [476]. However, "the knowledge generated from models" [433] or the derived insights provided by algorithmic concepts in the form of rules, statistics, or tables full of values is typically in the scope of visual analytics. Further it is important to allow to hypothesize about the data and combining whatever kind of technology is useful for the tasks at hand with the ultimate goal to explore the data. Put briefly, in a quote by Thomas and Cook, visual analytics is "the science of analytical reasoning facilitated by interactive visual interfaces" [495]. In their definition human users are not explicitly mentioned nor are the tasks and hypotheses involved in the whole process of knowledge discovery. The analytical reasoning [420] stage demands for many additional tools to effectively and efficiently support task solving while hypotheses are built, confirmed, rejected, or refined [277], typically involving human users playing the key role. The combination of computers with their analytical power and humans with their perceptual abilities and strengths make visual analytics a powerful tool which is described in a quote that goes back to Albert Einstein or Leo Cherne: "Computers are incredibly fast, accurate, and stupid. Human beings are incredibly slow, inaccurate, and brilliant. Together they are powerful beyond imagination" (see Figure 3.1).

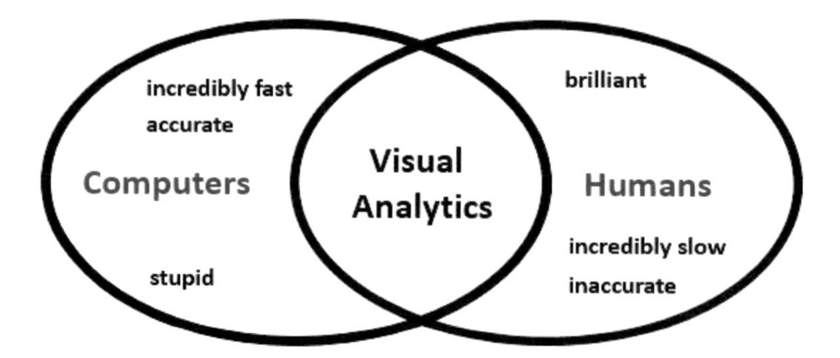

Figure 3.1 A quote by Albert Einstein or Leo Cherne describes the general ingredients of visual analytics: "Computers are incredibly fast, accurate, and stupid. Human beings are incredibly slow, inaccurate, and brilliant. Together they are powerful beyond imagination".

On the challenging and interesting side, visual analytics cannot be evaluated in the same way as visualization can be [306, 443]. Visual analytics is a combination of several processes that lead to a goal, defined, adapted, or modified by the users; consequently, it is not just primarily based on visual stimuli like visualization which could at least be shown in a user evaluation [397, 552] to investigate where and when people pay visual attention, to come closer to the solution of a task. Many of the cognitive processes happening in human brains, and also computational processes happening during algorithm executions, cannot be easily studied, but to understand whether a visual analytics system is well designed or full of design flaws, eye tracking is one way to, at least, get some insights into such complex processes composed of several repeating sub-processes with varying parameter settings. Eye tracking plays two roles here. It helps to record data about the spatio-temporal visual attention of several users of a visual analytics system, but eye tracking also generates a challenging dataset scenario worth investigating and creating a new application domain for visual analytics. From the perspective of visual analytics, eye tracking is used to evaluate its usefulness [306, 307], but visual analytics is again applied to analyze the recorded eye movement data [14] that might even be complemented by additional data sources like physiological measurements [44].

The goal to success is the sense-making skills and experiences of the data explorers who make use of both the interactive and linked visualizations as well as the computational power of today's computers. However, in most cases the human analyst has to guide the process of combining the machine with the provided algorithms and the implemented interactive visualizations. Hence, the humans with their tasks at hand seem to be the major ingredient in the visual analytics system. Those humans-in-the-loop, or sometimes described as "human-is-the-loop" [174], can serve as participants in user evaluations, with and without eye tracking, recording standard error rates, response times, and also visual attention strategies recorded by more and more advanced eye tracking systems [161, 235]. This kind of data about human behavior has great value for understanding how and if visual analytics systems function as expected or not. The real value becomes apparent when visual analytics is applied again to the recorded evaluation data, supporting hypotheses building, confirming, rejecting, or refining by taking into account analytical approaches as well as the perceptual strengths of the humans' visual systems. For visual analytics systems applicable to eye movement data this can lead to a dynamic visual analytics system since it allows the recording and analysis of the data in incremental and iterative processes.

3.1 Key Concepts

Visual analytics incorporates some major key concepts which are absolutely necessary to make it the powerful discipline it has become in these days with lots of contributors and a growing research community. This growth is caused by the various application fields [487] it is useful for, actually any kind of research area that generates data worth investigating, ranging from academia to industry. The success of visual analytics might be caused by the fact that it can try to tackle data problems that were not imaginable many years ago, that could not be solved by algorithms and visualizations applied separately. Only their combination with the power of today's computers and the strengths of the humans' perceptual abilities to quickly identify visual patterns paired with interaction techniques can solve several problems that already have a certain size and complexity that will even grow in the future.

Some of the key research areas involved in visual analytics focus on computer-related disciplines like data management, knowledge discovery in databases (KDD), data mining, machine learning containing several specific domain-relevant and efficient algorithms, general algorithmic approaches with computing power and the like, but also human-related ones like visualization, cognition, perception, individual as well as collaborative reasoning, and many more. The bridge between all these areas is built by human–computer interaction [19] in which the human factors play the key role. All of those interplays, benefits, and synergy effects have been and are still developed by the visual analytics community, for example, by communicating ideas at popular conferences such as the IEEE VIS with its well-accepted VAST challenge. However, evaluation of visual analytics [306, 307], with and without eye tracking, still remains a challenging issue, not only because all of those areas play a key role in the field, and this has made it what it has become.

3.1.1 Origin and First Stages

In several sources it can be found that visual analytics has some origins in the research areas of information visualization and scientific visualization [495, 529]. Actually, visual analytics more or less started to develop as a response to the terror attacks in the United States and the fact that the vast amounts of data have to be analyzed more rapidly for insights to support Homeland Security and to prevent further attacks. These cannot just be gained by pure visualizations or even pure automatic analyses. Their combination was the key to success as suggested in Thomas and Cook's book on "Illuminating the Path" [495]. In addition, the sharing of the results with experts and non-experts, all contributing to the analysis process in a collaborative manner, would be one of the goals.

Data is recorded at ever faster rates [3], but without proper analyses the data sleeps unused and the question comes up of whether the storage makes sense at all under such disappointing circumstances. The datasets have changed enormously in the days right before the invention of visual analytics into massive, dynamic, and partially incomplete datasets, stored in several data sources as heterogeneous data. This new complexity led to a rethinking of the technologies used these days to explore data, and finally brought the field of visual analytics into play. The power of human judgment played one key role in the development of visual analytics. In 2004, the Department of Homeland Security established the National Visual Analytics Center, led by the Pacific Northwest National Laboratory (PNNL). To reach the challenging goals that the new field of visual analytics promises, an agenda with a coordinated plan for governments and industry with investments was researched to guarantee that most of the developed advanced tools and successful technologies come to the attention of data analysts.

Searching the DBLP for the first archived paper that contains visual analytics in its title provides a result from 2002 on the topic of the semantic web [526]. This work combines visualization and analysis techniques while allowing to get web material that is of particular interest for the users. It seems that the term visual analytics has already been used before the famous book publication by Thomas and Cook [495], but what visual analytics actually really means has not been clarified in the beginning.

Nowadays, many research fields are included to make visual analytics successful and applicable to a variety of application examples. In their paper, Wong and Thomas [529] stated that "visual analytics will not likely become a separate field of study, but you will see its influence and growth within many existing areas, conferences, and publications." After several years of development we have to admit that it has indeed become a separate field of study, including more and more other fields, all giving benefit in their role to find insights in data. Hence, visual analytics can only survive if interdisciplinary research is done by several experienced scientists all over the world while their research outputs have to be presented and shared at famous international conferences like the IEEE VIS.

3.1.2 Data Handling and Management

Although the field is called visual analytics, expressing that visualization and the analysis techniques build the major ingredients, the real core of visual analytics is the data that has to be explored for patterns, anomalies, and insights. In many cases it is not the raw data that is used in the visual analytics system but the data is modified in some ways to make it readable by the system and to bring it into a suitable form for solving the users' tasks at hand. In this section we will describe and discuss some of the data aspects that are important in the field of visual analytics without explicitly focusing on the completeness of all the data-related aspects and operations.

The data-related processes can be divided into three stages focusing on preparing, checking and deriving, as well as advanced operations (see Figure 3.2). Some of those processes happen fully automatically by just using the computer, others require more intervention from human users to support an algorithm based on strong perceptual abilities, prior experiences, or the fact that humans can better understand the semantics of a textual description in a strange context. Algorithmic approaches are useful when the data source is clearly specified and structured in terms of parameters, rules, or patterns.

The first stage when working in a data-related research field typically comes in the form of data preparation. This means the data must be collected first, before it can be the focus of a visual analytics system. If the data consists of several data sources it has to be organized and each part has to be

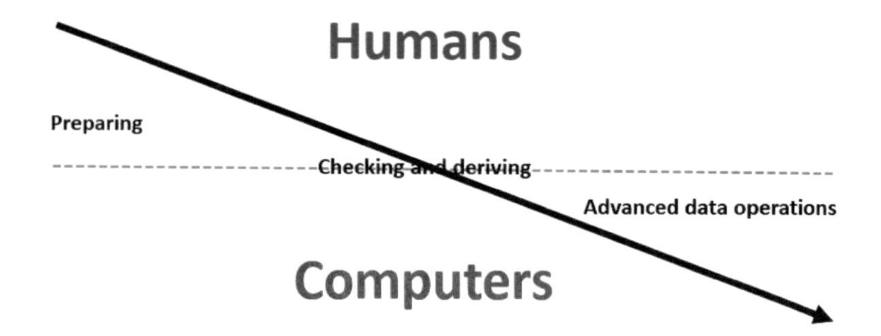

Figure 3.2 Data-related concepts may happen at three stages in a visual analytics system in the form of preparing, checking and deriving, and advanced operations. Humans and computers play different roles in these stages and are involved to varying extents.

checked for relevance concerning the tasks at hand. Data has to be annotated or even de-annotated for anonymization for example, which might also be done in rare cases by humans due to the lack of deriving the right computer-supported semantics. Exploring the correct format as well as the data types is as important as the linking between several data sources in case the linking is not explicitly given. In all of these steps the human plays a crucial role while the computer also supports most of the steps.

- **Data collection and acquisition.** Data can be collected in various ways. If the user is involved, this typically happens in a hand-written form with pencil and paper or directly stored in an electronic form, when the situation allows it. However, to make the data explorable by a visual analytics system it has to be brought into a computer-readable electronic format. Hand writing can be read by a computer, for example by a machine learning algorithm [366]. Normally, large data sources stem from device-supported recordings (like eye trackers, cameras, sensors, and so on) or dynamic and continuous data by simulations.
- **Organization and relevance.** In the case that a dataset consists of several heterogeneous data sources, those must be ordered by relevance. This decision is important to start with the most urgent data source first, for example in situations in which a quick response is required. The remaining data sources are read one after the other as soon as computation resources are available. Such an organization is also of interest if a primary data source has to be attached with additional attributes from secondary, tertiary, and further data sources.

- **Data annotation and anonymization.** Annotation can be done by the human users and is typically time-consuming, hence the best case is that the computer can do that automatically if it is instructed what to do. However, some datasets do not allow the computer to do that, for example, fixations in an eye tracking study while taking into account the semantics of a stimulus [370]. Also the opposite effect might happen, for example, if the data has to be anonymized which means that data elements have to be removed or encrypted. To avoid that humans read the data before it is anonymous, the computer should do this step.
- **Interpretation of data.** Identifying the format of the data is typically done by the human and then the computer is given some kind of template to apply that format to a dataset. This template describes which kinds of data types a dataset is composed of or if there are any unknown and non-identifiable data entries based on the given format. If the format is understood by the computer it can be brought in other specific formats to support rapid task solutions.
- **Data linking.** In some situations the data is not stored in one file in a clearly defined format but it is distributed over several files. Visual analytics systems should be able to link those datasets based on certain common keys. Also this process typically requires the human users in order to reliably merge several datasets into one. Sometimes, another external data source has to be considered to create the correct linking, while in some situations a linking might not be easily computable; however, it might be done by inspecting visual patterns observable when visualizing each of the datasets separately.

A second stage describes how the data is checked and additional values are derived. Included steps are the validation, verification, and cleaning of the data, the enhancement, i.e. which values can be removed, added, or have to be kept due to the fact that their deletion might lead to misinterpretations. If the data has to be presented in a textual form as well, apart from visual output, it must be decided which parts of the data are of particular interest, for example, as adequate labels. Further information about a dataset can be computed and attached to make it quickly accessible during the running tool. If the data is still not in a computer-readable and understandable form, this is the chance to get rid of pieces to avoid that. Storing relevant data portions to quickly providing them when the visual analytics system gets started again is also of particular importance. Most of the steps in this stage demand for the computation power of the computer and can be done quite automatically; however, in some scenarios the human users have to or can intervene.

- **Validation, verification, and cleaning.** It should be checked if the data in use is incorrect, redundant, incomplete, i.e. has a certain number of missing elements, where these elements are located, and if they can be replaced, interpolated, simulated, or just ignored. One goal of this step is to clean the data in a way that the number of errors or missing values lies below a certain, typically user-defined, threshold. Another question is, if there is an error or missing element, how serious this effect is and which impact it will have on the rest of the data. In particular, for eye movement data it could be important to clean the data from calibration errors [444].
- **Data enhancement.** In some cases the original data is not in its final form for loading into the visual analytics system. The reason is that the data has to be enhanced by removing superfluous elements, adding extra elements, or by even annotating the existing data elements with further add-ons. Those could come from the user by manually adding information, but also by the computer, for example, an algorithm could categorize data elements and tag them or compute even more advanced statistical or projected values and attach them to the corresponding data elements.
- **Data presentation.** It has to be decided which components of a dataset are shown graphically and which ones textually. For those shown graphically we have to find out what extra information is required to improve the interaction response of the visual analytics system. For the components shown textually we have to decide what part of the text is displayed. Text can be pretty long, even labels or whole text passages, meaning it cannot be shown completely. Only the relevant part of the text should be shown and the rest might be omitted.
- **Meta information about data.** Each process applied to data can generate additional information, for example, error reports, performance feedback, quality of results, trustworthiness of the data, uncertainty values, provenance and lineage of the data to provide information about the context in which the data was recorded. Also ethical, privacy, or security issues can be considered. Metadata can also be any kind of advanced computations on the data that are stored as additional attachments, i.e. data about data. Also units, whenever existing, are important metadata, even considering the change to another unit scale.
- **Data transformation.** A very general process is to bring a dataset into a computer-readable form. That means that the raw data is brought into a computer-supported format, an already preprocessed dataset is

adapted to a certain well-defined other data format specified by a visual analytics program, or a dataset is reduced by size and complexity to make it responsively usable in the visual analytics tool. For such data transformations the data has to be read and parsed which already means understanding the general format rules.

- **Data storage.** Either the whole dataset or filtered parts of it can be stored for further use, in a prepared format with data annotations. The storage format can be different from the internally used format during the execution of the visual analytics system, the format can be changed to allow faster access and better interaction responses.

Another stage consists of more advanced data operations which are typically done by the computer, not by the human since they demand fast computation and the rules how to apply them are clear in a way that an algorithm can quickly process the data. Interesting and popular operations in this context are finding better structures by ordering and clustering, reducing dataset sizes by summarizing, classing, classifying, aggregating, and projecting data, or by allowing fair comparisons, i.e. by normalizing data. The intention of such operations is to increase algorithmic, visual, and perceptual scalability, but it may be noted that such operations also typically lead to information loss. Hence, it should be decided whether or not they are adequate and do not introduce negative issues.

- **Ordering and sorting.** Quantities are typically ordered, i.e. brought into an increasing or decreasing structure by following some kind of order relation expressing which values are higher and which ones are lower. This approach can also be applied to n-tuples of quantities like pairs used for adjacency matrices, ordering each dimension in the same way to detect group relations. For multivariate data such a 1D ordering is useful to investigate the impact of one attribute to another one, i.e. if there are any easy-to-identify correlations among the attributes, positive or negative ones. Even lexicographic ordering for text is important. In general, the ordering could be a step towards applying classification and aggregation techniques, taking into account value ranges, i.e. classes, this is in particular useful if additional data is attached to each object corresponding to an ordered value.
- **Data clustering.** Putting data elements together in some way due to the fact that they share a common similar property can lead to reflecting on additional insights about a dataset that we would never get without this effect. Such a grouping of elements is typically shown

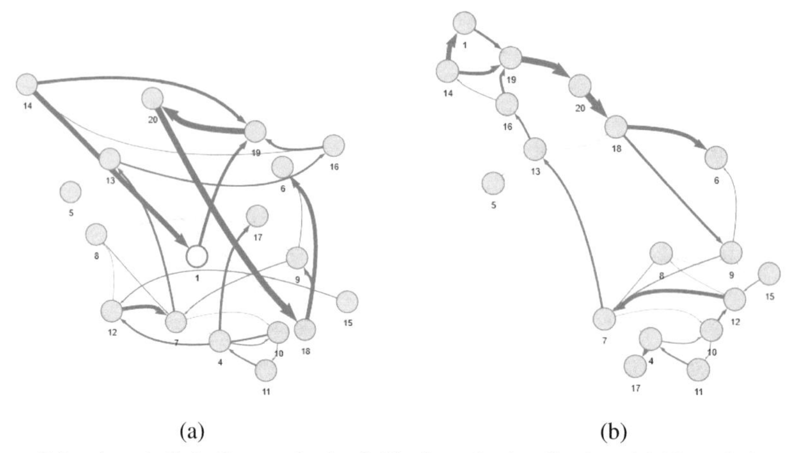

Figure 3.3 A node-link diagram in the field of graph visualization. (a) The relational data without clustering, just randomly placed nodes. (b) Computing a clustering of the same data as in (a) and, based on that, using a graph layout that takes into account the node clusters, encoded by spatial distances of the nodes.

by spatially positioning them close to each other, in case they are related, while the not-related ones are moved apart. Clustering is a typical example occurring in a graph visualization in which the relation among the objects is given by the graph edges connecting the objects (see Figure 3.3 for an example) or for fixations in eye movement data into areas of interest based on the spatial distance in a stimulus. In some data situations, a relation has to be computed first, for example by pairwise similarity values. Also a clustering could be a pre-step towards classification and aggregation techniques, to meaningfully reduce the amount of data to be visualized.

- **Summarization, classing, and classification.** Data elements or whole parts of a dataset can be summarized in a way that their size is reduced. In the best case several data elements get mapped to one representative element. To do this reliably without introducing too many interpretation errors those “new” representative elements have to be chosen carefully. This procedure is also useful for time-varying data in which whole time periods might be mapped to certain time-characteristic and well-described categories, for example, special events or attributes best describing what happens during a time period. A “rush hour” for traffic data would be a good example. Typical summaries for groups of data elements are based on a certain class or category in which the data elements fall in most cases.

- **Aggregation and collapsing.** Folding certain data elements that are contained or classified to belong in a container to a new aggregated element, due to the fact that they carry similar aspects, is a powerful strategy. This also brings along several negative issues like information loss and the problem of what this new element should look like. However, in hierarchical organizations of which many exist in the data domain such a data aggregation might be caused by collapsing sub-hierarchies to which certain objects with their attributes are attached. In such a case, as well as in many others, the represented value after the aggregation could be determined by minimum, maximum, sum, mean, average, standard deviation, or any other statistical value, maybe even shown as a box plot.
- **Data projection and dimensionality reduction.** If the data consists of various data dimensions, like eye tracking data [66, 79], too many to explore for patterns or to display visually due to visual scalability issues, the data might be projected from a higher to a lower dimension. The general goal behind a projection is the idea of preserving similarities and dissimilarities existing in the original data in the projected data as well. If this effect is not supported the projection method might fail in terms of leading to misinterpretations when exploring the data for insights. There are various dimensionality reduction methods typically differing in the fact if they are linear or non-linear [176].
- **Data normalization.** If several scales are taken into account while measuring data this might lead to unfair judgments of the data when graphically depicted. Normalization is a step to adapt each data element or data series to a common scale, for example, by stretching it to the largest value while adjusting all involved parameters. In the field of eye tracking this could be important for different scanpaths. Visual attention can vary from participant to participant in terms of exploration speed, although the order of visits in a stimulus might be similar. Normalization helps to identify those similarities.

Data is typically processed in a different visual analytics system-independent process, maybe on a different computer or server, to get the computing resources only for the purpose of displaying and interacting in the already processed data, allowing a responsive system. Apart from the running system, the remaining new, or modified data can be processed over longer time periods and when the processing is ready, this data can be interactively visualized and explored, not weakening the computing power of the visual analytics system. This separation strategy guarantees a user-friendly tool experience; no-one wants to wait for a long time for a data analysis result while wasting valuable time. A data processing step should be as independent as possible from the rest of the visual analytics system, in case the system and the tasks at hand allow such a scenario.

3.1.3 System Ingredients Around the Data

To explore the core ingredient in a visual analytics system, i.e. the data, various other techniques from several research disciplines have to be applied in combination. This interplay guarantees that the users of visual analytics systems can efficiently explore the data for answering their tasks at hand. On the negative, challenging, but also interesting and valuable side it also brings new research topics into the field that mostly come in the form of how to combine the techniques in a clever and effective way. This leads to steady progress of the field and a growing of the research community as well as powerful applications [278]. Visual analytics is actually composed of two major aspects, machines and humans, taking the best out of both worlds, and hence, the power of it is their clever combination while the human plays some kind of key role in the whole process.

Computers support efficient algorithms by their computing power and allow automated analyses in case it can be specified how such algorithms have to be applied. This means the human users just initiate the algorithms and then those run until they provide an expected or unexpected result. Statistical approaches to reduce a larger dataset into expressive numbers are typically computer-driven. Mathematics is a powerful field that comes into play here with a lot of theoretical concepts and disciplines. Strong research areas involved in the visual analytics process in which the computer is definitely required are knowledge discovery in databases (KDD) as well as data mining [186] to derive patterns in the form of association and sequence rules. However, although such rules bring some structures in the data and show static as well as temporal correlations between data elements, the output of the involved efficient algorithms is typically so large that further concepts have to be applied, with interactive visualization being among them. In particular, the visual depiction of data demands for appropriate devices and displays as visual outputs for guiding the input in form of interactions. However, although the displays are computer- or hardware-based, the decision which display(s) to use is still on the users' side, such as small-, medium-, or large-scale displays like high-resolution powerwalls or combinations thereof.

Humans with their perceptual abilities to quickly and effortlessly recognize and identify visual patterns are the driving force in the whole visual analytics process. Using their hypothesis building strategies they start some kind of reasoning to draw conclusions from what they see to guide and adapt the supported algorithms and concepts and to build the right mixture of ingredients with the goal to solve their tasks at hand. Humans use cognitive aspects to combine the facts in a clever way to refine the hypotheses, confirm, or reject them or even generate new ones. In some situations, not one individual person can solve the problem, but a quick or exact solution is oftentimes based on a collaborative strategy exploiting the strengths of many others. To make this even more advanced and powerful, people from different locations on Earth, in different time zones with varying technologies can step in the analysis process and can give feedback to the tasks at hand while the biggest challenge is to collect all the information and bundle it into something meaningful. The human users are also responsible for the dissemination of the achieved results, which means presenting them to a larger audience, for example, at conferences. Although the human plays the major role in these processes, computers and advanced technologies are required to allow all of this. Many years ago, due to a lack of technological progress, this was not possible, and hence, visual analytics as we know it today can and has to keep pace with the progress in technology.

The combination of computers and humans, i.e. the interplay as stated by Einstein or Cherne is partially realized in the research field of human–computer interaction (HCI). But this interplay exists in many more forms, for example, it also includes the algorithmic approaches that can be adapted during runtime by human intervention, i.e. changes and modifications of parameters, for example, if they have run into a local minimum where it is impossible to get out automatically. This requires the users' input to give an algorithm additional support during runtime to work more reliably and to search in a certain user-defined direction. But still, the computational power is exploited to allow running an algorithm automatically to a certain stage until the human users come into play. Hence, the decision-making process is guided by humans as well as the computers which are actually the strength of visual analytics, but the extent to which each part is involved is typically guided by the human. Although visualization is the means to provide insights in the final result or intermediate steps of algorithmic processes, it is still a combination of human–machine processes, the same holds for interactions although they are initiated by the human users. The generation, presentation, and dissemination of results is also mainly guided by the humans but the computer is required to prepare and display the results efficiently. Output devices are fundamental for visual analytics, but the users can adapt device parameters, and they could even make use of much more advanced technologies if they have access to them, for example, in a research institution or industrial environment.

To investigate if a visual analytics system is really useful or at least several parts of the provided functionality, evaluation can help to measure performance for the humans but also for the computer with all of its algorithmic approaches. Humans with their varying knowledge and experience levels as well as further specific properties can be evaluated in several user study settings, for example, in an eye tracking study to record visual attention behavior. The machine can be checked for runtime performances, reliability, trustworthiness, and soundness in the form of the numbers and complexities of errors it produces while working with the data. Also interaction techniques on the user side can be evaluated, on the computer side it must be checked for responsiveness, hence human–computer interaction demands two aspects, well-performing users with their tasks, and fast algorithms. If any side suffers from performance degradation, this might have a bad impact on the visual analytics system. In summary, the evaluation of a visual analytics system is challenging, but also interesting because it brings into play new research fields.

3.1.4 Involved Research Fields and Future Perspectives

Visual analytics is called an interdisciplinary approach [279] with many cross relations to and between fields that bring some benefits for the goal of finding insights in data by supporting analytical reasoning (see Figure 3.4). Without this linking strategy it could not tap its full potential as it does in these days with a variety of application fields, in particular, focusing on large, complex, heterogeneous, and time-varying datasets. To get the data ready and prepared for the analysis, visual analytics systems are based on a well-designed infrastructure with respect to data handling and storage as well as all other aspects making it a responsive system with a good user experience.

The number of involved research fields is growing as well as the interest which is reflected by the size of the community as well as the variety of corresponding research activities and events such as the VAST challenge;

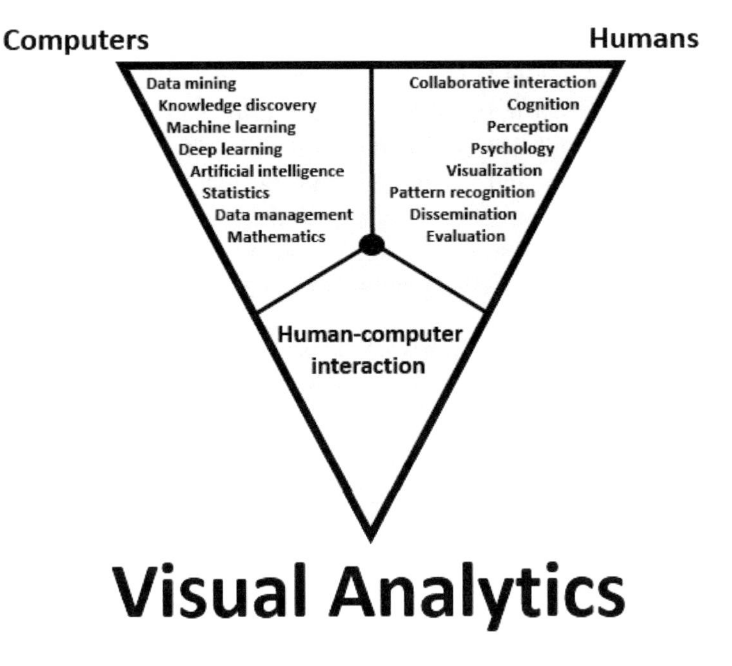

Figure 3.4 Visual analytics is an interdisciplinary field that makes use of research disciplines involving the computer, the humans, and also human–computer interaction (HCI).

there are too many to mention all of them here, but we will keep an eye on the most prominent ones which are also involved to the largest extent. For the algorithmic data processing we have to mention fields like data mining and knowledge discovery in databases (KDD). Also machine learning, deep learning, artificial intelligence (AI), or disciplines like explainable AI build the basis for many data analysis techniques as well as statistics and mathematics, in which the human user is typically not involved a lot. Data management is important because of diverse data sources, and is also available on the web existing in various and diverse research fields and applications.

From the human perspective, fields like human–computer interaction including individual and collaborative interaction, also over the web, play a crucial role for more and more data analysts spread all over the world with varying expertise. Cognition, perception, and psychology are important for enhancing and accelerating the decision-making process, in particular, if visualization is incorporated in the data analytics process, for example in cases where large data like time-varying high volumes or streams are visualized in a scalable form with a great overview first. Pattern recognition is used for starting further exploration processes, to build new hypotheses about the data, and to dig deeper in the already achieved insights. Presentation and dissemination of the results is typically mostly the job of the humans although computers are required to facilitate these issues. Evaluation is an increasingly needed research area that provides insights into the user experiences and can show design flaws and drawbacks that help to enhance the system. This brings into play novel problems like ethics and privacy, creating a new kind of data worth including in the analysis.

The field is still developing and has not reached the end of its progress. As long as datasets are generated and recorded, visual analytics will play a key role in order to process, analyze, and visualize this data with respect to the tasks of the human analysts. However, due to the steady progress in hardware and software technologies, visual analytics is also subject to a steady change to keep pace with more and more challenging problems. In particular, fields like machine learning, deep learning, or explainable artificial intelligence [123] bring new ideas and emerging topics into play, demanding the power of many involved aspects of this interdisciplinary field. Taking into account the opinions, real-time decisions which are difficult for visualization alone, and power of various users, maybe even in a collaborative manner, is a suitable concept but requires scalability in terms of client–server software architectures to reach out to the experts and non-experts in the world. Visual analytics on smartphones, although the display is small, could be of potential support, at least to access the people. Synchronizing, merging, and summarizing the inputs and outputs of the users is another challenge. Such apps would bring even more privacy issues into the field, a challenge that cannot be neglected in the future. Users could suffer from cognitive overload and hence, the problem might be split into parts and distributed among several people to reduce this cognitive burden for the individual person.

Algorithmic and visual scalability are not the only challenging problems, but also the general aspect of finding out if a visual analytics system is useful at all or for which tasks it is of particular use. Evaluation is a huge topic for the future development of such systems, specifically if millions of people could be included in the evaluation process. The number of people has to be large because visual analytics systems contain a variety of functions, algorithmic approaches as well as visual depictions in lots of parameter settings supported by interactions. Setting up user studies and recording the performance or visual attention data is not the biggest challenge here, but it is more the effective and efficient analysis of the recorded data with the goal to find or predict insights that help to improve the visual analytics system, maybe in real-time. Further, automatic adaptations based on users' eye movement behavior or general behavior like body movements and gestures, spoken words, interactions with the system but also between the users, facial expressions, and many more could be steps into the right direction. An exploration and interaction history, reflecting the steps and stages taken during an analysis process, which is not found in many visual analytics systems, can be of great support to achieve a faster way to find insights, specifically if users have to jump to and fro between earlier and later states of the system many times.

3.2 Visual Analytics Pipeline

Compared to the visualization pipeline (see Figure 2.26), the states and transitions in visual analytics contain many more concepts which is due to the fact that visual analytics is an interdisciplinary field combining the strengths of computers and humans. Figure 3.5 shows an illustration of the major concepts that are applied to start from raw and original data to finally generate insights that are either used to adapt the data aspects under exploration or that make the users confident and solve the tasks and help to confirm or reject their hypotheses. In this section the major concepts are described as well as their interplay and the transitions between them.

3.2.1 Data Basis and Runtimes

A good data basis builds the crucial aspect for an appropriate data analysis as well as an interactive visualization, which are important for the visual analytics system to generate insights based on users' tasks and hypotheses. Such data challenges can come in many forms, as described in Section 3.1.2. In nearly all cases, no matter from which application field the data stems, the visual analytics system cannot work with the raw or original data, hence it must be modified and adapted in a way that it meets the requirements of the system. This is described in the three stages starting with the raw data, preparing it, and finally checking it for its suitability for the system. The last transition is the deciding one, for which the computer is exploited with its computational power, i.e. the advanced data operations transform it into structures that reflect patterns, correlations, and rules.

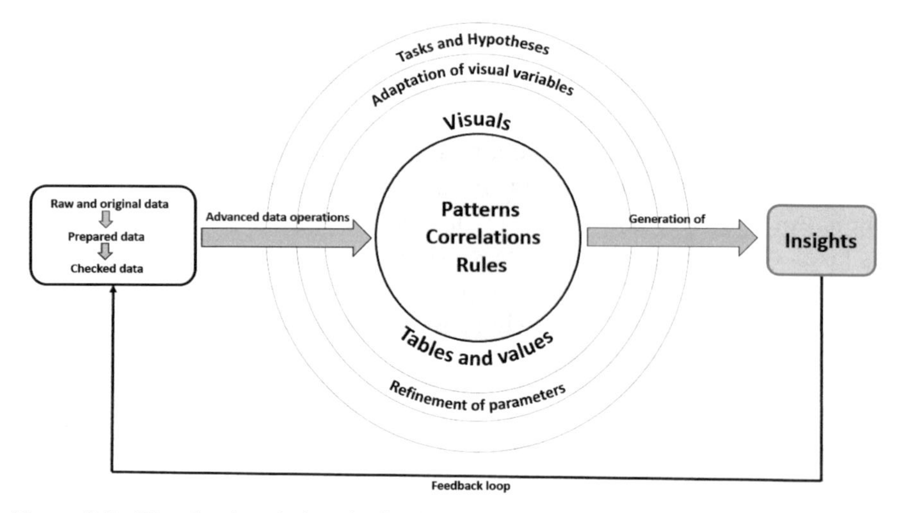

Figure 3.5 The visual analytics pipeline illustrates how data is transformed into patterns, correlations, or rules that can be regarded as tables filled with values or visual depictions. The users can adapt visual variables and refine parameters guided by their hypotheses and tasks, hopefully generating new insights that can be used to modify the data under exploration.

These transformations are defined by the users with their tasks in mind, hence due to the fact that the users can change their intentions quickly it is important that the algorithms run fast and adapt the outputs to the users' needs. This is actually the stage in which the responsiveness of the visual analytics system in terms of interactivity is based on. A poorly running algorithm cannot be mitigated by a fast interaction technique. The interactions do not only depend on the users' intentions but also on the algorithm performances. When working with algorithms it is crucial to have profound knowledge about data structures on which the algorithms are based and algorithmic runtime issues like NP-hardness [195]. In some situations, the algorithmic problem is so complex that we cannot expect an optimal solution in a reasonable time, hence a good heuristic approach is needed. This is actually the point where we need experts in the field of data structures and algorithms, typically needed to enhance the runtime and, consequently, the interactivity of the system. In the cases in which we really need an optimal solution for such a complex problem the users might be warned that a result is not expected for an indefinitely long time. Showing a progress bar in such a scenario is also not possible in most of the situations because we cannot predict how long the computation will take, i.e. we do not know after which time period the one hundred percent is reached.

In some scenarios it would even be a good idea to provide insights into the algorithm, which means showing the step-by-step iterations of an algorithm and how it processes the data. This would help to understand how long it might take until the algorithm terminates but, further, it would give insights into why an algorithm produced the wrong results or suffers from bad performance. Although such an approach would be suitable for the design of a visual analytics system, it is less useful for an end user who plans to just apply the system to find insights into the data based on the tasks and hypotheses. However, the algorithmic issues could be reported to the developer of the system while the users let the tool run without recognizing that such extra data is stored and transmitted. By such a strategy the system might be evaluated from an algorithmic perspective, but the end users' feedback could also be included and linked to the algorithms under different settings.

3.2.2 Patterns, Correlations, and Rules

The data gets its real value if it is processed by advanced operations such as ordering, clustering, data mining, projection, and so on, that finally brings it into the shape the users might want to see to get answers to their hypotheses and tasks. It should be mentioned that the result of such advanced operations first comes in a textual form, i.e. the transformed data is presented in tables consisting of values and numbers. Those could be read by humans but typically, it either takes a long time to understand them and to identify some patterns, correlations, or rules, or it is not possible at all to detect something relevant. This would lead to the fact that it is argued that the data does not contain any kind of pattern or a second, typically statistical or visualization tool is applied to confirm or reject this hypothesis.

Some experts working in certain application domains are trained to read and explore the data values without looking at corresponding visual depictions of it. One reason for that is that they typically do not like to learn visualization techniques since it would cost them a certain amount of time to get some experience to read them and they fear the cognitive efforts in the visual data exploration process. An example domain would be software engineering in which the programmers rather like to locate the bugs and performance issues in their code by reading the vast number of commands and instructions. They rarely use visualization tools [149] to debug their tool or to make it more efficient. Visualization can be of great help to quickly and effortlessly detect insights, in case the chosen visual metaphor is the right one for the task at hand. Hence, a visual depiction of the data is of particular importance to support the viewers and to guide them in the right direction as well as accelerate the insight detection process. This is, in particular, useful for applications in which a quick answer to the problems at hand is needed, like evacuation planning, epidemics, or preventing terror attacks [495].

A lot of experience is required to design a good visualization as well as to read it and to derive patterns, correlations, and rules. Moreover, it should be investigated whether visualization is really needed to solve the given task or if a pure algorithmic solution is strong enough to provide the right answers. Visualization is the means of choice if an algorithm cannot be clearly specified in terms of parameters and details about the computation steps. An example is the min–max search in a set consisting of quantities. This task can be solved on a sole algorithmic basis without asking a diagram for support. However, if we search for a pattern which is very vaguely defined, also depending on the dataset and domain, we might choose a visualization. But to choose that in the right way we need profound knowledge about the data types and structures involved in the dataset as well as the domain and environment from which it stems, also the user tasks are deciding factors in picking the right visual metaphor and right individual visualization techniques. In summary, applying visualization for a dataset to detect patterns, correlations, and rules is a challenging task, not only on the design level but also on the interpretation level, typically involving non-experts in visualization. This is the step in which user evaluation is important to understand if the designer was successful or not in supporting the pattern finding task which is interesting for the user.

A pattern is a user-defined order or structure in a dataset that follows a certain behavior, shape, model, outline, or template. It is something that is the opposite of random. A pattern carries some kind of user-defined meaning. In some cases it can be clearly specified in terms of parameters, but in most cases the users are able to define it as a pattern although they cannot specify it very clearly which makes it hard to be processed by an algorithm. Moreover, in some data situations it was not clear right from the beginning that a pattern existed in the data, hence an algorithm cannot be specified which refers to the "seeing the unseen" quote. If a pattern can be identified, also outliers and anomalies can be present which are points that do not fall into the pattern shape, but behave somehow differently. Outliers and anomalies can only exist if a pattern exists. A powerful visualization for depicting patterns is an adjacency matrix, if it is ordered in a meaningful way. Figure 3.6 shows

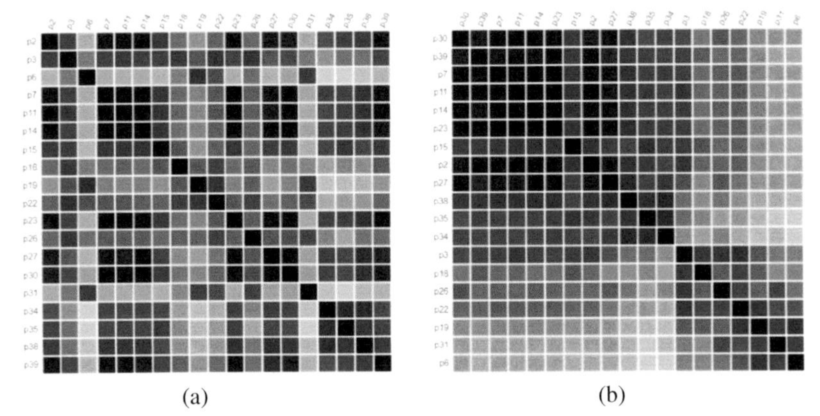

Figure 3.6 (a) Comparing the participants' scanpaths from an eye tracking study can generate pairwise similarity values shown in an adjacency matrix. (b) Applying a matrix reordering technique immediately shows a pattern in the matrix which is difficult to find by the pure textual values given in a table or 2D array [300].

two examples for the same dataset, an unordered one and an ordered one. The ordered matrix immediately reflects patterns in form of blocks along the diagonal.

A correlation is defined as two or more variables measured for the same observations standing in a certain related behavior. This behavior can be the same, similar, or one can be completely the opposite of other ones. An example would be a bivariate dataset, i.e. containing two variables for each observation, in which the larger values of variable A correspond to the larger ones in variable B and the lower ones of A correspond to lower ones of B, building a positive correlation. If the larger ones of A correspond to the smaller ones of B and vice versa we call this effect a negative correlation. If those values are stored in a table with two columns, an ordering of one column can help us to see the impact of this ordering on the second one. Scrolling down the columns can still be a solution to identify the correlation behavior but in many cases it is not that easy to identify in the textual representation. A visualization can help to rapidly judge which kind of correlation (apart from the standard positive and negative ones) exists. Scatterplots for bivariate data as well as scatterplot matrices and parallel coordinates plots [255] for multivariate data are standard and well-researched depictions for this kind of data focusing on visual correlation detection. Figure 3.7 shows an example for multivariate metric data from an eye tracking study [299]. The vertical axes encode the metric scales from top (large values) to bottom (small values)

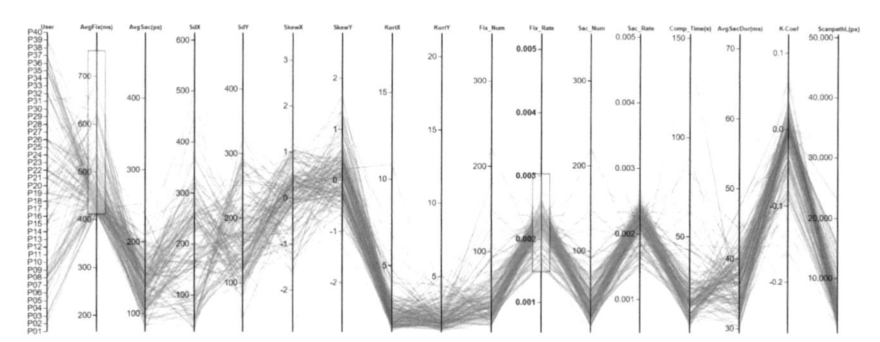

Figure 3.7 A parallel coordinates plot (PCP) for showing positive and negative correlations between pairs of metric attributes derived from eye movement data for selected study participants [299]. Axis filters are indicated to reduce the number of polylines. Image provided by Ayush Kumar.

and the polylines in between depict the values for each observation, in this case an eye tracking study participant. We can detect the positive correlations in the nearly horizontal more or less parallel running lines and the negative correlations in the crossing line patterns between the axes.

If data mining is applied we can generate more advanced rules that can exist in at least two major types denoted by association and sequence rules. Depending on the number of elements involved they can be binary or n-ary. An association rule expresses if two or more data elements are related at the same time to a certain extent, typically described in a percentage value or in natural numbers. This depends on the fact of whether the confidence or the support of a rule is considered. The support expresses the total number of occurrences of an element tuple in a rule while the confidence expresses the relative number of an occurrence of an element tuple to all occurrences of an element in that tuple. A sequence rule, on the other hand, contains a temporal aspect, describing an antecedent, the condition before, and a consequent, the condition afterwards. Also sequence rules exist with a certain confidence, i.e. probability, as well as a support value, even in multiple stages, if the sequence rule consists of several antecedents and consequents. Each following consequent typically lowers the probability because of the more restrictions due to the higher number of antecedents. Figure 3.8 shows an example from eye movement data [63] for which data mining rules have been generated and visualized. A typical visual aggregation for sequence rules can be achieved by considering common prefixes and put them together into a rule hierarchy, just like some kind of decision tree.

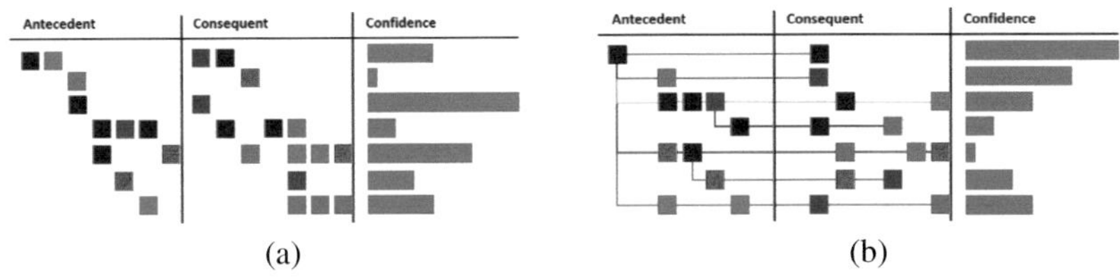

Figure 3.8 (a) n-ary association rules (b) and n-ary sequence rules generated from eye movement data can express which general relations exist in eye movement data [63].

It may be noted that apart from patterns outliers and anomalies can also exist, but those can only be detected if a pattern is known, i.e. the normal case from which an outlier can be distinguished. A similar effect holds for countertrend patterns that can only be detected if we know the trend pattern in the data. For example, if one time-series shows a growing behavior, a second one shows a decreasing one. However, it always depends on the perspective of the user what is defined as the trend and the countertrend. A trend is typically found in time-dependent data in which we can compare values over time. This means we identify a certain function that describes or models the trend behavior as well as possible.

3.2.3 Tasks and Hypotheses

Hypotheses guide the data exploration process. They describe insights that are to be expected by the users, for example at the time when the data is still existing in its raw form, before further adaptations have been made, the users do not know if those hypotheses can be confirmed or rejected. Furthermore, there are many more hypotheses which the users are not aware of before using the visual analytics system; they build new ones or refined ones during the exploration process. Each hypothesis involves one or a set of tasks that have to be answered in order to successfully confirm or reject it. A hypothesis can, for example, state that "the largest cluster of related people in a social network consists of 45 members". To find an answer to this hypothesis we need to know which people are connected to whom and how they are separated in individual clusters. This separation into individual clusters, if shown visually, is done pre-attentively [219] by exploiting the Gestalt laws of proximity and good form [292] without paying much attention and with fewer cognitive efforts if the visualization is perceptually well-designed, maybe as a node-link diagram.

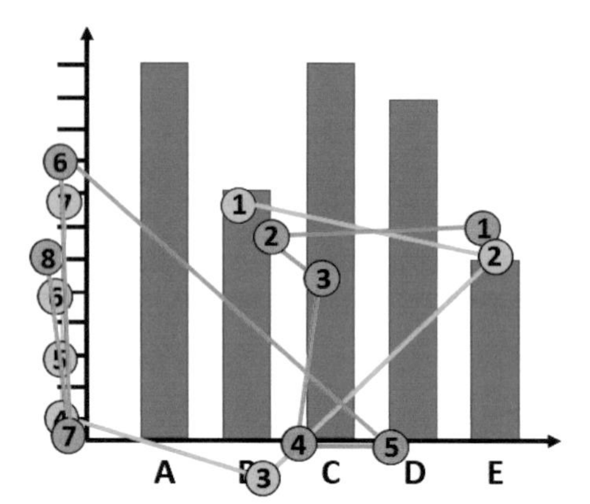

Figure 3.9 Confirming or rejecting a given hypothesis in a simple bar chart can generate a lot of simple tasks which might be recognized if the eye movements of observers are recorded and overplotted on the bar chart stimulus in the form of a gaze plot [203].

However, as a starting point, we need to answer some kind of estimation task to judge the number of people in each cluster as well as a comparison task to identify which cluster is the largest one. Finally, we might need a counting task to check the given number of 45. However, before we can start with answering the simple tasks and combine them to the answer of the complex task to find a solution to the hypothesis we typically make use of additional support. In our case, we could solve the tasks purely algorithmically. Visual analytics offers interactive visualizations as well, hence we could also try to find a solution visually if the right visual metaphor is chosen and if the preprocessing of the data computes a suitable structure, in this special case, a structure that moves the related social network members spatially closer together than the non-related ones. Only by this strategy we can start recognizing clusters of people.

Another simple visual example might be given as a bar chart. A hypothesis could be that "the value of the second smallest bar is 8 and the bars are labeled A, B, C, D, and E". If people's eye movements are recorded while checking this hypothesis we can get an idea of which simple tasks, and maybe in which order they have to be solved, to confirm or reject the hypothesis (see Figure 3.9). At least, without paying too much attention to the order, we can identify some simple tasks by looking at the visual depiction denoted by the term gaze plot [203]. The participants judge the heights of the bars (only two of them), compare those (otherwise they cannot find the second smallest one), follow a horizontal line (to get the scale of the vertical axis), read the number on the scale (to check for 8), and finally, read the labels. This visual stimulus also shows some extras, one is definitely the fact that not all bars are visually attended, a fact that might come from the peripheral vision, meaning in some situations we do not have to focus the eyes on a visual object to judge if it is relevant for a task to confirm or reject a given hypothesis.

We have to admit that the bar chart example is not needed to answer the given hypothesis, this could be solved by an algorithm while the labels could also be checked by just searching for them in the dataset. However, there are much more complex hypotheses for which visualization is required. For example, we might state that "there exists a periodic dynamic pattern in the data". This hypothesis could be algorithmically solved if we knew the period in the time-varying data, but this period might even change over time. Hence, a visual depiction would help and the perceptual abilities allowing fast pattern recognition would do the rest for us.

The examples above illustrate that a simple hypothesis requires a certain number of simple tasks that have to be combined in order to find a way to confirm or reject this hypothesis. In general there are various simple low-level tasks that can be combined in task categories, each describing the common procedure that is needed to solve each individual task contained in it. In a user study, with and without eye tracking, the participants are typically confronted by one or several of those tasks, simple ones or more complex ones. If the task is too complex the study participants start to subdivide the complex task into simpler tasks automatically. This subdivision can be identified if the visual attention behavior in form of eye movement was recorded and the analysis approach was sufficient to identify the task splitting strategy. In the following an overview is given about certain task categories without explicitly stating that the list is complete (see Figure 3.10). The tasks can be applied to both, textual representations in the form of tables and values, but also to visual depictions of data, probably with varying task performance in the form of task response times and even error rates, typical for user studies reflected in the dependent variables. Yarbus [539] has shown that the task even has an impact on the eye movements, hence it would be interesting to analyze eye movement patterns to identify the current user task in a visual analytics system, maybe to better guide a user.

- **Search tasks.** Locating textual or visual objects of interest builds the basis for nearly every other task. Search tasks can be quite time-consuming if the objects we are looking for are not quickly recognized, for example outliers should be visually highlighted and hence, be made pre-attentively detectable [219, 500].

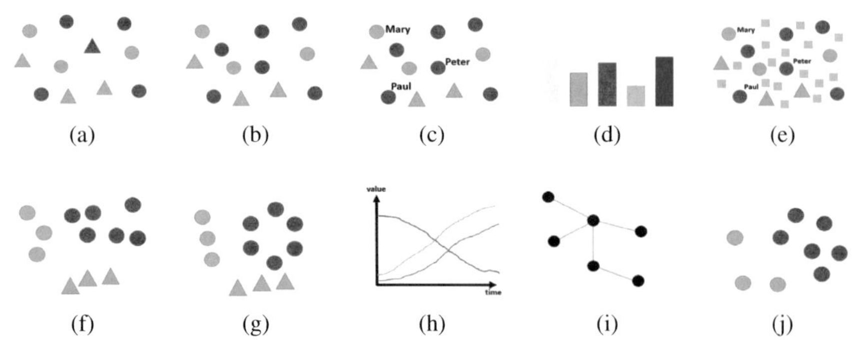

Figure 3.10 Visual scenes illustrating several tasks. (a) Searching a red triangle in a sea of distracting visual objects. (b) Counting the number of red circles. (c) Reading labels attached to visual objects. (d) Judging the smallest bar in a bar chart. (e) Estimating the number of green squares. (f) Comparing the red and blue circle clusters with the blue triangle cluster. (g) Identifying patterns from groups of visual objects. (h) Identifying correlations between visual curves. (i) Finding a route in a network. (j) Detecting communities by similarity and proximity.

- **Counting tasks.** If explicit values have to be checked, like a number of certain visual objects, we cannot just interpret them in a quick guess but we have to observe the objects one-by-one. A counting task [139] typically requires an object or pattern identification task prior to the real task.
- **Reading tasks.** Labels give extra details-on-demand or help to set the found insights into context to the seen objects. Also longer texts or text fragments give extra feedback on a visual scene in which the pure visualization is not exact enough, for example. Text can be an amplifier for the interpretation, but a visual depiction is in most cases more efficient to show patterns. Reading tasks have been studied a lot by means of eye tracking, also on small displays [382], with the major insight that the eye does not move smoothly but does several rapid movements as well as short stops [263].
- **Judgement tasks.** If a certain visual variable is used to encode a data variable it is important to allow judgments of the value visually depicted in this visual variable. For example, if a quantity is shown as a bar of a certain height or length it should be possible to judge this height reliably to avoid misinterpretations of the data. Color as a visual variable could be problematic for some people due to color deficiency or color blindness problems [348], hence the judgment process can vary

from user to user due to the impact of individual perceptual or vision properties [521, 522].

- **Estimation tasks.** If a crowd of visual objects is given that cannot be counted easily, but maybe a quick estimate as a first sufficient response has to be provided we typically do some kind of estimation task. In many situations when direct comparisons are required between two or more sets of point crowds we more or less automatically change from a counting to an estimation task, although the risk of inaccurate results may come into play. However, in estimation tasks we are less interested in response accuracy but more in a low response time as well as to reduce cognitive efforts that occur during counting and less during estimating.
- **Comparison tasks.** To set textual or visual objects in relation to others based on their properties or visual appearance they need to be compared to each other. This requires prior search tasks to locate the objects of interest, counting, estimation, or judgment tasks to make them comparable, hence a comparison is already a more complex task that also requires to remember visual objects in the short term memory [272, 410], at least if they are not both in the visual field at the same time, to allow a reliable comparison. Change blindness effects [376] may cause problems in this case.
- **Pattern identification tasks.** For these kinds of tasks we typically also use peripheral vision [120] to detect patterns that are not directly in the field of view. Pattern identification tasks are mostly based on user experience and exploit the general Gestalt principle [292] of "the whole is greater than the sum of its parts" to make the identification process cognitively less stressful. Moreover, patterns are in most cases not specifiable that easily by algorithmic concepts, hence visualization is an important ingredient in a visual analytics system.
- **Correlation tasks.** The impact of variables on other ones typically requires some kind of pattern identification task as a first step to identify correlations. This demands for having already some experience with those patterns and how they might look like, otherwise the pattern identification cannot work due to the fact that we do not know what to search for. Even more simple tasks are involved like reading tasks to identify the names of the variables that stand in a correlation behavior or judgment tasks to observe the extent of the correlation [223, 255].
- **Route finding tasks.** Following a single line or a sequence of connected lines is not an easy task if occlusion and clutter effects occur as in typical visual situations, for example, in graph visualization or route

network representations as in navigation systems. The law of good continuation [292] supports viewers in more easily finding a route, even in a complex network. Route finding tasks are a special example of following lines, which is a common task in many of the line-based diagrams such as node-link representations of graphs or line-based time-series diagrams consisting of several time-varying measurements.

- **Community detection tasks.** Identifying objects that belong together due to a certain criterion or relation is actually supported by the laws of proximity and good form [292]. If movement is involved for each object in a community separately, for example in animated graph visualizations [151, 190], then swarm-based behavior interpretation making use of the law of common fate is a quite successful strategy.

In general, a visual analytics system is needed for solving a series of complex tasks; in most cases the users are not aware of the fact that they subdivide such complex high-level tasks into simple low-level tasks, switching between top-down and bottom-up strategies, to find the right hints or solutions for confirming or rejecting a given hypothesis. The solutions to such simple tasks are the key factors in solving more complex analytical tasks, in particular if certain situations have to be understood quickly enough for timely reaction to urgent problems like epidemics, terror attacks, or other threatening situations. A reaction to unexpected events is as important as well-known events that require a fast response.

3.2.4 Refinements and Adaptations

In cases in which we do not exactly know if the visual metaphor or the visual variables have been chosen in a task-supporting way we are able to adapt those and watch the data from several visual perspectives. To reach this goal the users can interact with the visual analytics system until the visual representations meet the needs of the observers. A similar procedure can be applied to the algorithmic concepts for which the involved parameters can be refined as long as they have to. For example, changing a threshold value or a layout strategy (see Figure 3.11) might have a tremendous impact on the output of an algorithm, in particular in cases in which a probabilistic model or a randomized algorithm is applied, but also, in most cases, on the runtime, i.e. the time until an algorithm terminates with the result. One challenge for both algorithmic as well as visual refinements and adaptations can be the runtime until a new state is computed. Making these modifications many times in an exploration process can accumulate to a certain amount of time we have to

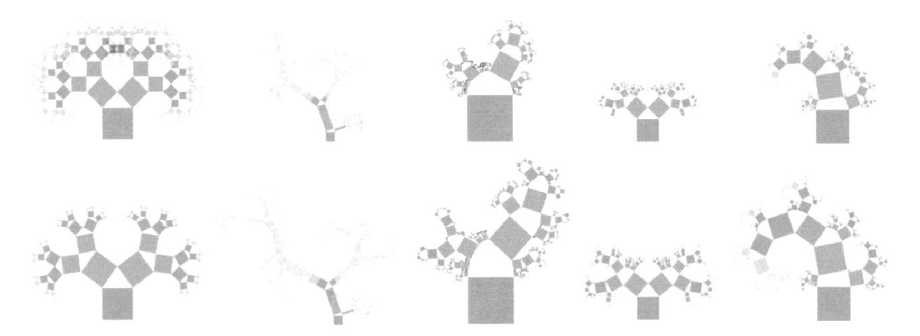

Figure 3.11 Changing the requirements for an algorithm can modify its output, which is seen in a visual result. In this case the layout algorithm for a generalized Pythagoras tree [31] for hierarchy visualization (top row) is changed to a force-directed one (bottom row) that creates a representation which is free of overlaps [363].

wait, valuable time that might be lost in which we might have done something else. If an urgent and timely solution is required, for example in an application scenario in which the users are under an extreme time pressure for making a decision, such an effect is not desirable and should be avoided when possible. However, some algorithms do not allow for a quick solution, even on the fastest computer [195], hence the user should be warned in such a situation.

Most modifications, either to the parameters of the algorithms or to the visual variables, have an impact on the other side, i.e. algorithms influence visualizations and vice versa. This means, for example, if an algorithm setting is changed, the new output will have an impact on the visual result, for example a refinement for the layout or arrangement of the visual objects. On the other hand, changing a visual variable might request an algorithm to compute a more efficient structure of a set of elements. This bidirectional interplay between algorithms and visualizations is important for a visual analytics system but also requires that the human observers are able to keep track of what has been changed and what stays the same or similar. This effect brings into play the mental map [356] which is a crucial aspect for keeping a system with visual and algorithmic ingredients usable and user-friendly. Such refinements and adaptations bring typical parameters in user experiments to check their impact on the user performance, but also on algorithmic performance without including humans. For eye tracking it is of particular interest that the visual analytics system can provide various types of visual stimuli to investigate the impact of parameter changes on them, and, further, on the interpretation of the users combined with the tasks. Only with these insights are we able to find out for which setting a certain design flaw occurs that needs to be removed. Moreover, with eye tracking we can record where (space) and when (time) the design problem appeared and who (participant) had bad experiences with it, reflected in the visual attention behavior.

The evaluation of such refinements and adaptations does not only depend on the shown visual stimuli but on even more factors. Interactions are supported by various input as well as output devices, can be done by individual users or in a collaborative manner, and the users themselves with their experience levels as well as tasks in mind build a crucial ingredient. Moreover, the eye can itself be used as interaction means, for example as gaze-assisted interaction [439], a concept that already records the eye movements during interacting with a visual analytics system.

3.2.5 Insights and Knowledge

Insights can be defined as any kind of extra knowledge that we did not have before letting a visual analytics tool run. Insight is the difference in knowledge after a run of a visual analytics system minus the knowledge before this run which requires that knowledge to be measurable as some kind of quantitative number. Insights can, and in the best case, are increased with each iteration for the user-in-the-loop who adapts and refines parameters over and over again to reach the goal of solving a given task. Insights depend on the task, i.e. how far we get to a possible solution; the more insights we get, the closer we are to the solution. It cannot be expected that the first insight solves our task at hand, but another run can provide us with even more insights, hopefully summing up, until the insight is so much that the users are confident. However, an insight could be as small as a new label that is successfully read as well as a complex link between two patterns that has been identified. Generally, and to sum up, insight is anything that brings us closer to a solution of a given dataset problem focusing on a task at hand. Gained insights can be applied to start the whole exploration, refinement, and adaptation process again under refined circumstances, under different viewpoints requiring that a user is able to remember already found issues and to strategically combine them.

In an eye tracking user evaluation it could be recorded how the user got the insight, i.e. what features have been used and what visual stimuli have been watched in a visual analytics system over time. This procedure generates a new dataset of spatio-temporal visual attention behavior together with the stimulus information that typically carries some meaning or inherent semantics, visually describing how a given task might be solved. If the insight is not enough at any stage the user does another iteration for which the visual attention behavior could be compared with the previous one to analyze which chunk of information was missing in the previous iteration. It can be evaluated if insights are as expected, good enough, how fast people came to them, if they made misinterpretations, and what they visually attended. Insight is much related to user confidence, meaning the more insight the users get, the more confident they are and the fewer eye movements are made from iteration to iteration, but as long as this is not properly analyzed the effect just serves as a hypothesis about the eye movement data. In any case the user getting frustrated while looking for more and more insights should be avoided. This effect might be detectable in the eye movement behavior, maybe by more and more chaotic visual attention patterns. On the positive side, the faster the insight is generated, with a minimum number of eye fixations, is a good sign for a well-designed system.

A maximum of insight finally leads to knowledge about a certain data scenario, aspect, or situation to allow making a decision. For example, in urgent cases for timely decisions with a lot of stress for the user, the eye movements might vary a lot [539] compared to a totally relaxed situation for which the rapid insight detection is not required but for which we have all the time in the world. Also for a real-time visual analysis in which the user cannot see the dynamic stimulus easily again and again, like in a replay, without losing real-time information, we could analyze the recorded eye tracking data to identify where the design flaws occurred for generating insights, hence real-time situations are challenging to evaluate by eye tracking. This is also problematic for a feedback loop allowing users to step back and adjust or adapt certain parameters, with the challenging issue that the real-time analysis is running at the same time as we wish to step back. For a non-real-time situation the feedback loop is a powerful concept since it allows us to go back to the data stage, but also to stay in the pattern, correlation, or rule generation stage, and refine and adapt it until insight is generated again. Visual analytics provides many opportunities to analyze data and to finally get knowledge about the data under exploration. To sum up, those insights are important to present, share, and disseminate to and with an interested audience.

3.3 Challenges of Algorithmic Concepts

Algorithms occur in many variations in a visual analytics system, be it for the problems in the original data (see Section 3.1.2) like fixing data capture errors, finding anomalies, or detecting noise, outliers, missing, or duplicated values, or for already processed and transformed data. Particularly for eye movement data, we could get low precision data in certain time periods, maybe caused by calibration issues of the eye tracking device. Moreover, if the data has to be transformed into a certain system-acceptable format we can be confronted by additional algorithmic problems like migrating or parsing the data, cleaning up the data, reducing, enriching, and up- or down-sampling it. If several data sources exist the data must be linked in an efficient but also meaningful way. More advanced data operations bring the data into a more structured and pattern-based form required to solve the users' tasks at hand, for example sorting, clustering, grouping, aggregating, classifying, categorizing, or projecting it. Even more, correlations and rules might be extracted from the data. Also most of the visualizations demand for efficient algorithms, for example, for computing suitable layouts that follow certain design and aesthetic criteria, like visual object placements without overdraw and occlusion effects or real-time animated diagrams that have to guarantee a high degree of dynamic stability to preserve the viewers' mental maps during the visual exploration. For real-time settings this cannot be pre-computed and it cannot be predicted where and when the user interacts with certain visual objects. A big challenge for algorithms in visual analytics systems is the fact that real-world problems cannot be defined in a proper algorithmic way, hence making it difficult to design, implement, or find the algorithm for a certain problem.

3.3.1 Algorithm Classes

There is no unique classification of algorithms, but the literature gives some explanations as to why a specific algorithm falls into a certain class [361]. Such a classification could be based on the computation strategy the algorithms follow during runtime like recursion, backtracking, divide-and-conquer, dynamic programming, greedy, branch-and-bound, brute force, and the like. Finding insights into the internal workings of an algorithm, typically supported by visualization or visual analytics, is described in Section 3.3.5. Moreover, a classification could focus on what the algorithms' results are based on, how their input is given, or what quality of results can be expected which is reflected in the classes denoted by deterministic, random, offline, online, optimal, or heuristic. In the following we discuss several of those classes and for which visual analytics scenario they are useful.

- **Deterministic or randomized.** We can distinguish between algorithms based on the result they generate. A deterministic approach follows a clearly defined computation strategy, meaning if the same algorithm gets executed several times on the same input data with the same input parameters, we will always obtain the same result. A randomized version can follow different rules from execution to execution due to the random effect, even if the same input data with the same input parameters is used. There even exist algorithm subclasses; for the randomized ones we could mention Monte Carlo and Las Vegas variants. This class of algorithms can be useful to consider in visual analytics if the runtime is important, i.e. randomized algorithms typically generate a result much faster than the deterministic counterparts, but the results have to be taken with care to not lead to misinterpretations.
- **Offline or online.** If the input is given, it only makes a difference how the input is given. This is reflected in the offline or online criteria in which offline means that an algorithm is aware of the input data and parameters all the time, which makes a solution more predictable. For online algorithms the input is typically not known, either completely or at certain stages during the execution which makes it sometimes hard to generate a good result due to the fact that a context information is missing or can change abruptly during the execution. Dynamic data problems, for example real-time data, are candidates for which a visual analytics system normally does not know what has to be expected.
- **Optimal or heuristic.** If the quality of the results and the time taken play a major role in the visual analytics system we should distinguish between optimal and heuristic solutions. This is in particular a good concept if the algorithm has a high runtime complexity, typically in the case of NP-hardness [195], for which we cannot expect a result in a reasonable time since no algorithm with polynomial runtime is known, while at the same time it might have large memory consumption, even for small data problems. An example would be asking for an optimal solution to a clustering or matrix reordering problem [34], maybe following the strategy of the minimal or optimal linear arrangement problem (OLAP). A heuristic approach can compute a non-optimal but still close-to-optimal and hence acceptable solution. This has the benefit that the solution is obtained much faster which means that we trade computation time for computation accuracy.

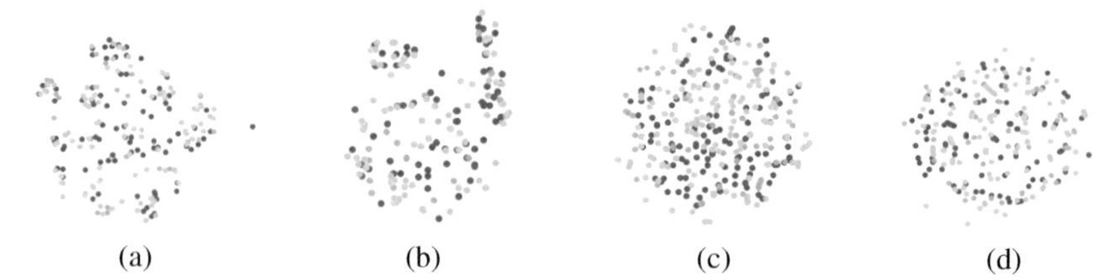

(a) (b) (c) (d)

Figure 3.12 Several dimensionality reduction algorithms applied to the same dataset consisting of eye movement scanpaths [79]. (a) t-distributed stochastic neighbor embedding (t-SNE). (b) Uniform manifold approximation and projection (UMAP). (c) Multidimensional scaling (MDS). (d) Principal component analysis (PCA).

The right choice of algorithms from a certain class is important for an interactively responsive system and as a designer of a visual analytics system we should have some expert knowledge about data structures and algorithms; even a simple algorithm such as a minimum or maximum search can end up in an inefficient end result. However, today's programming libraries mostly support efficient variants of the required algorithms, in the case that those algorithms are standard ones for a well-known algorithmic problem. The choice of the right algorithms depends on the data, its inherent properties, types, and structures, as well as tasks, and whether a fast or accurate/optimal solution is expected. In many cases there is a trade-off between runtime and exactness of the results. Visualization is important to show the output of an algorithm (see Figure 3.12 for examples of variants of projection or dimensionality reduction algorithms) but it is difficult to show the internal workings that could be done in an animated or static time-to-space mapping (see Section 3.3.5), in the cases that these internal operations can be tracked step-by-step.

The selection of an algorithm typically depends on the data and the patterns we plan to see in a given dataset. In the following we will describe several examples of data and tasks with corresponding algorithmic approaches being able to provide solutions to those given problems. How those are internally implemented depends on the decision of the designer, i.e. can it be based on a previously described algorithm class, as well as the way they process the data internally step-by-step (Section 3.3.5).

- **Counting.** Computing the number of elements is a basic algorithmic problem, but even that requires to understand what an individual unit is, like a number, a group of people, or the ups and downs in a time-series.

- **Ordering and sorting.** Bringing an ordered or sorted structure in a dataset seems to be a simple problem for one-dimensional data like a list of numbers, but if the to be ordered or sorted data has a higher dimension like a matrix [34] or multivariate data, we have much more variety and complexity in the repertoire of algorithms.
- **Clustering.** If a pairwise similarity relation is given among a group of objects or persons those can be clustered, meaning similar objects or persons are grouped together and non-similar ones are to be separated. Visually, this is typically indicated by the Gestalt law of proximity which leads to the pattern of clumped together visual elements.
- **Dimensionality reduction or projection.** In cases where the data has a high dimension, too high to easily inspect it visually, dimensionality reduction algorithms are able to reduce the dimensionality to a lower one, typically two dimensions, while the similarity and dissimilarity patterns in the data should be preserved as good as possible. However, such a strategy leads to some kind of information loss and computing such patterns can take a long time if a good result is expected (see Figure 3.12 for examples of t-SNE, UMAP, MDS, and PCA).
- **Route finding.** Identifying an efficient way in a network, based on some prior quality criteria, can be supported by algorithms. Popular examples are the finding of a shortest path in a road network, for example in navigation systems, or robot motion planning, i.e. efficiently guiding a robot through a complex environment following a certain goal.
- **Sequence comparison.** Data consisting of a set of sequences can be rather unstructured in terms of the order of the sequences or the alignment inside each sequence. In particular, for DNA or RNA sequences it is of interest to understand the commonalities of the sequences and even group or cluster them in a way that reflects these commonalities. This strategy is even applicable to general character sequences like texts or source code or even eye movement scanpaths, first translated into character sequences, in which each character represents a certain property in a scanpath.

Depending on the application fields there are various simple but also more complex algorithms consisting of several simple algorithmic routines that are linked in a way to compute solutions to the complex problems. Examples from this style are text or natural language processing algorithms like tokenization, stemming, and so on. Also graph- and network-theoretic problems have to take into account algorithms like computing a suitable layout for the users' tasks at hand, which gets even more challenging for a dynamic graph, i.e. one that changes over time for which the dynamic stability has to be taken into account to support the user as good as possible to preserve the mental map when using an animation to explore the dynamic data for trends, countertrends, or anomalies.

3.3.2 Parameter Specifications

Selecting the right parameters as well as the right values or value ranges is as important for an algorithmic solution for a task at hand as selecting the right class of algorithms and the proper way after which strategy an algorithm should process the data step-by-step during its runtime. In many cases it is quite hard to specify the values of algorithm parameters or the set of suitable parameters right from the beginning, which makes visualization a helpful concept to get the first visual idea about what an algorithm result might look like based on a fixed set of parameters and their values. Such a benefit of visualization is only useful if the runtime of an algorithm is not too complex, otherwise we would have to wait each time the algorithm is executed to see the result and, as a next step, adapt the parameters, and let run the algorithm again. This makes it important to have an initial feeling or experience about the class and type of algorithm as well as the size of its input parameters.

Although visualization can help to give insights into the algorithm results, we typically can only adapt one parameter at a time. If several parameters are modified at the same time it is hard, even impossible, to see which parameter caused the change in the results. On the negative side, changing one parameter after the other requires much more time to investigate the right parameter setting. A possible solution to this aspect would be to let the algorithm run several times, without seeing the results visually all the time in the end, but instead, statistical values describe the results. Those are not useful in understanding the details about a result but they can give a hint about the quality of the results depending on a certain value range for a parameter. Positively, this strategy can be used to reduce the search space for a good parameter configuration, serving as some kind of threshold value, but the statistical values are typically aggregations and averages of various results and, hence, it cannot be guaranteed that the best parameter setting is among the suggested ones. The selection process of the right parameter settings can be done manually or even another kind of algorithm could suggest and even adapt an automatically computed solution. If none of the parameter settings is appropriate to compute the patterns that help to solve the tasks at hand, we still have the option to try another kind of algorithm that might have a higher runtime complexity but successfully generates the patterns we wish for.

For algorithms from the class of online problems, we mostly have one chance to watch the data, because it is shown in real-time, hence the correct parameter setting has to be chosen in the beginning. This could be based on learning from previously seen scenarios or the data must be recorded and the important time periods will be observed in an offline setting after the real-time application has ended or even simultaneously, if the time permits. An offline inspection of the online data provides more time to explore and adapt the parameters to the users' needs, but in cases we have to react in real-time due to timely reactions, a visual analytics system must take into account such a complex scenario, even the combination of online and offline algorithms. It should even be possible to adapt parameters to modify the phase of an algorithm during runtime. For example, if visual output is given, even in form of a simple statistical value like a threshold number, we could initiate the algorithm to run in a new phase, for example from searching for an optimum to just a local maximum. This makes sense if the algorithm runs into a time-consuming routine with bad performance.

3.3.3 Algorithmic Runtime Complexities

Runtimes play a deciding factor for a visual analytics system in making it a live system and keeping it alive, avoiding waiting for a solution for a long time which is the worst thing that can happen if we are interested in a user-friendly system. If an algorithm runs only once, for example, to transform an originally not useful dataset into a usable form, we might accept that the algorithm takes a long time, but the user should be warned before running such an algorithm with a runtime estimation. In this case the algorithm might run on a different machine and inform the user when it is ready with its computation. In the meantime, the user can work at other tasks until the data is processed.

In a situation in which we have to use an algorithm all the time, after short visual exploration times, again and again, a badly performing algorithm is not a good idea. In such cases the algorithm should try to compute most of the needed steps beforehand, but this is more difficult than it sounds. In many situations we do not know what the user will be doing as the next steps, which makes it hard to predict what kind of information an algorithm should process beforehand to make the system a user-friendly and interactively responsive system.

A careful look at runtime complexities for a certain algorithm based on an input parameter $n \in \mathbb{N}$ brings us to the following classes of functions that describe the asymptotic runtime of a function based on the class it falls in [132]. Apart from this concept there exist many more runtime analyses like worst-case, average case, expected, amortized, or the analysis of competitiveness. The well-known Landau symbols [316] can help to give a mathematical abbreviation for expressing the runtime of an algorithm in terms of function classes an algorithm falls in. f and g are both functions from $\mathbb{N}$ to $\mathbb{R}_+$:

$$\mathcal{O}(f) := \{g : \exists c \in \mathbb{R} > 0, n_0 \in \mathbb{N} : \forall n \in \mathbb{N}, n \geq n_0 : g(n) \leq c \cdot f(n)\}$$

$$\Omega(f) := \{g : \exists c \in \mathbb{R} > 0, n_0 \in \mathbb{N} : \forall n \in \mathbb{N}, n \geq n_0 : g(n) \geq c \cdot f(n)\}$$

$$\Theta(f) := \{g : g \in \mathcal{O}(f) \wedge g \in \Omega(f)\}$$

$$o(f) := \{g : \lim_{n \longrightarrow \infty} \frac{g(n)}{f(n)} = 0\}$$

$$\omega(f) := \{g : \lim_{n \longrightarrow \infty} \frac{f(n)}{g(n)} = 0\}.$$

In a visual analytics system we would rather speak in terms of adjectives that describe the function classes which makes it more understandable for people not involved in algorithmic runtime complexities like non-experts in algorithms and data structures who are involved in the design process and just need knowledge about the adjectives but not the internal workings.

An algorithm has a constant runtime if the function modeling the runtime behaves as $f(n) \in \Theta(1)$, grows with logarithmic runtime if $f(n) \in \mathcal{O}(log\, n)$, grows linearly if $f(n) \in \mathcal{O}(n)$, and quadratically if $f(n) \in \mathcal{O}(n^2)$. Moreover, it has a polynomial runtime complexity if $f(n) \in \mathcal{O}(n^k)$, $k \in \mathbb{N}$, finally and worst for the runtime in these examples here, it grows exponentially if $f(n) \in \mathcal{O}(2^{cn}), 0 < c \in \mathbb{R}$.

3.3.4 Performance Evaluation

In cases in which the algorithms are too complex to estimate the runtime complexities using Landau symbols we need to measure the runtime in a

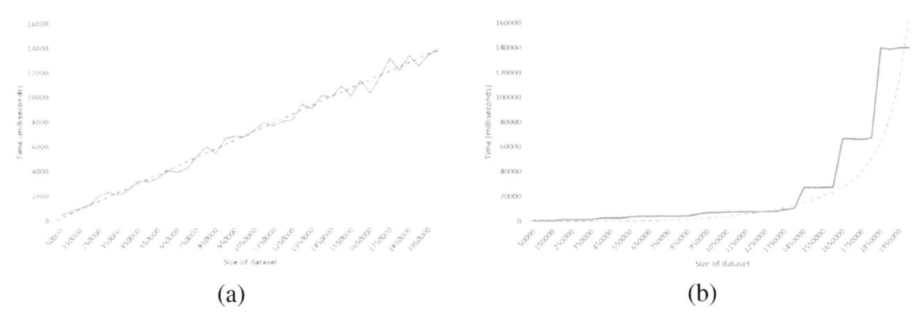

(a) (b)

Figure 3.13 Runtime performance chart for two algorithms generating visualizations. (a) A word cloud is generated for differently large dataset sizes, resulting in a linear-like runtime. (b) A pedigree tree is generated based on more and more people involved, standing in a family relationship, resulting in some exponential-like runtime.

different way. In such a scenario we could start the visual analytics system and for every computation that has to be made we could store the time it takes from starting the computation until the result is produced. This process could be done for several instances of the algorithm responsible for a certain computation by increasing the size of an input parameter, for example, the dataset size in terms of number of elements to be processed. Even an application on a different machine or a different environment can have an impact on the runtime performance. Plotting the measured performance values (see Figure 3.13), i.e. the time in this case, on the y-axis of a coordinate system and the size of the varied parameter, in this case the number of the data elements, gives a visualization of the runtime function. Having some experience with such mathematical functions and what they look like when plotted can give us an impression of which complexity class the algorithm falls in; however, it may be noted that this is not a mathematical proof, it is just a hint of what the runtime of an algorithm might be, which cannot be modeled in terms of input parameters and a function depending on these input parameters.

A performance evaluation might even be based on the memory consumption in addition to the runtimes. Negatively, if we use Landau symbols to express runtimes or memory consumption we only get the total aggregated performance of an algorithm after it has run. In a strange scenario, an algorithm might run very efficiently, maybe in $\mathcal{O}(n)$ time for most of the processing steps, but negatively, a few processing steps may take more time than expected, which would result in a much longer runtime. The aggregated performance is not able to explain this phenomenon, which makes it absolutely necessary to explore the internal step-by-step processes of an algorithm to investigate this negative issue. Such challenges are described in Section 3.3.5.

3.3.5 Insights into the Running Algorithm

An algorithm does not only generate results after it has finished running, it modifies a given data structure step-by-step based on input parameters typically following a certain well-defined goal. Only in rare cases does the output give a hint where the bug or error in the algorithm is located, and which line of code is error-prone and produces a wrong or misleading result. Further, it cannot tell us where a performance issue occurred and which program instructions and variables are involved. How the algorithm proceeds depends on the commands and instructions defined and implemented by the programmer of that algorithm. Visually representing the output of an algorithm is easier than showing its internal workings, which is due to the fact that the output is in most cases one instance of a problem solution, while the internal steps are several, not-finished instances, of the problem. Moreover, it might also be of interest to observe the variables in use, how they are modified, and how much time it takes to do such an operation. Even the memory consumption might be worth inspecting to identify another kind of possible bottleneck and shortcomings (see Figure 3.14). In a visual analytics

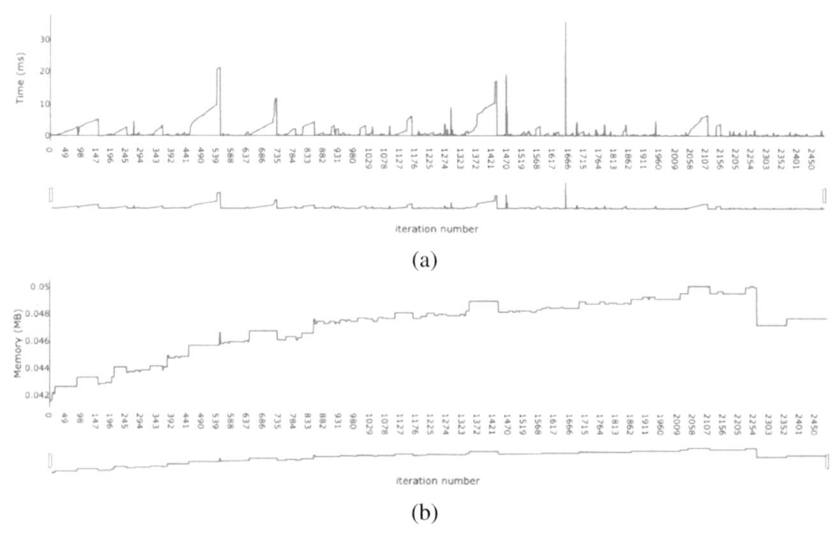

Figure 3.14 During an algorithm execution, the runtimes (a) as well as the memory consumption (b) might differ from iteration to iteration.

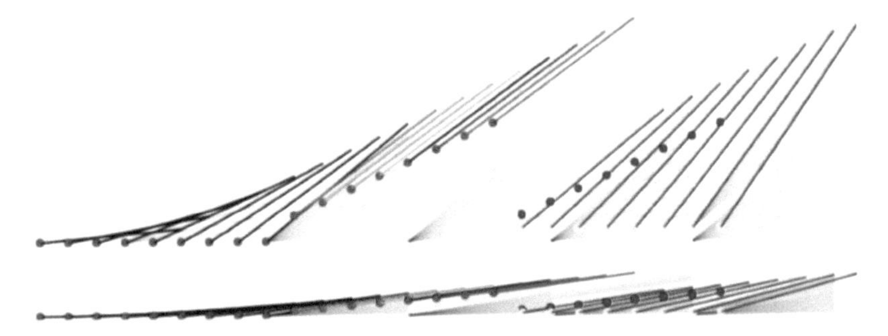

Figure 3.15 A time-to-space mapping of the vertices and edges processed by a Dijkstra algorithm trying to find the shortest path in a network is visually represented in a bipartite layout [75]. The time axis runs from left to right.

system it would be an extra insight if the users could choose to have a look into an algorithm and see how it works step-by-step (see Figure 3.15). Such an option would not be used by domain experts who would just explore their data, but by the designer of the system to understand the design flaws, not only on a visual basis but also on an algorithmic one, but supported by algorithm visualizations, typically as animated sequences consisting of the individual steps [59, 479, 480].

Before we can visualize the instances over time that an algorithm generates we have to access and store all the information that is required to get a suitable visual depiction of such a dynamic dataset in order to use it for insight detection. The tracking of the variables in use as well as the internal step-by-step states of the transformed data structure is challenging but can be of great help if visualized in an efficient way. We could even argue that the algorithms incorporated in a primary visual analytics system might be so complicated that another secondary visual analytics system is needed to analyze their internal workings for detecting design flaws in the algorithms of the primary visual analytics system. This procedure would make the primary visual analytics system a dynamic visual analytics system since the generated insights can be used to improve the algorithms, their runtimes as well as their visual representations, if eye tracking is applied as well.

To show the step-by-step execution of an algorithm [282] we have two major options which are denoted by time-to-time or time-to-space mapping. A time-to-time mapping also contains the concept of animation [505], which maps each time instance in a dynamic dataset to physical time, typically as a smooth animation with smooth transitions from one step to the next one to distinguish it from just showing individual time snapshots one after the other. In contrast, a time-to-space mapping is a static representation of the dynamics of the algorithm. The benefit of such a static representation is that it shows several instances at a time in the same display space which makes the dynamic data comparable over time. This is difficult, even impossible, for animated sequences. We have to stop and replay, but still comparison tasks are hard to solve successfully. A negative issue of time-to-space mappings is the fact that the display space has to be used with care to show as many time steps as possible, otherwise the individual time instances with their corresponding visualizations might get too small to be inspected visually. A concept that is between time-to-time and time-to-space mappings, some kind of hybrid approach, is rapid serial visual presentation (RSVP) [477]. A variant of it is called dynamic RSVP that shows a sequence of individual time steps as a time-to-space mapping and animates a sliding time window over the sequence to show the progress over time. This always gives an observer the chance to inspect a sub-sequence of a well-defined temporal length.

3.4 Applications

Visual analytics has existed for quite some time now which is the reason why it is applied in many application fields focusing on a variety of data analytics problems. The repertoire of examples for such tools is increasing day-by-day, the more complex ones focusing on heterogeneous data sources accessible online, while the most successful ones are typically published at renowned conferences or workshops or in journals and books. Some of the researched solutions even made it from academia to industry as a powerful concept to analyze specific datasets; however, for most of them, domain knowledge as well as some expertise in algorithms and interactive visualization techniques are required. From the various application examples we will just showcase a small selection for illustrative purposes.

The applications can also be distinguished by the size and the complexity of data they can handle, the types of tasks, the target users, like decision makers, who might have to react in real-time, or whether they have been designed for the expert or non-expert in a certain domain. Some of the prominent application fields are astronomy or astrophysics with continuously changing data streams full of noise, seismology with geographic information measured over several attributes, weather, climate, and meteorology focusing on predicting scenarios based on sensor and satellite data, security containing people networks and their interactions based on social media and global positioning system (GPS) data, or medical applications that are based on 3D human body data or DNA sequences as well as additional personal data sources including feedback and results of a therapy focusing on patients' formerly detected symptoms.

3.4.1 Dynamic Graphs

A dynamic graph Γ in one of its simplest forms is a finite sequence of n static individual graphs $G_i := (V_i, E_i)$, $1 \leq i \leq n$, $n \in \mathbb{N}$ that describe the time-varying behavior of relations between people or objects. The ith graph can therefore be based on actual time points, time periods, or just be a unique number expressing a certain order among the individual graphs. A static graph consists of a set of vertices $V := \{v_1, \ldots, v_k\}$, $k \in \mathbb{N}$ and a set of edges $E \subseteq V \times V := \{e_1, \ldots, e_m\}$, $m \in \mathbb{N}$ expressing which pairs of vertices stand in a relation and which ones not. If the weight or strength of a relation plays a role as well as the direction of the relation we typically denote that as a network instead of a graph. These data dimensions, i.e. vertices, edges, and weights build one of the simplest forms of a graph data type, but a dynamic graph adds one more data dimension to it, i.e. time or the sequential order.

Visual analytics of dynamic graphs [64] is a field that has been and still is the focus of research since time-varying graph data occurs in various application fields (for example flight traffic data as can be seen in Figure 3.16), in each scenario in which data elements stand in some kind of relation behavior. Analyzing the trends, countertrends, as well as anomalies over time between related persons or objects is of particular interest to identify group behavior, the spreading of disease as in contagion networks, or the message flow over time between people, to mention a few. Algorithmic approaches like generating task-specific layouts, clusterings over vertices and time, detecting shortest paths, or reorderings if adjacency matrices are

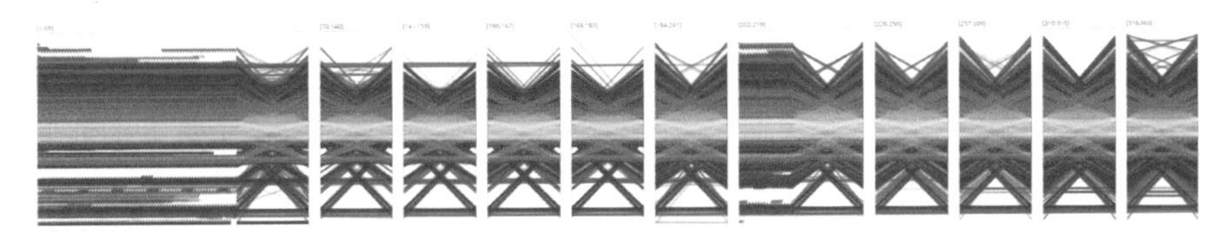

Figure 3.16 A flight traffic dataset taking into account temporal clusters while a bipartite splatted vertically ordered layout is chosen to reflect static and dynamic patterns in the time-varying graphs [2]. Image provided by Moataz Abdelaal.

chosen belong to the challenging aspects in this application field. From a visualization perspective [30] it is typically a user decision if time-to-space [75], time-to-time [151, 190], or hybrid approaches like RSVP [32] are applied. Apart from the visual challenges, algorithms are required to, for example, compute time-varying layouts focusing on a high degree of dynamic stability to help preserve the viewers' mental maps [18].

3.4.2 Digital and Computational Pathology

Another important field of research with a real-life background and application can be found in pathology (see Figure 3.17 showing an example of the PathoVA tool [134]). This field combines technological aspects from medical imaging as well as healthcare with the goal of improving the quality of diagnoses, i.e. to find the causes and effects of a disease, to predict its progress, and, based on the results, to find ways and therapies. In addition, based on various datasets, correlations between symptoms, patients' personal information, sensor data, and the outbreak of a disease can help to prevent others from getting sick. Although such a scenario sounds like science fiction

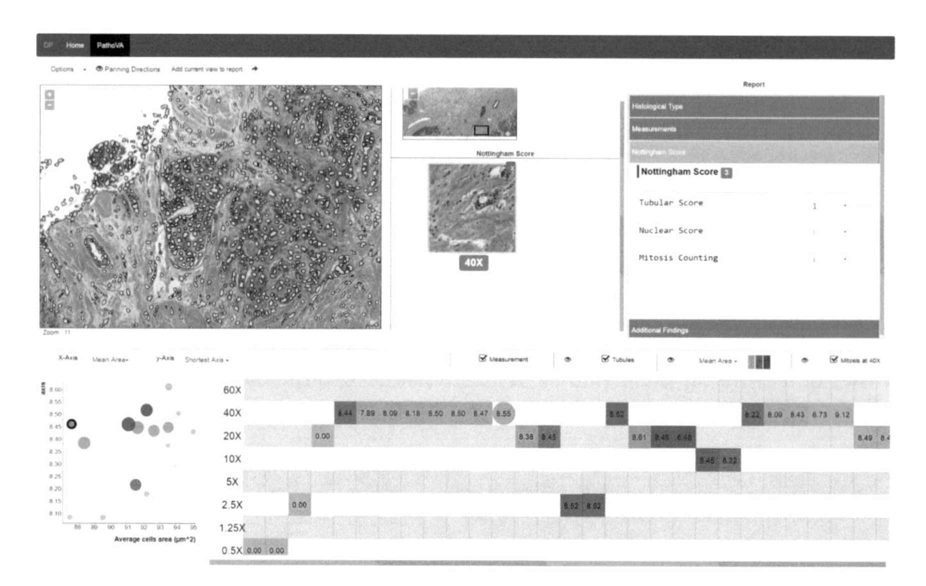

Figure 3.17 The graphical user interface of the pathology visual analytics tool with the image viewer, the image overview, a gallery with thumbnail images, a textual input to make reports, a scatterplot for showing correlations of bivariate data, and a view on sequential diagnostic data [134]. Image provided by Alberto Corvo.

or Utopian future wishes, visual analytics can at least support several, if not all, data analysis tasks for clinical researchers, given the fact that the data is available and ready to be used [135].

Typical tasks that are of interest come in the form of aggregating the data based on patient groups and their symptoms to derive a general kind of rule that holds if a certain subset of symptoms occurs for a special patient group. Based on such generated rules a prediction or therapy could be suggested while the new generated data and the reliability of the rule could be stored in the database with the goal of improving the accuracy of the set of rules. Data mining could be an important discipline here which shows again that visual analytics is some kind of interdisciplinary approach that has to take into account various technologies to be as powerful as possible. Insights from applied algorithms could be useful to find abnormalities in the data, facilitate the identification of a characterization that on the other hand helps to better group and categorize the data on several attributes and criteria, or, in general, information could be derived that is not directly visually observed from shown medical images for example. However, one problem this field has to deal with is of an ethical nature, i.e. the requirement for respecting individual human rights and privacy aspects. These challenges make data analysis and, in particular, the dissemination and sharing of the results to a larger audience difficult, although it would be a great benefit to find more reliable and accurate therapies given the fact that all of the patients' data is publicly available. The question is if the benefits outweigh the drawbacks and misuses of the data.

3.4.3 Malware Analysis

Activities in networks due to malware, i.e. malicious software like viruses, ransomware, or spyware, can have serious effects on the networks' performance, wasting time and causing massive cost to detect, identify, and remove the problem [142]. Data screening is a process in this domain that could be supported by visual analytics [106] in order to include the human users with their knowledge and expertise to accelerate the process of exploring the manner in which the malware behaves inside the networks (see Figure 3.18). Typical tasks in this context are discovering which areas of the networks are infected, how fast it is spreading, or where the leakage in the networks are. To find solutions to these tasks the data has to be transformed first; for example, it has to be projected to reduce its dimensionality, it has to be aggregated over time, or categories have to be derived, and the data has

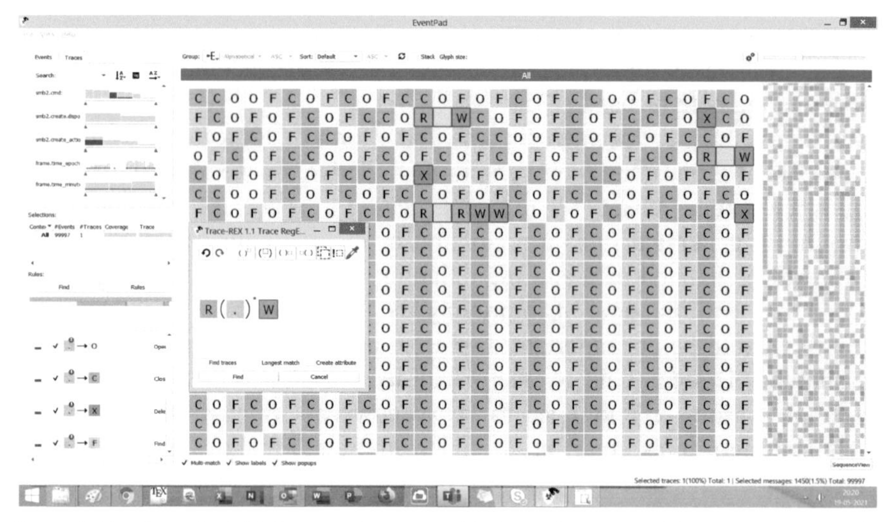

Figure 3.18 EventPad [106] is based on a graphical user interface with several interactively linked views to support data analysts at specific tasks to explore malware activities. Image provided by Bram Cappers.

to be classified to define events that are of special interest to support a data analyst.

In particular, in the cyber security domain, the data under investigation is oftentimes so complex and difficult to oversee, demanding some kind of external tool, like a visual analytics system, to get some insight into serious issues such as cyber attacks. However, even after applying such a tool it remains a challenging task to protect networks from such issues. In particular, the question comes up if such a visual analytics system cannot be used by the criminals to create much stronger, more difficult to detect, malware. This aspect shows that visual analytics can have benefits for both worlds, the good ones and the bad ones. On the one hand we try to find strategies and rules that describe malware attacks and their impact to mitigate the problem; on the other hand, it is a way to figure out how fast the malware can be detected and the network cleaned and freed from it, which shows the weaknesses of the malware and the ways to make it harder to detect.

3.4.4 Video Data Analysis

Video surveillance [232] generates a massive pool of data: individual snapshots of images that change over time, most of them containing

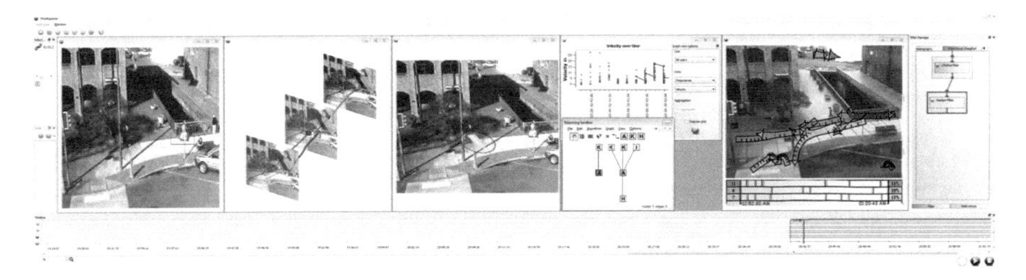

Figure 3.19 Visual analytics supports several views on the video data [232]: time navigation, video watching, snapshot sequence view, audio augmentation, statistical plots, graph views, schematic summaries, and filter graphs. Image provided by Benjamin Höferlin.

semantic information or visual patterns that might carry some meaning (see Figure 3.19). In video visual analytics [233, 234] we try to find reliable and fast answers to the tasks at hand, in particular, if the data stems from video surveillance to investigate criminal cases. Those tasks might be to compare video sequences to identify people who act in different scenes at different times, find analysis-relevant time periods in a video to reduce the time for browsing the video, or to detect certain well- or partially-known patterns, also known as weighted browsing. Visual analytics also takes into account additional information to augment and annotate a video for the observer, for example, based on object detection and tracking. Moreover, further derived information from the video sequence like color distributions might be of interest to apply some kind of temporal clustering or to faster compare several video sequences.

Watching the video to identify certain patterns that bring us to the solutions of analysis tasks would be one way but it requires to watch each video completely. For data that is produced from video surveillance systems this strategy is too time-consuming and would not quickly enough lead to a desired goal. Making a quick decision based on the gained insights is as important as to understand the relations between the environments, objects, and actors in a scene recorded as a video. Automatic analyses alone are a powerful approach to quickly examine the time-varying data but, negatively, the semantic information cannot be reliably extracted and judged by a pure machine-based approach. This is the point at which visual analytics comes into play since it combines the power of the machines with the perceptual abilities of the human observers being able to recognize patterns and to explore them in context to a semantic meaning that the machine is not able to derive.

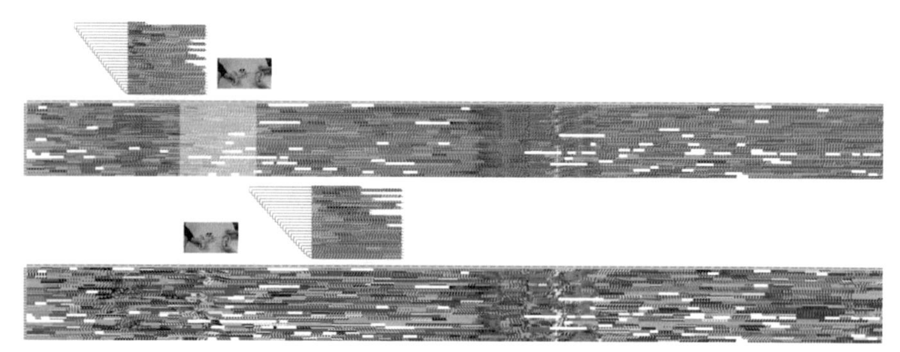

Figure 3.20 GazeStripes: visual analytics of visual attention behavior after several people have watched videos [309]. Images provided by Kuno Kurzhals.

3.4.5 Eye Movement Data

Eye movement data [161, 235] consists of space, time, and participant dimensions in its simplest form, but it can be augmented by additional data sources, for example, physiological data like EEG, blood pressure, pupil dilation, galvanic skin response, and many more [44], possibly being a candidate for big data analytics in cases where long-duration tasks have to be solved by millions of people like driving a car over longer distances. Visual analytics can be helpful to hint at design flaws in a static visual stimulus like a public transport map [372] or to understand the visual attention behavior in complex dynamic scenes (see Figure 3.20), given the fact that eye movement data is available for a certain number of people [308, 309]. This application is somehow related to the video application (Section 3.4.4) since the visual stimulus in an eye tracking study could be a dynamic one, meaning an accompanying video to the spatio-temporal eye movement data is required that has to be matched with the visual attention behavior to explore the visual task solution strategies applied by the eye tracking study participants.

Eye movement data cannot only be represented by a visual analytics system, but in addition, the visual analytics system itself could be controlled by eye gazes, leading to a field called gaze-assisted interaction [61]. For an eye movement data visual analytics system this would have two positive effects. On the one hand we can explore the recorded eye movement data while at the same time investigating if the recorded interaction strategies are based on a well-designed visual analytics system. This means each visual analytics system could be equipped with an option to explore the visual attention data as well as the data stemming from the gaze-assisted interaction. Another challenge in this line of research is whether the recorded eye movement data can be reliably analyzed to enhance a visual analytics system based on users' visual attention input, making it to a dynamic visual analytics system.

4

User Evaluation

Evaluation of an interactive visualization [185, 257] or a more advanced visual analytics system [307] can be done in several ways [315]. If we are more interested in the algorithmic runtimes (see Section 3.3.3) we might consider conducting performance testing that can give hints about the limitations in terms of processable dataset sizes or the step-by-step runtimes and memory consumption of an algorithm (see Sections 3.3.4 and 3.3.5). These limitations indicate whether the technique is still interactively responsive or whether we have to wait for quite some time to achieve a new result in form of a visual depiction, maybe a new layout of a graph or a set of points. This kind of evaluation focuses more on the algorithmic scalability of the visualization technique or visual analytics system.

On the other hand, we might consider conducting a study in which real users are involved, also causing ethical and privacy issues as a negative effect of incorporating real users (see Figure 4.1 for the most important study ingredients). Those could be laymen or domain experts. User studies can be done in several ways, with or without eye tracking, but their major intention is to get insights into the behavior of people while using a visualization or visual analytics system. The recorded data can be explored for finding insights like design flaws in the visual analytics system or which kind of visualization techniques are suited best for answering the tasks at hand. Eye tracking plays a crucial role in such a user evaluation because with this technology we are able to record the visual attention of study participants over space and time. Hence, a spatio-temporal dataset is generated which is much more challenging to be explored than the traditional measurements like error rates, response times, or qualitative participant feedback. When analyzing the eye tracking data in an efficient and effective way, we see the real value of eye tracking for visual analytics, the recorded eye tracking data alone is useless, only if visual analytics is applied on eye tracking data, the loop is closed and design flaws in space and time can be found and mitigated. This actually

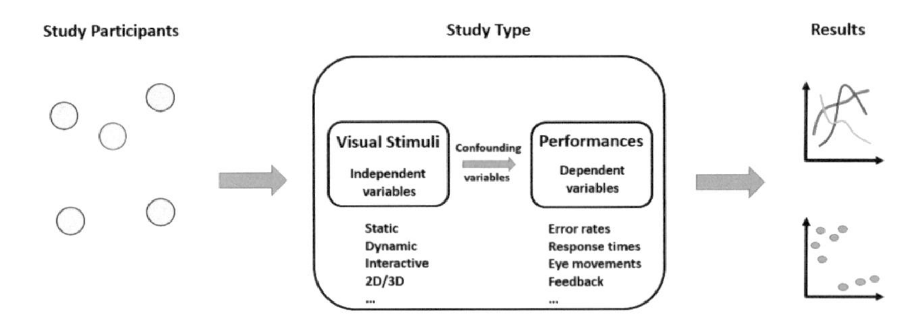

Figure 4.1 The most important ingredients in a user study are the participants, the study type with the independent, confounding, and dependent variables, and the results in the form of statistics and visual depictions.

requires some experience about the repertoire of existing eye tracking data visualizations [47]. However, both kinds of evaluations, performance and user evaluation, provide a wealth of insights for enhancing a visual analytics system, algorithmically or perceptually.

Evaluating visual analytics systems and analyzing the recorded performance data can have several benefits, for example, to obtain measured values instead of subjective feedback and personal judgments, but in case that spatio-temporal eye movement data is incorporated in the analysis process, we have to tackle many challenges before insights about the user behavior can be detected. One benefit is definitely that design flaws can be identified over space and time, even for certain specific participant groups, which leads to an improvement of the quality of the visual analytics system, maybe only for certain groups of users. This quality can be measured during the development or after the visual analytics product is finished, making it a problem- or technique-driven evaluation [450]. In the case that there are serious design flaws which lead to a degradation of performance at some task [426], the system might be redesigned in some way and evaluated again to check whether there is any kind of improvement. Even before starting the design and implementation phase, a user study, in this case more in the form of interviews, could help to guide the development process in the right direction. If several system variants are available and there is no clarity as to which one is better for certain tasks and user group, evaluation can help as a means to get a qualitative or quantitative comparison between the variants [397]. Moreover, not only can the visual analytics system be the focus of an evaluation, but also the users, allowing categorization into user groups based on varying performances or visual attention behavior. The obtained evaluation insights provide useful information to make the designed system more acceptable to a user group if the evaluation results convince them.

4.1 Study Types

Depending on several criteria we have to base a user study on one or several classes of study types. Those might differ in the way the data gets recorded, where it gets recorded, which kind of data gets recorded, the number, properties, experience levels, or quality of study participants involved, the technologies applied to record user performance and behavior, how the evaluation data gets evaluated, which tasks are involved as well as the actual or expected duration of the tasks, and many more. The human user is a crucial ingredient in a user evaluation, in contrast to a pure algorithmic performance or runtime evaluation that, on the other hand, is similarly important for a visual analytics system and its user-friendliness. Moreover, the type of stimuli shown in a study make a difference for the type of the study, for example, if the stimulus is static or dynamic as in interactive, animated, or video scenarios which mostly build the basis for studies in visual analytics. The bases for the visual stimuli are built by the data types and where they stem from, i.e. either real-world data or artificially generated synthetic data.

The study type also decides the preparations that have to be taken into account before the users can be recruited and confronted with the tasks to measure their performances. An appropriate study setup is required to guarantee proper conduction of the study with as few negative issues as possible to avoid erroneous study results. Many study designs are first evaluated themselves by so-called pilot or feasibility studies [160] to check whether the procedure has to be adjusted or not, or whether and how the real study can be conducted. For example, not only prior explanations and questionnaires, but also the execution of the study should be tested. Before having seen users in real action in a study we cannot really predict how they will behave, in most situations; they do not exactly follow the working plan and time schedule as we expected before running the real experiment. Moreover, the type of study typically limits the concepts that go hand in hand to finally get the best out of the study. In particular, for visual analytics systems equipped with lots of visualization techniques, interactions, and algorithms, i.e. supporting many modifiable parameters, several study types have to be considered to identify design flaws based on the opinions and performances of real users.

4.1.1 Pilot vs. Real Study

Before starting a real and oftentimes expensive study in terms of money or human resources it is a good idea to test the setup of the actual study in a pilot study that runs some time before, to leave room for possible adjustments. The pilot study can be regarded as some kind of feasibility experiment. However, a feasibility study is more than a pilot study since it asks whether a study can be conducted at all and in particular, how it can be conducted. A pilot study adapts the role of a preliminary study, but we should take into account that it already costs some resources. Generally, it follows the same design as the actual study but on a much smaller scale with fewer participants and, if the scenario allows it, also with a smaller number of trials. In the best case the number of trials should be the same as in the real study to check how much time the study takes for one participant and if learning or fatigue effects might occur and, in particular, after which trials.

The general goals of a pilot study are to test and check for aspects like strangely unexpected events and study design issues, also technical problems, missing study elements, additional costs, the study duration that causes fatigue effects in case the duration is not well-chosen, hence also the feasibility of the study. The participants from a pilot study should be different ones from the real study to avoid learning effects, but they should belong to the same population based on the same recruitment criteria as in the real study to avoid different characteristics which could be a confounding variable. The results obtained by running the pilot study should not be merged with the results generated from the real study to avoid grounding the insights on two study setups, which might cause misinterpretation problems and which might add some bias to the study. Moreover, adapting formerly built study hypotheses or research questions after the pilot study has been conducted should be avoided.

Since a visual analytics system contains a variety of functions with lots of interactive visual output it is quite a problem to guess how the study participants will behave in the given task situation and in the visual analytics environment. For this reason it cannot be expected that all functionality can be tested in one user study due to the fact that there are just too many parameters that can be varied. A pilot study can help to test if the study setup works as desired or if modifications have to be made. However, even before running the pilot study we have to reduce the number of functions to be tested. Moreover, if eye tracking is applied to record the visual attention behavior the equipment should be tested to make sure it is running properly, i.e. if the eye tracker records the data over space and time accurately. In addition, it should be checked whether the participants who are not familiar with eye tracking can effortlessly work with the new technology without being too stressed and feeling uncomfortable. Eye tracking studies are very different from regular studies without eye tracking.

4.1.2 Quantitative vs. Qualitative

Whether a study is based on a quantitative or qualitative setup makes a huge difference in the study design as well as the reliability and expressiveness of the generated results, for example, the instrumentation, population size, or even the analysis as well as the visualization techniques of the recorded data to mention a few [111]. Quantitative evaluations are typically conducted in a laboratory environment and allow a quite high precision for the setup, control, as well as the results, which are important criteria for generalization aspects. A quantitative study follows a strict plan including stages like hypothesis building, finding independent, dependent, as well as confounding variables, the study design is based on within- or between-subjects procedures, the evaluation and statistical analysis of the recorded values, the presentation of the results, and so on. To run such a quantitative study in a reliable and smooth way the experimenter has to control a lot of aspects to guarantee a successful study. This means that the independent variables are varied and their impact on the dependent ones, typically error rates and response times, is measured. The more independent variables are modified at the same time the less can be said about this impact since we do not know which variable caused the change. Compared to a qualitative study the results of a quantitative study have a high degree of certainty due to the controlled setting.

In a qualitative study we do not ground the results on explicit measurements with error rates and response times. This makes it more applicable to complex visual stimuli equipped with interactions like visual analytics systems. The benefit of this type of study is the fact that they can easily be included in a design process of a visual analytics system, for example, by interviewing the participants from time to time and by asking for qualitative feedback which can be used to improve the system after each development stage and hence, guide the design and implementation process. The participant feedback is mostly written down by taking notes or recorded by audio/video to make it more accurate when summarizing the content later on and to mitigate the experimenter's challenge of remembering all the important aspects that have been said during the experiment. Think- or talk-aloud studies might have an impact on the participant's behavior as well as writing down notes. This is even more problematic in remote eye tracking studies in which the participants should not move around too much to avoid calibration errors, and hence inaccurate scanpath recordings.

For eye tracking technologies it makes no difference if a quantitative or qualitative study type is used. The eye movement data can serve as additional input for either enriching the dependent variable response time with temporal visual attention information to identify the causes of response time variations between stimuli or when the time is wasted, why an error has been made, and which visual region in a stimulus might be problematic and worth enhancing. For a qualitative study we might enrich the verbal feedback by visual attention data to understand why, where, and when a participant made the comments. This gives a bit more insights from the cognitive processes that are not recordable by eye trackers. However, linking eye movement data with other data sources like qualitative feedback is a challenging task and likely producing interpretation errors. For visual analytics, both, quantitative as well as qualitative study types are useful. Quantitative studies can help to compare simple stimuli like static diagrams to find out which one would lead to better performance measures for a specific task at hand. Qualitative studies, on the other hand, might be useful in more complex scenarios, even during the design process, but the generated results are less exact due to the missing quantitative values.

4.1.3 Controlled vs. Uncontrolled

Having the control over a study design with all its independent and confounding variables is of particular interest to obtain reliable and trustworthy results. However, such a high degree of control comes at a high price. First of all, it takes a longer time to setup the study due to the fact that control must be given in many respects. Moreover, controlled experiments that typically take place in a laboratory, measure the performances of the study participants one after the other, not as in uncontrolled settings in which many participants can take part at the same time, for example, in an online setting. Although every effort is made to make an uncontrolled study as controlled as possible, there is no guarantee that the control always gives reliable results and the participants follow the required rules and procedures. Uncontrolled study types mostly measure the performance of hundreds or thousands of people [150] as in a crowdsourcing online experiment

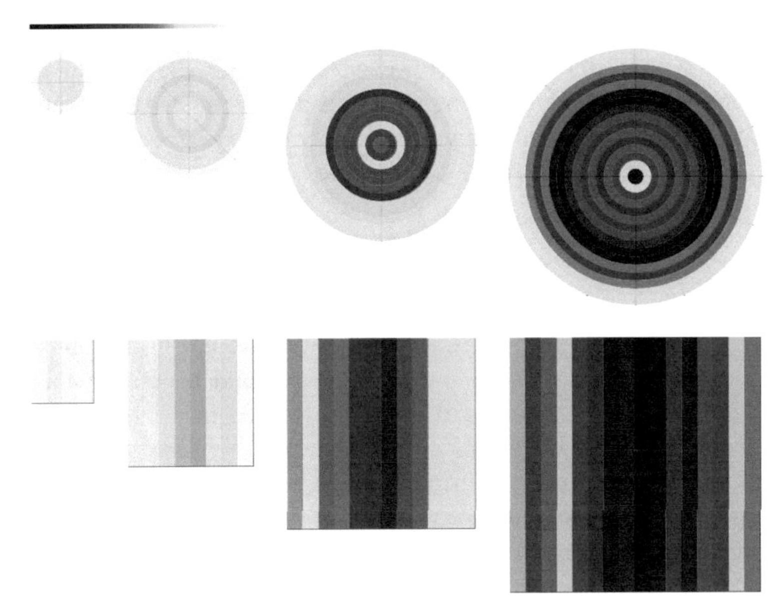

Figure 4.2 A comparison between Cartesian and radial diagrams in an uncontrolled user study recruiting several hundred participants in an online experiment [150]. Image provided by Stephan Diehl.

(see Figure 4.2), mostly being of a short duration of a few minutes to attract as many people as possible, sometimes offering money or an additional gift for an elected winner. If there seems to be a common pattern or strategy, this typically shows an impact from an independent variable on a dependent one, but still there is no guarantee for it, even if thousands of people show the same behavior.

We might say that the more control is forced on a study the smaller the population in the study. The question is, how we can put as much control as possible in a study while at the same time recruiting a large number of study participants which would allow us to increase the number of independent variables to be varied? This limitation of controlled studies has an impact on the visual stimuli, being rather static and trying to focus on comparing two or more visuals based on the same task. This might result in an assumption that one technique is better in terms of error rates and response times than another one under certain circumstances. For uncontrolled studies the parameter space can be larger, i.e. showing an interactive stimulus like a visual analytics system and allowing many people to experiment with it, maybe asking for a certain well-defined task and measuring the performance, or trying to get qualitative feedback from the people. However, explaining the task in a controlled setting for static technique comparisons is mostly much easier than if a more complex interactive stimulus with a variety of flexible parameters is given. Negatively, if the technique is too complex, the population size is reduced due to the fact that we need to recruit more domain experts to solve the task reliably. Asking non-experts would require a longer training phase for which we cannot guarantee, in an uncontrolled setting, that the training is done successfully.

Eye tracking brings some kind of control in a study design. It demands for calibrating the system to make it produce reliable and accurate data. Moreover, the eye tracking equipment is typically located in a laboratory, asking study participants to meet an experimenter who is familiar with eye tracking and the setup of the study. Currently, most people are not familiar with eye tracking as a technology, hence it might be challenging to let them participate in an uncontrolled setting, maybe in some kind of online eye tracking experiment. A future scenario, however, would be to allow many people to take part in a study to produce massive amounts of this type of data [44]. For example, head-mounted or wearable eye trackers might be an option to record eye movement data for a driving task which is some kind of longitudinal study type, typically in an uncontrolled setting since the experimenter cannot play the role of a co-pilot all the time. But, positively, experimenters could try to control such a study by video surveillance and give instructions via audio to not distract the participant too much. For visual analytics, this could also be a suitable scenario, with the difference, that the stimulus is always at the same place, i.e. shown in a display, not like a 3D dynamic scene as in car driving [387] or plane landing [431] experiments.

4.1.4 Expert vs. Non-Expert

Conducting an expert study requires recruiting people that have special knowledge or expertise of a certain domain. The recruitment process can be difficult and time-consuming since by only asking people whose knowledge level is high we limit ourselves to a small population, i.e. to only a few possible candidates. Moreover, spending the time of the experts might cost a lot of money since they can no longer concentrate on their daily job. To reduce costs and increase the value of a study it is unavoidable to ask experts who really enjoy to take part in such a study and who do not see this as a waste of time, i.e. just being interested in the money that can be earned. For example, medical scientists working as doctors in hospitals might be good candidates to evaluate a visual analytics system for analyzing patient data [134], but in reality they do not have much time and are typically very busy doing their daily jobs. This situation is particularly even more problematic if we are interested in a problem-driven evaluation that demands meeting the doctors on a regular, maybe weekly, basis to give expert feedback and to guide the development of the visual analytics system.

Another question arises when determining the quality of the experts. Each expert behaves differently, has more or less interest in the application, understands the visual analytics system differently, and provides feedback at different levels of granularity. Deciding whether a person is an expert or not might also be done after the study when evaluating the results. But this procedure should be done with care and the rules about classifying people's performance to identify experts should be given beforehand, i.e. prior to the study execution. This mitigates the problem of adding a bias to the study results, meaning the experts could be classified in a way that the results hold for confirming a certain research question or hypothesis. Also the design and presentation of the stimuli in a study should take into account which kind of experts are taking part. The domain experts are typically neither visualization nor algorithm experts, hence the ingredients from a visual analytics perspective should be easy to explain, to understand, and to use. On the other hand, if only the visual analytics system should be evaluated separately from the application domain, we might ask visual analytics experts, but then the application domain knowledge might be missing. In the best case, we need experts from the application as well as the visual analytics domain, a fact that again reduces the number of possible expert participant candidates. This is a general problem due to the fact that visual analytics combines many fields making it an interdisciplinary approach.

Finding an expert for participating in an eye tracking study can be even more problematic. First of all there is some kind of reluctance with respect to eye tracking, also for non-experts, due to the fact that the technology is not well-known for people unfamiliar with usability research. The first step when evaluating visual analytics by applying eye tracking is to explain the technology to possible participants and that it will not harm the eyes. Moreover, they might even be convinced by the new technology and be curious to participate. If someone is finally convinced to take part in the study, it is important to record if an eye tracking system has been used by this participant before since the familiarity with this technology can have an impact on the study results. A performance issue or visual attention strategy might not be caused by the task applied to a visual stimulus but by other influences, like the technology, which is another confounding variable in the study. It is always a good advice to conduct the study with non-experts first, and then refine the tasks, stimuli, and independent variables to reduce the study complexity. As a second stage we can use the adapted and refined version for the few experts that are available, hence we should not waste our valuable resources on an ill-designed study. As another interesting insight to explore the expertise of a participant, the scanpaths can be compared which could reflect expertise changes over time, i.e. with the progress of the study, but this insight could even be used to classify expert and non-expert users.

4.1.5 Short-term vs. Longitudinal

Traditional studies in visualization or visual analytics are based on a single study session that might last an hour per participant, letting people watch a stimulus or interact in a visual analytics system while performance, visual attention, or qualitative feedback is recorded. This one session setting makes sense since the fatigue effect is quite high, leading to performance drops after some time. In a longitudinal study setup [460], the participants can be invited to several sessions on different days during which a lot of data in a multitude of forms is recorded. That means they can relax a bit between the sessions to reduce the fatigue effect while they participate again and again in subsequent sessions. However, such a study setting brings some negative issues that have to be considered in the evaluation of the recorded data. One problem is that people typically forget some of the required aspects from session to session that are needed to successfully take part in the study. Hence, they have to be instructed and trained every time to bring them to a similar state each time. Another problem is that people might exchange their opinions with other study participants, introducing another kind of bias, comparable to learning effects. The study participants also do not behave the same in all of the sessions, making it hard to interpret and compare the recorded data between the sessions.

For visual analytics evaluated in an eye tracking study it can be quite challenging to recruit a lot of participants that spend their time on several sessions. The chance is high that not all people will finish all sessions, making the recorded data difficult to compare. However, in the case that enough participants can be found to solve a task while their eye movements are tracked, it can be interesting to research their visual attention strategies with respect to changes over time as well as comparisons between groups of participants having different properties. For example, short-term and long-term memory could be compared with respect to the recorded eye movement data, giving hints about how much can be remembered in a visual analytics system. The easier the longer complex scenarios and visualization setups can be remembered the more user-friendly such a system is, reducing the burden of cognitive effort when starting the system again and again. Also a mental map is important in cases where human users need to quickly get started from session to session. Modifying a setup of the visual analytics system between subsequent sessions can lead to confusion and frustration, caused by having to learn new aspects again and again, and not allowing the participants to always use the same setup every time, making them faster and faster and, hence, more efficient while solving a given task.

4.1.6 Limited-number Population vs. Crowdsourcing

If a lot of participants are required in a user study there is the option of conducting a crowdsourcing experiment [52, 194] which is only possible in the modern day due to the internet making it possible to quickly reach out to many people. To recruit a large number of people in a rapid manner we can make use of platforms such as Mechanical Turk (MTurk) [221], or invite people via mailing lists or advertising on social media like Facebook, Twitter, or LinkedIn, which can lead to a quite large number of people being attracted, given the fact that the study setup and visual stimuli are easy to understand. Otherwise, the number of people giving up will be quite high which is one of the biggest issues when conducting a crowdsourcing experiment. Hence, the more people can be accessed, the more people will complete the experiment, even if there is a high number of dropouts. To prevent a lot of incomplete answers, a pre-test can be conducted asking certain study-related tasks. Only the participants who have already invested much time in solving these tasks and paying attention seem to likely complete the real crowdsourcing study. Although there is no guarantee, it can at least serve as some kind of participant filter before running the actual experiment.

Crowdsourcing experiments bring into play some challenging issues that limited-number population studies conducted in a laboratory do not have. The biggest issue is definitely having no control over the participants, i.e. some kind of screening procedure like in controlled lab experiments is not given and one must more or less rely on the personal details the study participants provide, leading to an anonymity problem with lots of limitations due to the confounding variables that cannot be controlled. However, even if there is some kind of uncertainty in the recorded performance data, crowdsourcing experiment results can be exploited as some form of pilot study to investigate which parameters are important and worth further investigation [150], i.e. a crowdsourcing study can serve as a way to improve or optimize the setup of another more controlled smaller-scale experiment. In such a follow-up experiment the relevant parameters can be fixed and only a limited number of participants has to be recruited to generate performance values. Another problem in crowdsourcing experiments is that it can be quite challenging for longitudinal studies to match the participants and to get reliable results due to the several sessions and prerequisites to every session. In general, the tasks in a crowdsourcing experiment are performed with less attention than in small-scale experiments in a lab due to the missing experimenter.

Crowdsourcing experiments have to be carried out with care for visual analytics and, in particular, in combination with eye tracking to investigate the usefulness or the development of a system. Due to the missing control of the study participants and the task difficulty as well as the complexity of the stimuli it cannot be expected that most of the recorded eye movements are reliable results. Moreover, these days not many people own an eye tracking device, although they are getting cheaper and cheaper, making them affordable for the everyday user. In the future, standard computers might be equipped with eye tracking technology which would make crowdsourcing experiments of this new style imaginable. However, if such data was available it is questionable how accurate the recordings are for each individual user. Although it is still hard to think of such an advanced experiment, technically it might be possible in a few years' time, given the fact that the eye tracking hardware will become well-accepted by many people and, hence, cheap enough for mass production and integration.

4.1.7 Field vs. Lab

Field studies are conducted in the natural environment of the participants and are not like an artificially generated study setup in a laboratory, for example. This helps to understand causes for certain correlations or impacts from independent to dependent variables that could not be found in a lab. To obtain similar characteristics in a lab the experimenter has to simulate many aspects without ever reaching the same circumstances as in a natural field environment. In most of the cases in a field study it is difficult to measure the variable values as in traditional lab studies due to the fact that the study must be invasive to obtain good performance from the study participants which should be avoided as much as possible in a field study. Most of the results are hence received by pure observations of the participants while several environmental parameters might be varied to understand their impact on performance, behavior, or qualitative feedback. Interacting with the participants is challenging since this might lead to a confounding aspect to the natural environment.

Field studies are typically time-consuming, are expensive, and only a few participants can be recruited. Also the collection of the study data is more difficult as well as the evaluation of it. Moreover, due to the low number of participants it might be hard to generalize the found insights to a larger population. The collected data consists mostly of qualitative aspects like observation data to reduce the effect of being too invasive. However, if the participants are equipped with adequate sensors, we could even get quantitative time-varying performance data that produces very exact traces of the people's behavior, in the case of eye tracking the visual attention paid to a given static or dynamic scene. Without the right sensors the experimenter has to be as close as possible to get information while at the same time being far away enough to not distract the participant. This situation might cause a trade-off in a field study.

Whether a field or a lab study type is chosen makes a huge difference for the application of eye tracking technology. In a field study we might consider more head-mounted and wearable eye trackers since they must provide some freedom and room for the participant to move while not feeling distracted or uncomfortable. In a lab study we can also use head-mounted eye trackers but even remote eye trackers integrated in a computer monitor might work in this setting, but the freedom is limited a lot, requiring a person to sit rather quietly on a chair during the experiment. Actually, there are four different general scenarios for an eye tracking experiment given the stimulus variants and the flexibility of the study participants. The stimuli could be static or dynamic while the participants could be fixed to a location like a lab or freely move around, maybe in a field study. The participants might sit in front of a monitor with a static stimulus, they might move around to inspect a static stimulus when displayed at a high-resolution powerwall, the participants might sit in a car and the stimulus changes dynamically over time, or the participants freely move around while the stimulus is dynamically changing. The last scenario is probably the most challenging one for evaluating the recorded visual attention data. A visual analytics system is typically used while sitting still on a chair, while the stimulus itself is quite dynamic.

4.1.8 With vs. Without Eye Tracking

Many user studies in visualization and visual analytics do not make explicit use of eye tracking technologies [161, 235]. They more or less record user performances in terms of response times and error rates as well as some qualitative user feedback. While some kind of think-aloud protocol [199] during a task solution goes toward getting insights into the time-varying visual attention behavior, although described verbally, the recorded protocol is not exact enough to make reliable claims about visual attention. Think-aloud is also said to distract a study participant during the task solution [111] because some amount of cognitive capacity is dedicated to the verbal feedback all the time and hence less to the actual task solution. Moreover, to evaluate a think-aloud protocol the content has to be either written down during the study run or it has to be recorded as audio or video before its results can be processed, aligned, and prepared to be included in the study results.

Eye tracking does not lead to such a distraction. The viewer can completely concentrate on the task solution while inspecting the stimulus and, at the same time, the eye tracking device is recording the visual attention in the form of fixation points with their fixation durations. Although this spatio-temporal data contains more detailed information about the user behavior it also brings new challenges into play in the form of more advanced technologies to record the data, more complex and time-consuming algorithmic analyses and interactive visualizations for this technique, and the question of how reliable the recorded data is in terms of tracking accuracy as well as the relation to cognitive processes [268, 269]. Chapter 5 gives more in-depth details about eye tracking.

4.2 Human Users

The human users play a key role in a visual analytics system. They are responsible for modifying the views, interacting with the visualizations, or refining algorithm parameters with the intention to explore a dataset for insights and knowledge. Hence, they decide if a visual analytics system is well-designed, understandable, and created for the tasks at hand. On the challenging side, the users have various properties like being experts or laymen, belonging to certain age groups, having cultural backgrounds, being of certain genders, or even suffering from color deficiencies, color blindness, or having visual acuity problems, to mention a few. All of those aspects have to be taken into account when recruiting participants for a study, no matter which type it is. This requires that the users-in-the-loop, i.e. the study participants, have to be tested for a variety of aspects and properties before running the study, even before thinking about the final design of the study. Those human aspects definitely play a major role in the study setup, for example, they might even serve as independent variables to measure their impact on the dependent ones like error rates, response times, or visual attention strategies recorded in an eye tracking experiment.

The human users can play different roles in a study, typically described by the purpose of the study or the goal it is based on, like exploratory, predictive, formative, or summative [12]. These goals might be used to build another kind of study type classification in which the study participants are more closely taken into account than in the study types discussed in this book (Section 4.1). The individual users with their roles can have any kind of background knowledge but before starting a study, or even before designing it, the experimenters typically try to find the best possible way to recruit study participants [171]. In the recruitment process there is some kind of trade-off since the expertise of the participants stands in a negative correlation behavior to the relationship between the experimenter and the participants themselves. This means that the more expertise the population taking part in a study has to solve a given task, the farther away this population is from the personal environment of the experimenter. For traditional small-scale and simple-task studies this trade-off might bring some kind of bias in the study if this problem is not carefully taken into account.

4.2.1 Level of Expertise

The level of participant expertise plays a crucial role in the conduction of a study which might also be given by the environment in which a study is conducted, for example in a large company with the employees as participants [449]. This expertise is already important in the first stage when introducing the participant to the study setup as well as the technology and instrumentation used to measure the performance. The more expertise the users have, in particular in visual analytics, the less training has to be provided in a practice runthrough. The trade-off between the level of expertise and the accessibility of the people as well as the study costs typically decides the population that finally takes part in a study. For example, a lecturer at a university might first start by recruiting the students from a lecture since they have a closer relationship to the lecturer than experts from other universities. However, if the level of expertise has to be higher to solve the

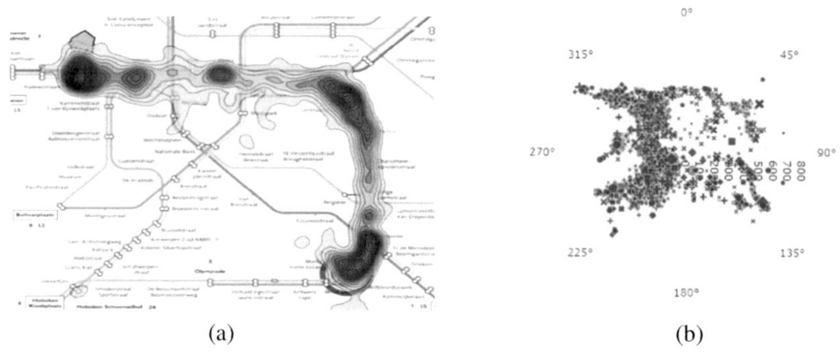

Figure 4.3 Examples from a visualization course at the Technical University of Eindhoven educating students in eye tracking and visual analytics [70]. (a) A visual attention map with contour lines. (b) An eye movement direction plot.

task and to understand the visual stimuli, like a visual analytics process, the students might not be the right choice for a study population. To raise them to a certain level of expertise they are educated, in the best case, in eye tracking [71], and hence, trained during the lecture to finally serve as some kind of experts in the study. After some training of about eight weeks the students are able to design and implement their own interactive visualizations for eye tracking data (see Figure 4.3), enriched by algorithmic concepts to transform the data [70], showing a steep learning curve, which means they obtained profound knowledge of this specific application domain that they did not have before attending the course.

The training has an impact on the individual performance, possibly introducing some kind of bias. It is important that the individual participants' expertise after the training session is checked by test questions and a thorough practice runthrough. All of the participants should be on a similar level after some time to take part in the study. Randomization, replication, and permutation are powerful concepts to average out certain differences in the expertise and performance that might even vary during an experiment, typically reflected in learning or fatigue effects. Measuring the expertise is difficult but not impossible. It can be based on the accuracy of the answers to, and response time to, test questions in the practice runthrough since only the expertise relevant for the particular study is important, not the expertise stated by a participant. People could even be excluded from a study based on the level of expertise, but this should be done after the participant has completed the study. Moreover, the rules for such an exclusion must be defined beforehand, prior to the execution of the study to avoid introducing biases that manipulate the study results.

4.2.2 Age Groups

We might tend to say that the expertise grows with age but there are specific age-related tasks that can be solved faster and more accurately by infants than by adults. Similar to this, old people might have much more experience to solve specific tasks that young people cannot solve. In most cases, studies are designed in a way to focus on a certain age group requiring that the participants in this group are able to provide answers to a given task in a reasonable time with an acceptable accuracy. For sure, there are tasks that can be solved more accurately and faster with growing age, independent of the expertise level. However, the training stage is typically more difficult for infants since they do not have the global understanding of certain aspects that we need to explain the connections between scenarios in the right way. If infants under the age of six have to be recruited we have the additional challenge that those cannot read text properly and quickly enough, hence giving them textual tutorials is impossible. On the other hand, their verbal feedback is on a different level than the one given by adults which makes the analysis step a bit more time-consuming, but given the fact that the information contained is less granular than for adults due to the limited vocabulary and semantic expression, this can also be a chance for more efficient procedures to find patterns and insights in the recorded verbal data. Children tend to focus more on the relevant and simple aspects while adults try to interpret more due to their life-long experiences and hence more complex dependencies might be mentioned.

From a technical and technological aspect, younger people might have fewer problems than old people since they have grown up with all of these issues, practicing them on a daily basis, for example when using mobile phones, apps, or certain public services like ticket machines. Perceptual and visual issues (Section 4.2.4) can occur in any age group but older people tend more to vision-related and cognitive diseases affecting memory and thinking abilities, for example eye cataracts or Alzheimer's disease. Amnestic diseases tend to cause problems with the memory while non-amnestic diseases tend to cause issues with decision processes, both of which are important when taking part in a user study involving visual stimuli. Also senso-motoric issues tend to grow with the age, making it hard to solve certain visual analytics tasks, in particular if it comes to complex interaction techniques for which several senses have to be used in combination to make the best of the system. But such diseases can be a great opportunity for user studies since they can serve as a way to understand the difficulties this population group has in daily life, hence evaluating a system for insights can be a useful strategy to improve the daily living environment for these people. The biggest challenge, however, is to get participants for a study related to such issues, also for the age group of young children, maybe under the age of six. Ethical issues also build a limitation in this research field and hinder the development of appropriate tools and systems.

For eye tracking, age plays a crucial role, not only due to reasons of the study setup like the sitting position's height and distance to the monitor to allow a smooth and comfortable calibration and tracking phase for a remote eye tracking device. Moreover, also many instructions are dependent on the age of the participants in an eye tracking study; infants must be controlled a lot more than adults, while elderly people may need many more details explained on the technological side. This is the case for controlled studies in a lab, but even more for uncontrolled settings in the field. For visual analytics it could make sense to include very young people, in case they serve as domain experts, for example, for analyzing data stemming from a certain kid-like environment like a kindergarten in which data about strategic game playing is recorded. A study setting could make sense in which a nursery teacher and a child watch such scenes on a monitor while the kid is eye tracked and the teacher tries to figure out the visual attention behavior of a kid in a specific scene. In this case eye tracking is required to produce another kind of data to the given video stimulus data as well as visual analytics to let the experimenter, here a nursery teacher, visually explore the data. For elderly people we could find similar useful scenarios, focusing on understanding the problems they might suffer from in their daily lives with the intention of enhancing their situation based on the insights found by applying visual analytics.

4.2.3 Cultural Differences

The internet gives access to people all over the world, in different time zones and various cultures. These cultural differences are also problematic in user studies [259] and can introduce a bias. A famous example in this context is the text reading direction. For people from a Western civilization a left-to-right reading direction is appropriate whereas in Asian countries like Japan the reading direction can be vertically, i.e. from top-to-bottom, and even right-to-left as in Arab countries. In a controlled experiment in a lab this effect can be checked by reading tasks and asking the recruited participants about their cultural background. In uncontrolled crowdsourcing experiments which might be run all over the world the reading direction must be taken with care, for example in a study setting in which text labels play a crucial role in understanding a certain correlation. One might say that the text orientation can be based on the reading direction of the participants but adapting the labels by their orientation can bring layout problems into play which did not exist in another text orientation, hence introducing a confounding variable that is hard to control. Moreover, translating a text into a participant's native language can bring semantic meaning issues that are again hard to control.

There are several visual variables or habits that differ from culture to culture. These differences can lead to misinterpretations in a user study for both sides, the study participant as well as the experimenter. This does not only hold for the visual stimuli or the verbal output in form of conversations or qualitative feedback, but also gestures might be problematic due to their different meaning among cultures. A famous example in this respect is color which can have a tremendous effect on the meaning, also depending on the context. Color is one of the most exploited visual variable in a visual analytics system and hence its associations have to be understood to avoid misleading results. For example, in a visual analytics system for medical data we might choose the color black to indicate that a person has died, but this can cause interpretation problems for people who interpret the color black as health. A color should be intuitive and quickly understood, and not lead to debates about its meaning.

For eye tracking studies such interpretation issues might be detectable in the visual attention behavior. Longer fixation duration can be a sign of unclear meanings caused by a variety of aspects, with cultural ones mostly not being on the list. Also knowledge about eye tracking technologies might vary among cultures, leading to some kind of bias in the study. It is a good advice to record people's cultural backgrounds and add some extra questions in a tutorial before running an experiment. In a laboratory experiment, for example, at a university with students stemming from many cultures, such background information can be collected by asking the student participants. In a controlled study setting the chances are smaller to run into such problems, but in an uncontrolled crowdsourcing experiment it can become a big problem one has to be aware of.

4.2.4 Vision Deficiencies

One important aspect in a user study is to understand the deficiencies that the participants might have. These can come in various forms, too many to be checked before running the experiment. But at least the most crucial ones related to color vision and visual acuity [314] should be taken into account. Doing the check absolutely right is nearly impossible unless we have the expensive equipment that an eye specialist would have as well as a lot of expertise that one gains after having studied this field for several years. The human eye is very complex and can be regarded as the window to our soul. This means it does not only support the seeing tasks for the study participants but it can even be observed by the experimenter for emotional states that might hint at other issues apart from the traditional user performances, visual attention behavior, or qualitative feedback. However, such emotional aspects are hard to measure and might even be related to cognitive processes. Emotional states can have a big impact on the study results, hence they create some kind of bias in the study.

The health of the eye can at least be checked for visual impairments that influence the performance of a participant. For example, visual acuity is important if texts of certain font sizes have to be read (see Figure 4.4(a)) which can be checked by a Snellen chart [470]. These well-known

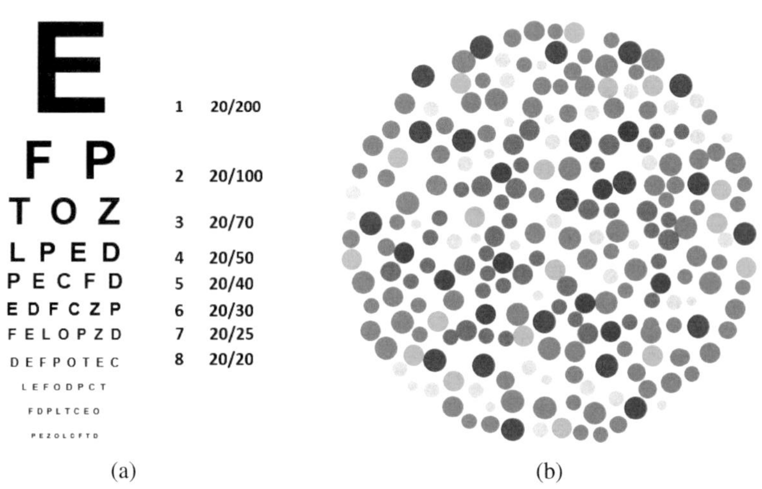

Figure 4.4 (a) A Snellen chart [470] can help to identify visual acuity issues. (b) An example plate of an Ishihara color perception test consisting of several pseudo-isochromatic plates [258].

alphanumeric capitals appear in different but well-defined sizes and should be read while the participant is standing at a given distance. To compensate such visual acuity issues the study participants typically wear glasses or contact lenses, i.e. their visual acuity is corrected-to-normal. Another problem is color blindness, mostly red-green, which is one of the most occurring color vision deficiencies among the population. In particular, this negative issue can affect a study related to visual analytics in which color scales play a crucial role to encode various kinds of data. The famous Ishihara tests [258] (see Figure 4.4(b) for an example plate) provide a number of plates that focus on asking people to read a number or pattern contained in a certain color noise pattern focusing on a mixture of colors to check for a specific kind of color blindness.

Color deficiencies are not a problem for the tracking of the data in an eye tracking study, but more for the conclusions that can be drawn from the recorded eye movements. People might fixate much longer on a certain region in a stimulus if they are color blind since they try to interpret the visual stimulus with respect to color coding and try to infer some meaning from the perceived color. This means that their scanpaths might vary from participants with normal color vision. Moreover, similar challenges might occur for visual acuity which change the reading behavior of the people in case they are not able to clearly see a written text and hence draw the wrong conclusions. Braille technology [490] is a way to support people with visual impairments but the question is to what extent such a support can be incorporated in an eye tracking experiment. Further artificially created negative issues that cause visual deficiencies are deep eyelashes that partially cover the eye which might lead to tracking problems in an eye tracking study.

4.2.5 Ethical Guidelines

Ethics describes aspects related to morality and tries to define a border between concepts that are either right or wrong or good or bad, ending in a crime if the ethics rules are not followed. Such criteria also hold in user evaluations, taking into account several perspectives like the study participants, the experimenter, but also the results related to the participants and existing in a data-like form. Recording personal data is one thing, sharing it with others is a different one. Before the data recording starts we should ask each participant for permission to use, i.e. record and share the sensitive data, maybe in an anonymized form. It is important that pressure is avoided when asking the participants for their agreement. From a study setup view it is advisable to respect the people involved in the study; this holds for the study participants as well as for the experimenter who can be mistreated without respect, for example, if people just take part for the reason of earning money. Everybody should be treated similarly, no matter which gender, age, culture, religion, and so on this person is associated with.

For the recorded data there could be the general problem that the results are not convincing enough or not convincing at all. In these days with various other research groups working in the same or similar area it is tempting to manipulate the results in a way to obtain more convincing results. Such fraud must be avoided, but it might be quite challenging to detect it. For this reason it is desirable to share the recorded data to make it publicly available for the research community to check for any inconsistencies in the analysis steps. However, the data might already have been manipulated before the sharing, which is hard to avoid. Some kind of mechanism would be beneficial that directly shares the recorded data in its raw form which would minimize the chance of modifying to get more convincing results. However, at least in the visual analytics field, such an approach is hard to realize and might contradict the goal of data protection and data privacy. At least, in many institutions there is some kind of ethics committee that checks the rules before running an experiment. Negatively, it can take quite a long time until permission is given to start a study, i.e. an ethics committee can sometimes slow down research, but it is definitely needed.

People tend to be more under pressure in an eye tracking study than in a traditional study without eye tracking from the perspective of ethical reasons. This is due to the fact that people might fear that they are more observed because of the general opinion that the eyes tell more about a human than any other organ. The eyes are some kind of window to the soul [246] and it is said that emotional states can be recognized [332] by reading the eyes as well as if somebody is telling the truth or lying. Ethics can be a problem in eye tracking research [320] because it might be unclear for the ethics committee to assess how invasive the procedure will be, if eye damage can occur, and which additional body-worn sensors are used to record additional personal and private data, which can cause a delay for the confirmation to start a study. This delay is typically caused by missing information about the relatively new technology, in particular when evaluating visual analytics systems. For example, the types of visual stimuli and the way they are presented plays a crucial role, also taking into account that people might have health issues when exposed to a flickering stimulus, maybe causing epileptic problems. People should be informed about the option to give up, even if they are forced to take part in the study with the money as some kind of pressure. Moreover, the data recorded in an eye tracking study contains much more confidential and privacy information than the data from a traditional user study without eye tracking.

4.3 Study Design and Ingredients

A technique- as well as problem-driven user study has to take into account several necessary ingredients to make it successful and to obtain results that are based on a reliable and correct setup. Any kind of erroneous procedure in the study execution can make the results useless and hence a lot of time is wasted and sometimes even money. For this reason it is important to make sure that the most crucial aspects have been incorporated in a proper and structured way [450]. For a technique-driven study we need more controlled aspects to compare the technique with the state-of-the-art while for a problem-driven study we mostly need to record or note down qualitative feedback given in interviews to improve the design of a developed tool. Starting from the fact that the participants are already recruited and the stimuli can be generated in the required parameter settings to model the independent variables for a technique-driven study or the system snapshots for a problem-driven study, we can think about how to create a study design that allows us to record the dependent variables as well as qualitative feedback and eye movements in a way that they can either be statistically evaluated for confirming or rejecting a formerly given hypotheses or research questions, or to qualitatively improve the current state of a visualization or visual analytics system. The stimuli generation process is driven by a list of user tasks [389] that have to be responded to during the running experiment to check the hypotheses or research questions based on recorded performance measures, qualitative feedback, or visual attention patterns.

All of this guides the major design of the study, but during this process it also has to be considered if the participants have to be split into several groups to avoid learning effects, for example, as in a between-subjects study design in contrast to a within-subjects study design. Moreover, the number of participants and the number of trials for each participant can be fixed after the coarse-grained design has been created, combined with a detailed plan describing how the individual steps are executed and how they link to each other. For example, if several tasks are evaluated, a certain permutable order of the task blocks has to be defined while inside each block another kind of permutation is important to compensate learning and fatigue effects that typically happen over time. The number of replications decides the coverage of certain configurations of the independent variables. A pilot study is an appropriate way to figure out if all the study parameters are well chosen or if some have to be modified and adapted to guarantee a smoothly running actual non-pilot user study. For eye tracking studies, the technology also has to be included in the study design process in an adequate way. Actually, in most cases, if we can run the study without eye tracking we can also run it with eye tracking by just using suitable eye tracking technology, i.e. either remote or wearable eye tracking systems. The downsides of adding eye tracking as an opportunity to record additional performance measures in the form of visual attention can be tested in a corresponding pilot study.

4.3.1 Hypotheses and Research Questions

The hypotheses or research questions build some kind of starting point to initiate a user study. They more or less guide the whole study design process and force us to follow certain rules since they describe the needed insights that we hope to get from the users and the performance they show while solving given tasks, focusing on responding to the tasks corresponding to the given hypotheses or research questions. Hypotheses can state that something is better or worse to a comparable variant of a similar style focusing on a specific task or they can claim that something is useful in a way that human users can understand which typically also includes some kind of user task. The terms "better/worse" and "useful" include some kind of performance that has to be recorded, evaluated, and compared, in most cases errors or response times in quantitative study setups and verbal feedback in qualitative studies. Varying an independent variable can give insights into the impact it has on the assumption that a hypothesis is true or not, i.e. can be confirmed or rejected. For example, a hypothesis might hold for small datasets but for larger ones it can no longer be confirmed. In some situations it is even unclear if a hypothesis has to be confirmed or rejected due to missing statistically significant evidence.

Hypotheses express a stronger claim than research questions; however, they also need more grounded statistical approaches to be accurately confirmed or rejected than research questions. In typical quantitative user studies, a formulated hypothesis might describe the fact that variant A is better than variant B with a special focus on a user task and the measured performances. In qualitative user studies we do not have the performance but we have to rely on verbal feedback to check the hypotheses. For example, we might argue that variant two is better than variant one of a visual analytics system by investigating the judgments or personal opinions of the study participants. However, we could also measure performance but in general in such a complex visual stimulus and study setting it is hard to compare performance due to the fact that the people start interacting and follow different exploration strategies. This makes the recorded values incomparable due to the fact that the concrete task is split into different subtasks, each participant uses a different subtask organization which is one reason why a hypothesis is typically not checked with raw numbers in a statistical evaluation.

Eye tracking can give insights into the visual scanning strategies and allows comparisons between the participants, even if they followed different subtasks to respond to the main task. Hence, eye tracking is a suitable way to confirm or reject hypotheses focusing on such viewing behavior, i.e. over space and time. However, the hypotheses are difficult to statistically evaluate, but visual analytics can serve as some kind of evaluation since it allows analytical reasoning incorporating algorithms, interactive visualizations, and the human user (Chapter 6). As a challenging aspect, the hypotheses in eye tracking studies can be much more complex due to the spatio-temporal nature of the recorded data, even combined with extra data sources [44]. Hypotheses might be built that refer to the space and time dimension in the data at the same time. Moreover, the semantics of a (dynamic) stimulus can be taken into account, like getting some information in a certain region at a specific time point that is applied later on in a different region in the dynamic stimulus. Such a scenario brings challenging issues related to cognitive processing, for which we cannot easily find answers.

4.3.2 Visual Stimuli

The visual stimulus plays the role of the interface between the independent variables and the visual output, i.e. what the study participants can see or interact with to solve a given task. If an independent variable is modified this has an impact on the visual stimulus, for example varying the size parameter of an underlying dataset on which a visualization is based has the impact that more visual elements have to be included in the visual output, hence maybe affecting the performance for a task solution due to visual clutter. This means the independent variables first have an impact on the visual stimulus which again may have an impact on the dependent variables that come in the form of performance measures or visual attention strategies in eye tracking studies.

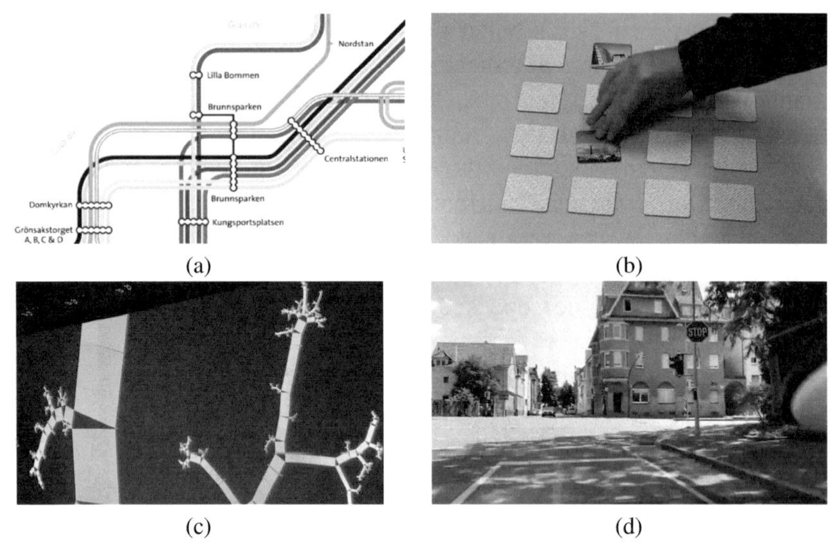

Figure 4.5 The way a stimulus is presented and the degree of freedom of the participant's position has an impact on the study design and the instrumentation. (a) A static stimulus, like a public transport map [372], inspected from a static position like sitting in front of a monitor. Image provided by Robin Woods. (b) A dynamic stimulus, like the game playing behavior of people recorded in a video [71], inspected from a static position. Image provided by Kuno Kurzhals. (c) A static stimulus, like a powerwall display [441], inspected from a dynamic position, allowing movement to change the perspective on the static stimulus. Image provided by Christoph Müller. (d) A dynamic stimulus, like driving a car with many other cars and pedestrians crossing our way while dynamically changing our positions [44].

However, the impact on the dependent variables is decided by the strengths of the human users when solving a task. In summary, we can say that in a visualization or visual analytics study the independent variables affect the appearance of the stimuli while the dependent variables are affected by the human users with their task performance.

A visual stimulus can be of several types, including the way it is presented and the degree of freedom of the participant's position which have a direct impact on the study design and the instrumentation (see Figure 4.5). We can distinguish between static and dynamic stimuli as well as static and dynamic participant positions. A static stimulus like a public transport map might be inspected from a static position, for example when sitting on a chair in a laboratory experiment [372]. A dynamic stimulus could be some kind of video that shows people playing cards while a study participant is watching it and sitting still [71]. A static stimulus could be a powerwall display and people walking around dynamically to watch the visualization from several perspectives [441]. Finally, a dynamic stimulus occurs when driving a car while many other cars and pedestrians are crossing our way [44]. The drivers sit still but move their heads dynamically to visually observe the dynamic scene while they actively change their positions by navigating the car dynamically.

Moreover, apart from a stimulus being static or dynamic it can be either 2D or 3D while a dynamic 3D stimulus seems to be the most complicated one for conducting a study since many parameters are flexible and hence uncontrollable. The dynamics can have several forms, i.e. a video or animation is just one linear sequence of static frames or time steps while interacting with a stimulus means changing the states on user demand, typically generating related static stimuli forming a graph structure [68]. The type of stimulus can even be classified by the way it is generated. For example, real stimuli are based on real-world datasets while artificially generated ones are mostly based on a stochastic model or manually created, sometimes in a time-consuming process, to guarantee similar characteristics of the underlying data allowing fair comparisons later on. The presentation of a stimulus and the flexibility of the study participants typically decide which kind of eye tracking device to choose. For visual analytics we are confronted by a dynamic stimulus equipped with various interaction techniques. In most cases the data analyst sits on a chair in a laboratory, but with the growing fields of immersive analytics [347] and augmented, virtual, and mixed reality [303] more advanced technologies are incorporated in the data analytics process and hence more flexibility for the study participants has to be guaranteed.

4.3.3 Tasks

The tasks in a user study can come in various forms, ranging from simple ones for which just one static diagram has to be inspected to very complex ones demanding for a variety of interaction techniques including gestures, touch, or gaze [413], applied to several visualizations, changing parameters, letting run several kinds of algorithms, and understanding the components, their interplay, and how they affect the dynamic user interface like buttons, sliders, menus, and the like. Simple tasks are mostly required in controlled studies in a lab for a technique-driven setting to evaluate if one static visualization is better than another one [389]. The complex tasks, however, are more interesting in visual analytics systems in which complex relations, correlations, rules, patterns, and models have to be understood and brought in context to others to finally derive some knowledge. The measurement of performance like error rates or response times does not make that much sense for complex tasks as it makes for simpler ones because the study participants automatically split the main task into several simpler subtasks to reduce the cognitive effort. These subtasks do not even vary in the number and function but more in their order.

Each participant has a different understanding of the main task and splits it into subtasks as required, and as the visual stimulus is understood. To respond to the main task it is required to merge the solutions of the simple subtasks in the end, which can be a challenging problem and requires some complex and well-structured cognitive processes [305]. Hence, in such a user study scenario in which the participants are exposed to complex tasks it is a good strategy to use a think-aloud setting with a lot of qualitative feedback, as well as eye tracking to better compare the visual task solution strategies [539] of the individual participants, maybe to classify them. Moreover, if a combination of interaction techniques is allowed, for example, in a 3D walking scene, video recording is a suitable choice to better analyze the gestures, body movements, and facial expressions of the participants. Complex tasks in complex interactive systems turn a study into some kind of behavior-driven experiment instead of the standard technique- or design-driven settings. Tasks target finding answers to the previously formulated hypotheses or research questions about a dataset. However, the various types of recorded data also challenge the analysis of the data and the verification of the formulated hypotheses also gets more challenging, maybe requiring visual analytics as a means to find insights in the heterogeneous data. More explorative types of tasks go in the direction of asking whether a tool or system is usable or whether the design of it has to be improved in some way. Those tasks are mostly relevant in a design-driven user evaluation and are asked after each development stage.

A task can also demand combining certain aspects from a visual stimulus to respond to a simple question. The visit order of the regions in the stimulus is important to solve the subtasks step-by-step that lead to a solution of the complete task. For this, the semantics of the visual scene has to be understood and applied to another region. Such a task can be as simple as “why is the road wet?” (see Figure 4.6). Without eye tracking it is unclear how the participant combined and linked visual objects in the scene to provide the final answer, given as qualitative feedback. Moreover, the response time might support as additional performance measure to compare

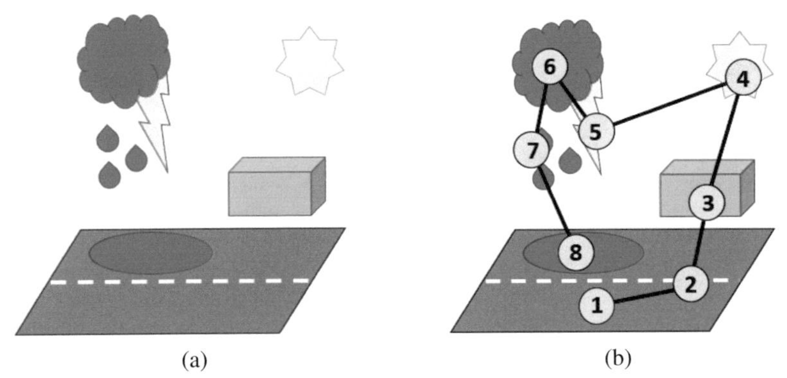

Figure 4.6 "Why is the road wet?" is a task that can be solved by watching a given abstract visual depiction of a scene (a). The visual scanning strategy to solve this task has to follow a certain visit order to grasp the information subsequently to solve the task (b). Eye tracking can give some insights into such viewing behavior [416].

scenes and explore their complexities in terms of a quantitative measure. Such simple tasks could also be important for visual analytics to understand if there is a design flaw in a certain visual or more complex process, i.e. the order of visual attention is not the optimal one to solve a task, for example the order of interaction techniques or diagram types. The task and even the task category [304] has an impact on the scanning strategies applied to the same visual scene as illustrated in an early work by Alfred Yarbus on a painting called "The Unexpected Visitor" [539]. For visual analytics, other tasks might ask whether the system is effective, efficient, user-friendly, engaging, easy to understand and learn, or usable by different user groups, all of them including real users who generate scanpaths worth analyzing.

4.3.4 Independent and Dependent Variables

With the independent variables in a user study we guide the possible impact of a visual stimulus on the dependent variables. While we do that we come across many confounding variables, those that also have an impact on the dependent variables, but which should be avoided or controlled whenever possible to increase the reliability of the results, in terms of the real value of this dependency given by the performance variation of the participants when responding to a certain task. Since we cannot see the independent variables, which are the factors to be studied, and hence, can neither react

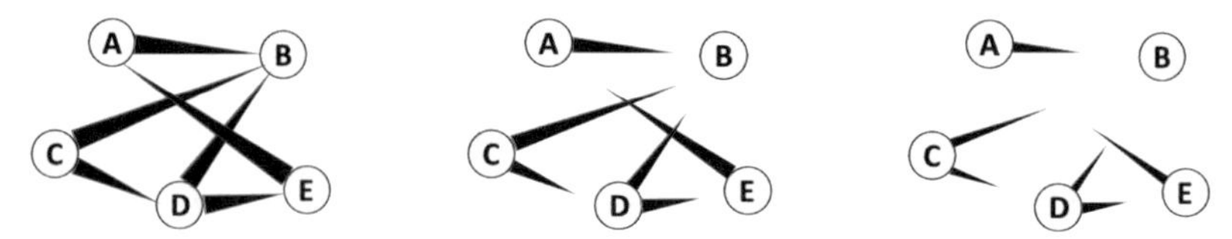

Figure 4.7 Varying the independent variable "link length" can have an impact on the dependent variables error rate and response time for the task of finding a route from a start to a destination node in a node-link diagram with a tapered edge representation style [97].

on them nor measure their impact on the dependent variables, we need some kind of extra means to show the change to the human user, i.e. the visual stimuli which serve as visual output. Those reflect the modifications of the independent variables and get perceived, observed, and visually processed by the user in order to efficiently solve a given task (see Figure 4.7 for a change of the independent variable "link length" in a partial link study [97]). The user efficiency comes as a challenging issue and it is impossible to measure whether a user performed as efficiently as possible. However, to achieve a real value for the performance measure we should trust the participants and tell them to either respond as fast as possible or as accurately as possible. Reaching both goals at the same time are typically two conflicting situations, i.e. the faster we respond the more errors we make and the more accurate we are the slower we are with the response. Most study setups ask for responding as fast as possible while still keeping a high degree of accuracy, in cases in which we are interested in the response time as a dependent variable reflecting a performance measure to compare two or more visual variants for example. In a comparative user study, for example, it typically does not matter if the participants behave similarly for both settings, hence replication, permutation, and randomization of the task blocks and trials should average out this effect.

In a visual analytics system the independent variables can at least be based on properties like the data, visualizations, interactions, algorithms, or displays.

- **Data based.** Modifying properties of the underlying data in a visual analytics system can show the impact on the dependent variables. For example, given the fact that the data structure remains unchanged – only its size in terms of the number of elements or the granularity gets changed – might have an impact on a certain well-defined user task. Another data aspect to be tested could be the completeness of the data in terms of missing values and errors it contains. An independent variable could vary the extent of such data gaps, maybe with artificially generated data based on a stochastic model.

- **Visualization based.** There are various visual variables that can be varied to test their impact on user performance. For example, data elements can be visually encoded differently focusing on lengths, sizes, areas, shapes, colors, and many more. Moreover, layouts and arrangements, visual complexities, compactness, or sparsity/density properties can be adjusted to understand how they affect the user performance. Even the positions of several views in a graphical user interface can be modified which seems to be a meaningful independent variable for a visual analytics system to be tested.
- **Interaction based.** If user interactions are allowed those can come in a variety of ways [544]. Understanding the difference between two or more interactions that focus on the same effect in a visualization can help to pick the most effective and efficient one based on the user performance. Moreover, input devices like a mouse, gesture, gaze, voice, keyboard, or many more might be compared, or even a combination of them to find the best way to interact in terms of user performance. However, testing different input devices typically demands completely different study designs and setups, hence the results must be taken with care due to the fact that the impact might not be caused by the input device but rather by the different study setup.
- **Algorithm based.** A visual analytics system provides lots of algorithms to process, transform, aggregate, project, or modify the given data. Supporting several algorithms that actually produce similar results, or one algorithm that produces several different results based on the same input parameters, like a stochastic approach based on a random function, could serve as an independent variable. It may be noted that the runtime performance of the algorithm is not investigated here, but rather its visual output that can be perceived and explored by the user based on a certain task.
- **Display based.** The output device can also serve as an independent variable. For example, small-, medium-, or large-scale displays provide more or less space for the visual output, hence they might have an effect on the user performance. Typically, a certain repertoire of interaction techniques is required depending on the displays like mobile phones, standard computer monitors, or large-scale high-resolution powerwall

displays, even AR/VR environments are suitable concepts, in particular in immersive analytics applications [347].

If we vary independent variables, only a few variations are possible, otherwise the parameter space of the study explodes and many more participants are needed, in most cases too many to successfully conduct the study. Consequently, testing the independent variables should be done with care, by investigating only the most crucial ones in a clever way. If a between-subjects study design has to be chosen we need many more participants due to the splitting into separate groups. A within-subjects study design tests all independent variables with all participants, hence no splitting into groups is required and, consequently, fewer participants have to be recruited. What both study designs have in common is the fact that we investigate the relationship between independent and dependent variables, sometimes leading to false positive (type I error) and even false negative (type II error) effects. A type I error describes the effect that a relationship is found although none exists while a type II error describes the effect that no relationship is found although one exists.

In an eye tracking study we generate many more dependent variables than in a traditional study without eye tracking [44]. The standard performance measures like error rates and response times are generated as well as in a standard study without eye tracking. Further metrics from a long list that are recorded in an eye tracking experiment are, for example, spatial fixation coverage, temporal lengths, fixation durations, saccade lengths, saccade orientations, scanpath lengths, Euclidian distances between start and end points, and many more [299], leading to multivariate data. This data is typically stored in a tabular form with rows and columns, while the rows represent the observations or cases, i.e. the study participants, and the columns represent the attributes, i.e. the measured dependent variables. Statistically evaluating the data is possible but visual analytics combined with machine learning concepts can provide even more insights than pure statistical approaches, for example, correlations between the dependent variables. Furthermore, in an eye tracking study we can even collect qualitative verbal data stemming from interviews, think- or talk-aloud protocols, open- or closed-ended questions, questionnaires, and even participant behavior, emotions, gestures, body movements, and so on, which might be of interest and worth analyzing. Finally, ratings focusing on aspects like satisfaction or frustration might be requested, mostly by showing a Likert scale ranging from 1 to 5.

4.3.5 Experimenter

In a controlled study setting the experimenter guides the participants and hence, plays a bigger role than in an uncontrolled setting in which the study participants more or less have to guide themselves through the study trials. The degree of control is higher if the experimenter is present and sitting next to the participant, eager to help, like in a laboratory experiment. However, the experimenter has to behave as constantly, smoothly, and similarly as possible for each participant to avoid biases. Moreover, the experimenter can be a confounding variable for several reasons, for example, just the fact that a person has a different effect on another person makes the mimicking of a similar behavior a difficult, nearly impossible challenge. Some study setups even try to hide the experimenter from the participants, for example, by a glass pane that is transparent on one side, but still the voice of the experimenter can have an impact on the performance of the study participants. Furthermore, even if the experimenter tries very hard to behave the same for every study participant, it is just not possible due to mood swings and personal reactions to certain participants. Experimenters are human beings and hence prone to feelings, errors, and misunderstandings. The only way to mitigate this situation is to average out the effects in the performance measures by testing many participants while, hopefully, the experimenter does not have a positive or negative impact on all of them in the same way.

Also in an uncontrolled study the experimenter has an impact, for example when interviewing people later on. However, in running an uncontrolled experiment, the study participant is typically left alone, for example in a crowdsourcing experiment, and the experimenter has to rely on the participants' best performances without giving them further instructions or giving them the chance to ask for help. The experimenter has neither the opportunity to intervene, in case the participants do not follow the study plan, nor does the experimenter have a good way of observing the participants or to take notes, unless the session is recorded as audio or video. In an uncontrolled setting it should be avoided that the experimenter is seen. Normally, it is not needed because all the instructions can be given in written form that have to be read by the study participants. However, also the writing and explanation style might have an impact on the participants, but at least it is the same style; however, the interpretation is based on the personal attitudes and understandings of the study participants. In summary, the experimenter plays a role to different extents when taking the perspective of before, during, or after a running experiment.

The experimenter plays a crucial role in an eye tracking study since a participant is typically not able to put on the eye tracking device or to calibrate it in a proper way. This challenge is due to the typically unknown technology to the laymen, i.e. non-experts who are not familiar with eye tracking. Another challenge for the experimenter is to explain the required features and functionalities a visual analytics system can have. For a controlled study, to which eye tracking studies typically belong, this can be managed by explaining the most important functions and by letting the participants do a practice runthrough and a test session in which they are guided as much as possible. For an uncontrolled eye tracking study, if this is possible at all, the biggest challenge is to setup the study for each individual participant and to explain all of the required features in a written form. Positively at least, we might have the recorded eye movements to check for the problems that occurred to improve the study setup for the next study participants in line.

4.4 Statistical Evaluation and Visual Results

Only storing the user performances or scrolling through the textual information does not really help since the data already has a certain size and is composed of various attributes. Statistically evaluating the recorded data [506] helps to aggregate or summarize it to a valuable claim but although statistics is a powerful concept it might even not show all the insights in a similarly powerful way as a visualization, diagram, plot, or chart can do. Statistics can hint at similar properties although the data, if visually depicted, shows a completely different phenomenon, for example illustrated by the Anscombe's quartet (Section 2.1 and Figure 2.7). It is a good idea to evaluate the usefulness of statistics first, before running the study, i.e. to explore whether it is worth the efforts. In many cases, a qualitative feedback can show more insights than any statistical evaluation could do [447]. A mixture of statistical analyses with its summarized values as some kind of data aggregation and a visualization that shows visual patterns from which a human can derive insights that a statistical number cannot give could also be a suitable alternative. The power of both, statistics and visualization, is the fact that the results from a user study can easily be presented and communicated to a larger audience, people who are able to derive knowledge from the results or to even generalize them. There exist various free and commercial tools to support statistical analyses, typically also integrating visual output in the form of standard plots like bar charts, histograms, box plots, scatter plots, or even parallel coordinates. Those tools are, for example, GGobi, JASP, PSPP,

R, RapidMiner, MATLAB, PRISM, SAS, SPSS, Stata, or STATISTICA to mention a few.

Although statistics provides a summary in form of values about a certain recorded performance measure in a user study, it has to be taken with care since it gives no absolute guarantee about those facts and how certain they are. Statistics aggregates the values while incorporating some kind of vagueness since there is always the chance that the statistical result may err in some situation. However, this error is kept as small as possible so as to only have a few cases in which the computed result does not hold, hence the statistical values have probabilities that give the likelihood that the results could really occur or not by chance. If qualitative feedback or eye movement data is recorded, the statistical evaluation cannot be applied directly, meaning that the textual and spatio-temporal data has to be transformed into suitable numbers first before statistics takes its part. Qualitative feedback might be projected to some kind of Likert scale first while eye movement data could be split into x-, and y-coordinates, fixation durations, or saccade lengths, to mention a few useful data transformations. Based on those numbers we might build statistical plots, however, researchers more and more make use of more complex visualizations (see Figure 4.8), in particular, to show eye movement patterns consisting of data dimensions like space, time, and participants [298].

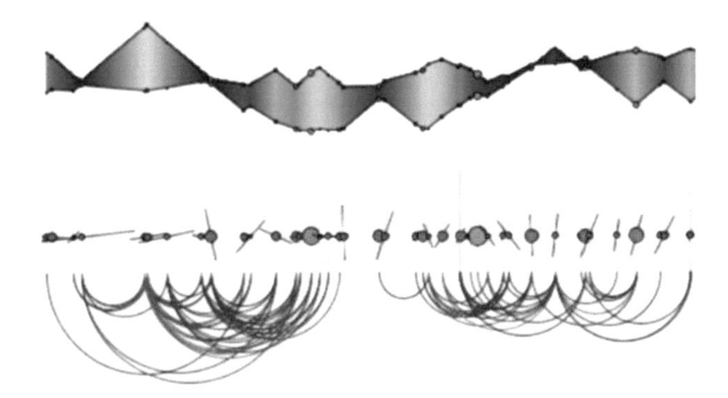

Figure 4.8 Since eye movement data is composed of at least three data dimensions like space, time, and study participants, the visual representations also get more complex with many aligned and linked visual components supporting pattern identification in the data [298]. Here we see the x–y positions in the top row, the saccade lengths and orientations in the center row, and the filtered pairwise fixation distances in the bottom row while time is pointing from left to right.

4.4.1 Data Preparation and Descriptive Statistics

There are various statistical tests, and all of them have certain goals and have their benefits and drawbacks. Moreover, their application depends on certain circumstances and data properties, i.e. the quality, distribution, and structure of the dependent variable to be evaluated. Fortunately, software packages, freely accessible or commercial ones, are a great support, but profound knowledge about the existing tests and their differences is required to know which test to successfully apply. Doing statistics correctly is a challenging task since most visual analytics experts are no experts in statistics. However, interpreting qualitative feedback or using a visualization technique to identify and explore design flaws might be a better option than relying on pure statistical numbers that might reflect a mixture between the good and the bad and hence cannot give enough details about serious problems in a system. However, if statistics is a like-to-have feature we can give some basic rules for applying it to performance measures.

First of all, the data to be statistically evaluated has to be prepared for the analysis. For this reason, obviously erroneous data has to be removed from the repertoire of the performance datasets as well as participants who had vision deficiencies and obviously could not solve the given task in any way. Removing or cleaning data are serious actions that should be taken with care. Never just clean a dataset just because it does not seem to fit into the final result. The rules about which data to be cleaned or even removed should be clear right from the beginning, i.e. before the first participant started to take part in the study. Otherwise, it might be tempting to remove participant data that leads to better results, for example, to confirm or reject a specific hypothesis. In any case, there must be a pre-defined rule about the reliability and usefulness of datasets that make it into the final round, ready for statistical evaluation and this rule should not be modified or adapted during the study. In any case if the recorded data is modified due to whatever reason it should be mentioned in the study report to recap why it was done and which data was under investigation in the final statistical evaluation, i.e. on which data the results are based.

Normally, performance data consists of a list of quantities, numbers or values with which we can do arithmetic operations. This fact brings simple descriptive statistical values into play, for example the minimum or the maximum of a list of numbers. Moreover, to make it a bit more complicated we can compute the mean or the median of all values. The minimum and maximum can be real outliers of a list of values, which don't tell us anything about the major distribution of the numbers. The mean value, i.e. the average value, takes into account all numbers, however, it can be a value that is far from being a representative value of the distribution, for example in a case, in which the numbers are distributed at both ends of a scale and not in between. On the other hand, the median value, which could be explained as the value in the middle, could also not tell anything about the distribution of the list of values. However, it is nice to have such descriptive statistical values that give an impression about a central tendency of a value list. They already give some very general hints about the data, for example, when inspecting the minimum and maximum in combination which provides insights about the range of the values. Negatively, they do not tell us enough about a distribution of the values and can lead to wrong conclusions, like in the example of the Anscombe's quartet.

To provide even more insights into a list of user study performances, statistics is equipped with further expressive values, for example, indicating the distribution or spread of a list of values around a certain point or in a certain value range, but telling us more than just the standard minimum, maximum, mean, or median values. The variance, for example, takes into account each value by including the difference to the mean value which is important for the spread of the values in a list. To reach this goal, the sum of the squared differences of all values to the mean divided by the size of the value list gives the variance (Var) which is at the same time also the squared standard deviation (SD), a more commonly used term in a scientific report or research paper. If $X := \{x_1, \ldots, x_n\}$ expresses a list of performance values, $\overline{x} := \frac{\sum_{i=1}^{n} x_i}{n}$ the mean value, then the variance and standard deviation are defined as

$$\mathrm{Var}(X) := \sigma^2 = \frac{\sum_{i=1}^{n}(x_i - \overline{x})^2}{n}$$
$$\mathrm{SD}(X) := \sigma = \sqrt{Var(X)}.$$

Generally, the variance describes the spread effect of the values in a given list from their mean value. It may be noted that if we use a population for our performance value list, we use n to divide, while for a sample, i.e. a smaller part of a population, $n - 1$ is used.

4.4.2 Statistical Tests and Inferential Statistics

Apart from descriptive statistics we can have a look into inferential statistics that tries to generalize the insights gained from a small population to make it applicable to a much larger population. One goal is to generate statistically significant results to reject the null hypothesis (H_0) which serves as a default, expressing that no statistically significant relation occurs between two value lists, typically focusing on differences between performance measures in user studies. H_0 remains true until it can be rejected by statistical means, for example using a p-value that expresses the probability that the null hypothesis is true. This brings the term significance into play which has been defined as the fact that the probability p should be smaller than a given value of $p = 0.05$, also denoted as a significance or α level which can be understood as the probability of a type I error, i.e. the fact that we support an alternate hypothesis in a case in which the null hypothesis is true. Generally, the lower the value of p, the more significant is the obtained result. There are several tests that support this significance aspect in inferential statistics, for example the t-test [551] like Welch's or Student's t-test, analysis of variance (ANOVA) [183] like one-, two-, or three-way as well as factorial ANOVA, with the Kruskal–Wallis [296] or Friedman tests [187] as special cases, just to mention a few.

- **t-test.** With the t-test we are able to compare two mean values of two given performance value lists, i.e. distributions of dependent variables. The result of the test expresses whether the values indicate a certain difference to a certain significance level. To reach this goal the variance of the value lists is taken into account.
- **One-way ANOVA.** Comparing three or more mean values of performance value lists, i.e. some distributions of dependent variables in the study, to test them for differences, is more difficult than just two. The F distribution is considered to do this reliably in a one-way ANOVA, hence using F-tests to statistically evaluate equality of mean values. Moreover, a post hoc test after the ANOVA test is required to check between which pairs of mean values a difference exists since the ANOVA can only tell us that there is a difference between all the means.
- **Two- or three-way ANOVA.** If two or three independent variables come into play we rely on the results of a two- or three-way ANOVA which is some kind of extension stage of the one-way ANOVA. These kinds of ANOVAs have the additional benefit that they can find results about interaction effects between two or more independent variables.
- **Factorial ANOVA.** Again, if more than one independent variable is under investigation we typically apply two- or three-way ANOVAs, or in the case there are even more of them to be analyzed, we refer this to as

a factorial ANOVA. However, a four-way or even more-way ANOVA is not found very often due to the challenge of understanding the generated results and their reliability.

Although these tests are in most cases applied to performance measures from a user study, they might even be applicable to eye movement data. To make them useful we have to derive some kind of quantitative values for the dependent variable coming in the form of eye movement patterns. For example, the fixation durations or saccade lengths could serve as standard dependent variables, but it may be noted that those do not express the visual attention behavior in terms of spatio-temporal behavior, they are more like a separate form of information, having lost the context to the shown stimulus, i.e. the spatio-temporal information.

4.4.3 Visual Representation of the Study Results

For descriptive statistics there mostly exist several impressive visualizations that are easy to understand and hence useful for non-experts in visualization, researchers who just want to get an overview of the recorded study data, the distributions, or correlations between two or more dependent variables. For individual participants or participant groups we could use bar charts (see Figure 4.9 (a)) to visually explore the performance differences while pie charts (see Figure 4.9 (b)) might be useful for a part-to-a-whole relationship, i.e. how much time each participant spent for a certain task with respect to the total time spent by all participants for the same or for all tasks. This might give a hint at the percentage of the time for each participant and could help to identify the slow ones and separate them from the time-efficient ones.

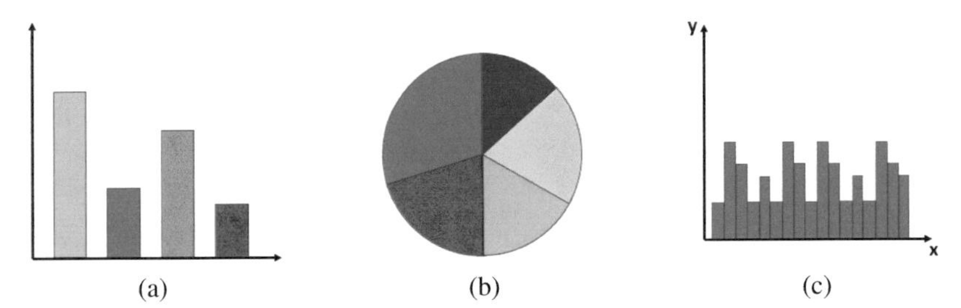

(a) (b) (c)

Figure 4.9 Easy-to-understand diagrams are often preferred for depicting the results of a statistical dependent variable in a visual form. Such a variable could, for example, be a performance measure like response times or error rates or participants' individual feedback in the form of values indicated on a Likert scale: (a) a bar chart. (b) a pie chart. (c) a histogram.

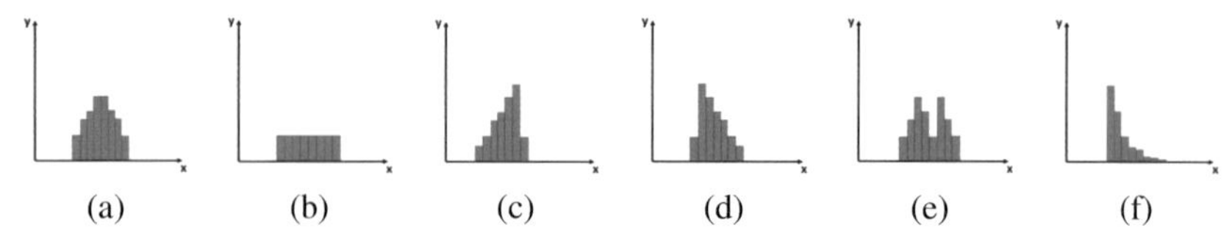

Figure 4.10 A histogram can contain various patterns indicating a property of the distribution of the dependent variable under investigation. (a) Bell-shaped or normal. (b) Uniform. (c) Left-skewed. (d) Right-skewed. (e) Bimodal. (f) J-shaped.

A histogram (see Figure 4.9(c)), on the other hand, can be used to show the distribution of a population with respect to the performance measure, i.e. a quantity is mapped to the x-axis instead of a categorical information, the participants, as in the case of a bar chart. In a histogram we typically encode the number of people falling into a certain bin, i.e. a value range on the x-axis representing the corresponding performance values. The height visually depicts this number and hence, gives an easy-to-understand overview of the bins that are frequently hit and those in which not many values fall. The shape of the resulting histograms can be interpreted for patterns indicating a property of a certain distribution (see Figure 4.10).

If the evolution over time of a performance value is of interest we might use line charts that connect the points indicating a certain value at a time point (see Figure 4.11). This mostly results in visual shapes that help to identify trends or countertrends, in case several of those temporal variables are plotted, for example, focusing on fatigue or learning effects that might have an impact on a performance measure over time. Moreover, temporal data might include anomalies or outliers, i.e. effects that do not follow the overall trend pattern. It may be noted that line charts should be taken with care if discrete or categorical values are depicted at the x-axis since the lines would reflect some kind of interpolation effect that might let us perceive non-existing values between the discrete time steps or even between two neighbored categories. However, the shape created by adding lines to the points is perceived as some kind of closed curve and hence comparisons to other such curves are perceptually easier than if only the point set would be visible. This is due to the strengths of the Gestalt laws of good continuation and closure [292].

Depending on whether univariate, bivariate, trivariate, or multivariate data is measured, we have to rely on different types of visualizations to show patterns in the data. For example, for univariate data, i.e. data that is composed of just one variable, like the response time, we could show a

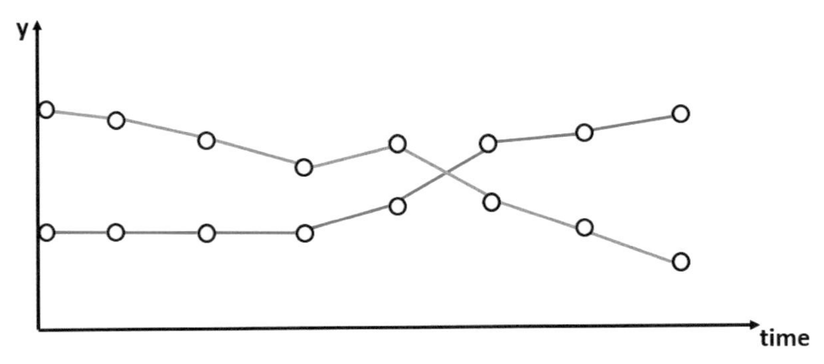

Figure 4.11 A line chart is useful to depict several time-varying performances to identify trends as well as countertrends and to compare them for differences over time.

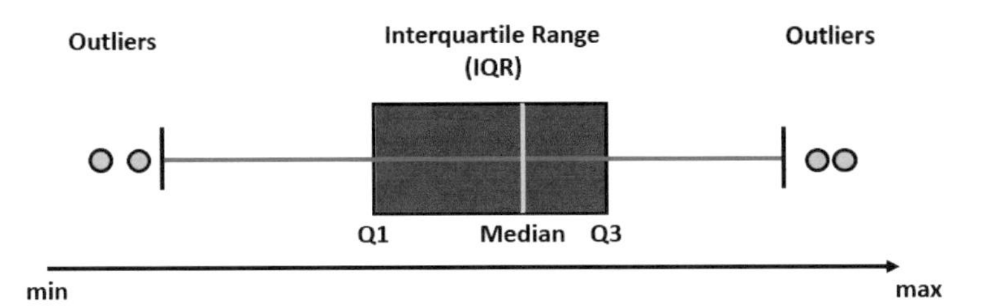

Figure 4.12 A box plot can show the distribution of a univariate dataset, for example the performance measure of the response time or the error rates.

histogram, or a box plot (see Figure 4.12). The box plot is useful to show the general spread of the univariate data on a quantitative scale. It indicates data values like the median, which is the middle value of the dataset, the minimum, the maximum, the first and third quartiles which are the medians of the lower and upper halves of the dataset. Moreover, if correlations of two variables are of particular interest, for example, between an independent and a dependent variable or between two dependent ones like the response time and the error rate, we typically show this bivariate data in a scatter plot (see Figure 4.13(a)). Correlations between more than two variables, for example tri- or multivariate data, are depicted as scatter plot matrices (SPLOMs) (see Figure 4.13(b)) or parallel coordinates plots (see Figure 4.13(c)). The most prominent patterns to be visually derivable are positive or negative correlations, for example, a longer response time might cause more (positive correlation) or less (negative correlation) errors.

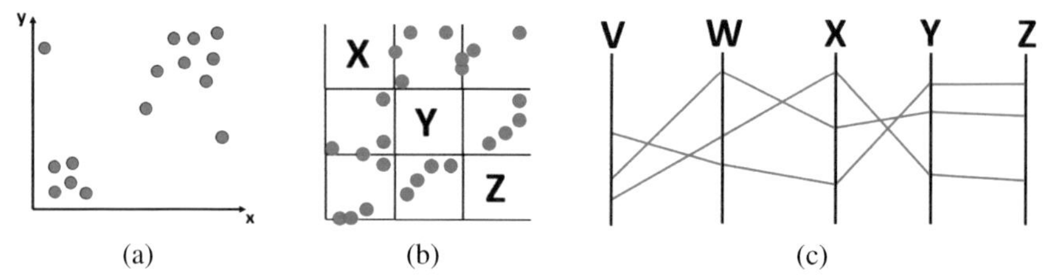

(a) (b) (c)

Figure 4.13 Visual depictions of bivariate and multivariate data. (a) A scatter plot for showing the correlations between two variables. (b) A scatter plot matrix for depicting more than two variables. (c) A parallel coordinates plot (PCP) as an alternative to the scatter plot matrix for representing more than two variables.

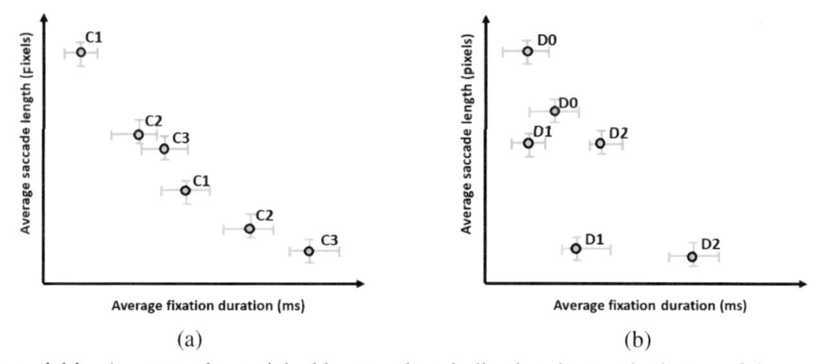

(a) (b)

Figure 4.14 A scatter plot enriched by error bars indicating the standard error of the means (SEM). The average saccade length is plotted on the y-axis while the average fixation duration is shown on the x-axis. (a) The complexity levels. (b) The task difficulty.

In an eye tracking study such bivariate data could be given by the two variables fixation duration and saccade length (see Figure 4.14 for a scatter plot for this kind of data). For even more variables, like derived metrics from the visual attention behavior, we might show the correlations among them by a parallel coordinates plot (see Figure 3.7). However, depicting the results of an eye tracking study, in particular, the visual attention over space and time, requires many more complex visualization techniques. If a spatial start- or target-oriented task is asked we might provide an overview of the temporal distance to either the start or target or both in form of a line plot (see Figure 4.15). Several visualization examples will be explained in detail in Section 6.4. The simple statistical plots like histograms or box plots are useful to give a first impression on the distribution of the recorded eye movement data but such diagrams are not able to provide deeper insights into the spatio-temporal data.

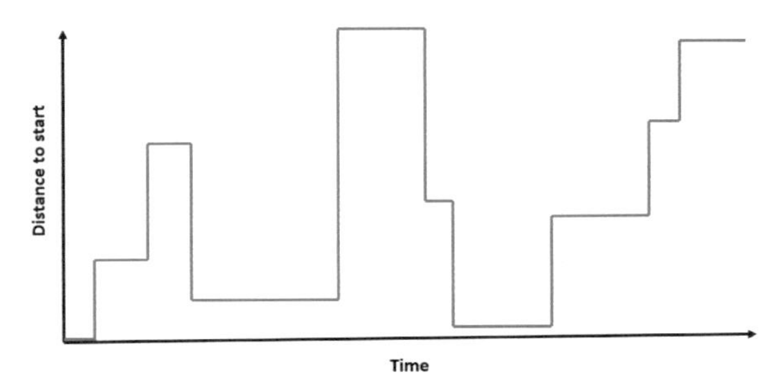

Figure 4.15 The Euclidian distance to the start is plotted over time to show the progress of visual attention with respect to such a relevant point of interest in a visual stimulus.

Moreover, if even further patterns are algorithmically computed, for example, generated rules by data mining techniques [63], we have to select suitable visualization candidates from the large repertoire of information visualization techniques.

4.5 Example User Studies Without Eye Tracking

Many more visualization techniques and corresponding interactions have been developed than we can test for usability, either as a stand-alone concept or in a comparative study to check the differences of the approaches. Moreover, due to the various parameters to be modified for each of the techniques, the design space, and hence the number of independent variables, explodes to such an extent that we can never check all of them. For this reason, it is a wise decision to first focus on the most crucial variables to not waste valuable resources in the form of user study participants. In addition, the number of tasks can also be quite large, adding one more dimension to the repertoire of possible user experiments. However, in this section we are going to look into typical examples that take into account a certain visualization technique, an interaction, or a visual analytics system while asking users to perform tasks to record performance measures, qualitative feedback, as well as in some cases even more.

The user studies surveyed here are far from providing a complete list since there are various ones already just in the very specific field of graph visualization [76]. The explained studies just serve as illustrative examples to show the reader in which direction such research is pointing and which ingredients are typically incorporated, for example, the visual stimuli, independent variables, the number of study participants, and on which type a study is based. Moreover, the output in terms of statistical results, how they are computed, and how those results are visually depicted will be discussed. It is worth noting that in this section we only focus on user studies without the explicit use of eye tracking as a technology to record visual attention which would go beyond the scope of this section, but those will definitely be discussed later in Section 5.4.

4.5.1 Hierarchy Visualization Studies

Hierarchical data exists in many application fields and a visualization of it is useful [446] to explore the hierarchical structure but it also serves as a means to provide suitable interactions, for example, to collapse or expand the data on certain levels of hierarchical granularity. Hierarchy visualizations can come in at least four major visual metaphors like node-link, nesting, stacking, or indentation approaches. All of them have their benefits and drawbacks, but if it comes to a specific task, like identifying the least common ancestor of a given list of relevant nodes, some hierarchy visualization techniques might be preferred due to the fact that the users perform the task faster and more accurately with this technique than with any other. This is the starting point for a user evaluation since the statement that one visualization is better or worse than another one is some kind of hypothesis that can be confirmed or rejected, based on the performance of real human users. Figure 4.16 shows some examples for hierarchy visualizations designed and developed by students which indicates that hierarchy visualization can be learned and applied quite quickly.

In two controlled and qualitative lab studies with 15 and 20 participants, respectively, it was investigated which one of four combined tree visualizations like RINGS, radial tree, treemap, or hierarchical is useful under

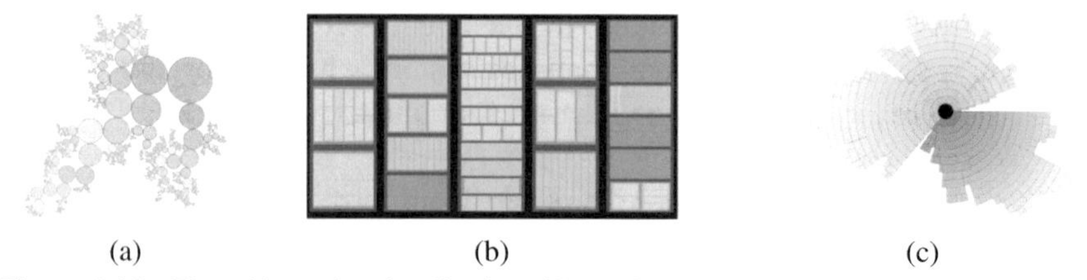

(a) (b) (c)

Figure 4.16 Three hierarchy visualizations illustrating examples from a huge design space for depicting hierarchical data. (a) A bubble hierarchy. (b) A treemap. (c) A sunburst visualization. Images provided by the students from a design-based learning course in 2018 at Eindhoven University of Technology.

the assumption that the provided tree visualizations complement each other since no individual one can focus on all aspects in the hierarchical data such as depth, size, or branchings [494]. Moreover, it was partially investigated whether several of the techniques in combination can be more powerful than one technique alone. A year before, only three tree visualizations, RINGS, treemaps, and Windows Explorer were compared [520] by asking 18 participants. Qualitative ratings as well as task completion times were measured. Another comparative study using the hierarchical visualization testing environment [13] also looked into four hierarchy visualizations by recruiting 32 participants answering eight tasks. The visualizations under investigation were Windows Explorer, the information pyramid, the treemap, and the hyperbolic browser while the response times and subjective ratings were recorded. In a similar direction we find a study focusing on six tree visualizations [291] with 48 participants. This study uses Windows Explorer as a baseline comparison system and quantitative as well as qualitative measures were recorded. Only three tree visualization techniques were compared for hierarchies with large fan-outs [471] by recruiting 18 participants. Response times, error rates, and verbal feedback was recorded for various hierarchy-related tasks. There are many more hierarchy visualization user studies, focusing on different independent variables checked for a variety of tasks. Such studies focus on hierarchies in source code [21], treemaps vs. wrapped bars [536], 2.5D treemaps [333], progressive treemaps [425], as well as combined treemaps [331], node-link trees [398], a space-reclaiming variant of the icicle plots [509], indented trees [192], or H-tree layouts [436], to mention a few.

4.5.2 Graph Visualization Studies

Empirical user evaluation in graph visualization has become a prominent research field due to the various visual variables useful to encode one or more properties of graph and network data like vertices, edges, weights, temporal evolution, and additionally derived quantitative or qualitative metrics about the underlying graph data [76]. Graph visualizations can give a great overview of the relational aspects among objects or humans stored in a dataset, quickly reaching a situation in which too many data elements and relations among them can lead to occlusion and visual clutter [426] effects in cases where the visual variables are not well chosen. Hence, it makes sense for user studies to investigate which technique is most suitable for certain tasks and what impact the variation of an independent variable has on the dependent ones.

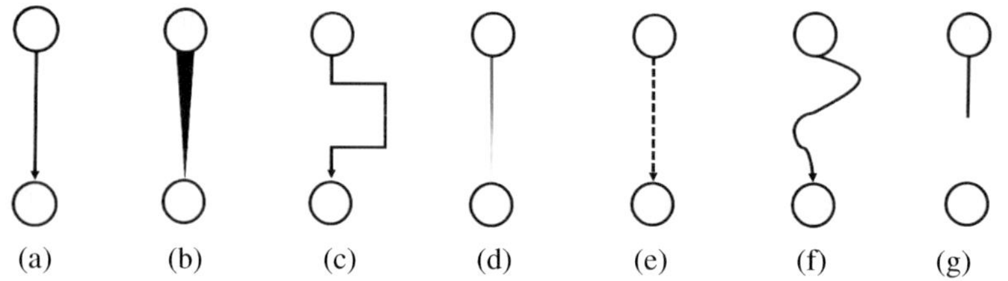

Figure 4.17 Seven edge representation styles: (a) standard with arrow head; (b) tapered; (c) orthogonal; (d) color gradient; (e) dashed; (f) curved; (g) partial.

The visual metaphor for graphs can be investigated, for example, asking if node-link diagrams or adjacency matrices are better in terms of performance measures. One study asks for typical tasks [200] focusing on understanding which one of the aforementioned approaches produces better results in a comparative study. Thirty-six participants were recruited while the error rates and response times were measured as performance indicator. Another comparative study [238] tries to understand which kind of edge representation styles like animated, tapered, curved, or standard with arrow head (see Figure 4.17 for a few examples of edge representation styles) is suitable for the task of finding connected nodes in a network. Twenty-seven participants answered various trials while their response times and error rates were recorded. Also qualitative feedback was investigated to understand people's preferences. Moreover, partial links [97] were researched to understand the shortest link length that still gives good performance results. Forty-two participants answered path finding tasks in a controlled laboratory study while error rates and response times were recorded. The crossing angles effect was checked by 22 participants in an uncontrolled online study setting [248]. The response times were recorded and those of the correct answers were analyzed. Brief interviews provide some qualitative feedback. Finally, the dynamics of graphs with respect to memorability tasks was investigated [17]. Error rates, response times, and the 25 participants' ratings were measured and recorded. There are many more studies related to graph visualization, some even with eye tracking, which is less prominent in hierarchy visualization, in particular focusing on aesthetic criteria [29, 407] like the impact of link crossings [247], the layout [400], or edge representation styles such as curvature effects on link interpretation [136, 534]. Moreover, some studies focus on connectivity models with respect to node-link diagrams and adjacency matrices [6, 280], while others take into account the mental map in dynamic graph visualizations [16, 408],

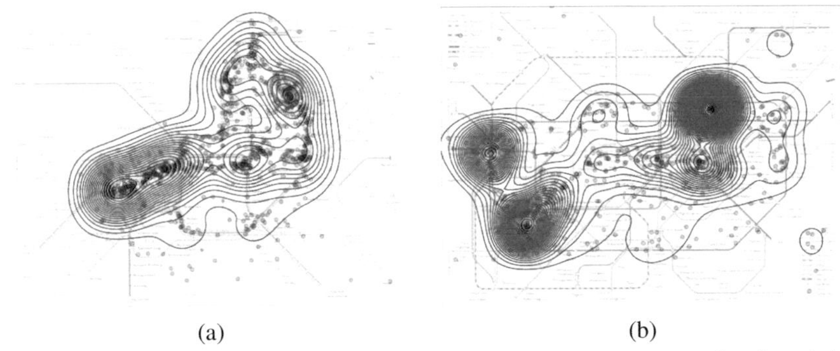

Figure 4.18 Using interaction techniques to adapt parameters in a contour line-based visual attention map. The public transport maps of (a) Zurich, Switzerland and (b) Tokyo, Japan.

however the dynamics effect is less studied in hierarchy visualizations for example.

4.5.3 Interaction Technique Studies

The way we interact with a visualization or a visual analytics system is important for efficiently exploring the represented data (see Figure 4.18 for an example of interaction techniques). It turned out that there are at least seven major categories in the field of information visualization [544] worth investigating in a user study. On the challenging side, it is quite difficult to compare interaction techniques that are applied differently but focus on the same effect, in particular for visual analytics that is equipped with a large repertoire of interactions [396] to make it a powerful concept for data exploration. There are too many parameters to control to achieve a similar scenario to make two or more of the interaction techniques comparable. Also the display types, like small-, medium, or large-scale displays, as well as the environments, like using a standard computer monitor or walking in a virtual reality scene, make a difference to the way in which we apply an interaction and the repertoire of interaction methods. However, instead of comparing interactions, it is always possible to test if users understand an individual interaction technique and to apply it in a reliable way to visually analyze an underlying dataset, for example for finding insights and knowledge in it.

Interactive public transport maps provide a way to find insights into routes, in particular, focusing on additional passenger information [96]. The usefulness of such incorporated interaction techniques was evaluated by recruiting 20 participants, split into two groups, to avoid learning effects between two different study settings, i.e. with and without interaction in a comparative study setting. Moreover, other tasks focused on understanding if the interactions are useful at all and can be applied by the participants in a suitable way. Response times and qualitative feedback were recorded and evaluated. In another study focusing on interactions for exploring dynamic graphs [69], 20 participants were recruited to examine if the interaction techniques incorporated in a dynamic graph visualization tool are intuitive for the study participants to solve a given task. This task was too complex to be solved by the static version of the tool without interactions. Response times were recorded and qualitative feedback was requested. In the same line of research focusing on dynamic graph visualization we can read about the evaluation of two interaction techniques [181]. Sixty-four participants answered tasks based on interaction in a controlled study while error rates and response times as well as qualitative feedback was recorded. Also eye movement data visualization is in the focus of user evaluation, in particular the combination of several views [100] and interactions for modifying the individual views as well as linking them together. An uncontrolled experiment showed the usefulness of the interactivity of such an eye movement data visualization tool while the qualitative feedback of 10 participants was analyzed for insights on the usability of the tool and its functionality with respect to find patterns and strategies in the visual attention behavior of people when inspecting public transport maps [372]. Interactive timeslicing, animation, and small multiples were investigated in several user studies for visualizing dynamic graphs on large displays [322]. For all study setups the same 24 participants carried out tasks while their response times, error rates, and qualitative feedback were measured. There are many more studies with a clear focus on interaction techniques in a visualization- or visual analytics-related context. Those range from preferential choices based on a set of interactive visualization techniques [28], Fitt's law with respect to the design of interactive user interfaces [350] based on pointing devices, or navigation tasks for off-screen targets on mobile phones for which the display space limitations generate challenges for standard visualization techniques [27].

4.5.4 Visual Analytics Studies

Evaluating the user behavior in a visual analytics system [46] is a challenging problem [514] due to the many linked ingredients like algorithmic concepts, visualizations, and interaction techniques. Consequently, the list of user studies focusing on an entire visual analytics system only contains a few examples. Moreover, most of these studies rather focus on qualitative feedback instead of measuring response times and error rates. The reason is that in many cases the developer of such a system is more interested in the development phases and whether the users are still confident with the system after certain stages. Evaluating a visual analytics system after the final stage of development might be a good idea to get extra feedback, but actually, visual analytics systems are much more complex than just interactive visualization techniques, hence it is a wise idea to evaluate them from time to time, typically with domain experts because they mostly focus on very specific and complex dataset scenarios stemming from a well-defined application domain. Eye tracking might be a good concept here to evaluate a visual analytics system [307] since the eye movement data can give us many insights into the linking of the system's components which cannot be found if just standard quantitative performance measures are recorded or even qualitative feedback is provided.

The exploratory strategies of 24 participants with respect to visual causality analysis were investigated [541] in a controlled experiment. The reasoning performance, strategies, and pitfalls were measured and analyzed while the participants also provided qualitative verbal feedback during the exploration stages based on the experimenter's requests. To answer the given tasks reliably the participants had to use many of the provided functionalities of the analysis tool, hence such a study involves more processes than a standard visualization tool or a simple interaction technique. Another study investigated how well people can recall their reasoning of visual analyses [334] by recruiting 10 participants using the WireVis system for visual exploration of financial transaction data [117]. A mixture of qualitative and quantitative data was recorded to get insights into such complex visual processes. Another follow-up study by the same authors tried to figure out if the recall of findings and strategies is possible and how this is done. Again 10 participants were recruited. A user study focusing on the ability of semantic interaction by using a visual analytics system to support sensemaking [173] was conducted in a laboratory setting by means of a high-resolution display. Six participants solved such interactions to understand how the models benefit the participants based on a majorly qualitative feedback. Twenty-seven participants investigated human factors of the confirmation bias in the field of intelligence analysis [131]. Qualitative feedback and quantitative measures were recorded in a laboratory setting. Also the priming and anchoring effects in visualization and visual analytics

Table 4.1 Examples of user studies focusing on aspects in visualization, interaction, and visual analytics: comparative (CP), laboratory (LB), controlled (CT), qualitative (QL), quantitative (QN)

Scenario	Participants	Study type	Literature
Hierarchy visualization	15 + 20	CP, QL, LB, CT	[494]
Hierarchy visualization	18	CP, QL, QT, LB, CT	[520]
Hierarchy visualization	32	CP, QL, QT, LB, CT	[13]
Hierarchy visualization	48	CP, QL, QT, LB, CT	[291]
Hierarchy visualization	18	CP, QL, QT, LB, CT	[471]
Graph visualization	36	CP, LB, CT, QT	[200]
Graph visualization	27	CP, LB, CT, QT, QL	[238]
Graph visualization	42	CP, LB, CT, QT	[97]
Graph visualization	22	CP, QT, QL	[248]
Graph visualization	25	CP, LB, CT, QT, QL	[17]
Interaction technique	20	CP, LB, CT, QT, QL	[96]
Interaction technique	20	LB, CT, QT, QL	[69]
Interaction technique	64	CP, LB, CT, QT, QL	[181]
Interaction technique	10	QL	[100]
Interaction technique	24	CP, LB, CT, QT, QL	[322]
Visual analytics	24	LB, CT, QT, QL	[541]
Visual analytics	10 + 10	LB, CT, QT, QL	[334]
Visual analytics	6	LB, CT, QL	[173]
Visual analytics	27	LB, CT, QT, QL	[131]
Visual analytics	726	QL	[507]

were investigated based on five individual studies, even including a series of Mechanical Turk experiments [507]. A total of 726 people participated in the studies, while the results were mostly based on Likert scale-based participant ratings. Some more examples related to visual analytics and user evaluation focus on think-aloud protocols targeting the design of user interfaces [262], i.e. the link between the human, the visual, and the analytics parts in a visual analytics system, adaptive contextualization [207], reasoning processes [159], modeling of user interactions during the data exploration process [141], or visual analytic roadblocks [313].

Table 4.1 provides a summary with some user study examples from the application fields of hierarchy and graph visualization, interaction techniques, and visual analytics. The additional study information in the table is based on the number of participants, the most important study types as well as the reference to find more details about each of the studies for the interested reader.

5

Eye Tracking

Eye tracking has become a well-studied field these days [161, 235], not only because of the progress of the hardware technology and the reduced costs for eye tracking devices, but also due to the various application fields that benefit from it. Such fields are numerous with varying backgrounds like marketing that is interested in the user behavior to improve the selling strategies, visualization and visual analytics which are focusing on understanding the interplay of their algorithmic, visual, and interactive components to identify design flaws and to build a starting point for enhancements, or software engineering trying to figure out how software developers produce source code and how efficiently and effectively they implement or debug such code while they interact with a software development environment or while they collaborate and communicate with other software developers. Moreover, fields like neuroscience, cognition, human–computer interaction, medicine, sports, or the automobile industry all benefit from eye tracking, given the fact that the data is measured accurately and analyzed afterwards.

To efficiently and effectively record eye tracking data, a profound knowledge of the anatomy of the eye is needed. Moreover, the internal processes and concepts involving facts about light, vision, perception, cognition, and psychology have to be researched and understood in order to obtain reliable and significant results based on eye tracking data. The human eye is very complex but to build a simple and cheap eye tracking device that is powerful enough to measure and record eye movements, even on a very coarse-grained and not very accurate level, at least the major components of the eye have to be studied as well as their connections, impacts, and causes for certain effects. Such effects also include diseases that might have a negative influence on the way we measure the eye movements but also on the trustworthiness of the results. Hence, the eye is involved in the field of eye tracking in many respects, it is not just "the window to our soul", it builds some kind of interface between the visual stimuli and the cognitive processes happening in the brain.

The field of eye tracking has existed for many years but the technology as we know it today is not comparable with the early attempts to observe people's eyes while they were solving a certain task. Although there was a general impression about what eye movement looked like, it was pretty challenging to note down the various numbers of movements, i.e. just very general observations could be made. The invention of the computer and the progress in hardware technology in general has led to the eye tracking devices that we know today with the ability to record eye movements very accurately as well as additional physiological measures. Moreover, the steady progress in software technologies, in particular data analysis, visualization, and visual analytics, give great support in helping us to identify patterns in the eye tracking data, even in real-time. Compared to the early attempts we can say that we are eye witnesses to the great advancements in the field; however, other, even more challenges, occur these days, moving away from the data recording to more data analysis challenges. Describing the many concepts that exist for analyzing eye tracking data is beyond the scope of this chapter, hence we move that to another part of this book which is located in Chapter 6.

Moreover, interacting with the eye, instead of using devices like the mouse, joystick, keyboard, or further concepts based on touch, gestures, voice, and the like, brings into play another kind of giving user input to an interactive visual stimulus. Such gaze-assisted interaction is researched as a novel discipline but although it seems to be promising, it also brings new challenges into play, for example the well-known Midas touch problem that describes the effect of making everything that is visible in a visual stimulus interactable by focusing on it with the eye. However, this effect is counter-productive since it generates some kind of over-reaction in the users' visual attention, hence good ways out of this dilemma have to be developed. This issue requires special gaze-assisted features which are not known in the same way for other interaction devices and methods, but if several interaction concepts are combined in a clever way, they seem to create a possible solution for this problem. However, user evaluation is required that brings another kind of data to be analyzed into the field.

In particular, applying eye tracking to visual analytics systems requires the knowledge of many concepts due to the interdisciplinary character of such systems ranging over, and including, general fields like algorithms and data structures, human–computer and gaze-assisted interaction, visualization, as well as cognition, perception, or psychology, to mention the most important ones. The interdisciplinary fields are one reason for the generated eye tracking dataset coming with a certain degree of complexity. No matter how complex the data is, it is worth analyzing and visualizing since it hides interesting and valuable patterns, correlations, strategies, rules, or associations that correspond to a certain user behavior. The understanding of such visual task solution strategies depends on how well we are able to translate the abstract eye tracking datasets into a language that easily explains the human behavior while solving a task. These insights can finally help to identify negative issues like design flaws to hopefully enhance a static, dynamic, interactive, animated, 2D, or 3D stimulus based on such advanced analyses.

5.1 The Eye

We primarily discuss the human factors of the eye since the humans are most relevant for visual analytics systems with their perceptual abilities to rapidly detect patterns and to apply the found pattern-based insights for refining parameters, algorithmic as well as visual ones. Without the functions that the eye is providing or just some limitations of them, our vision and perceptual abilities would suffer a lot which would have many negative consequences for the exploratory strengths and strategic behavior while using a visual analytics system.

Eyes are considered organs that are relevant for the visual system [522]. Effectively perceiving and processing static and dynamic visual objects in the brain requires unlimited vision for which the eyes play the deciding factor. To reach this goal, the eyes transform light into impulses that occur electro-chemically in the neurons which build the basic units of the humans' nervous system and hence are responsible for any kind of information processing. The majority of such neurons are in contact with hundreds of other neurons and are consequently able to transmit information further and further.

For eye tracking the eyes logically play an important role since they are obviously the core ingredient for this emerging technology. However, it is actually not of primary interest how the information is processed in the brain, but for the tracking we are more interested in how the eye functions and how to extract, measure, and record the valuable information that is required to reliably and efficiently track the eye. For this to be achieved we need profound knowledge about the eye movements, how they are initiated and controlled, and which features of the eyes have to be taken into account to reliably record the data we need for our eye tracking data analysis. A second, but still important and challenging aspect is the way in which the information is transmitted from the eye into the responsible brain regions and vice versa back from the brain to guide our reactions and interactions which brings

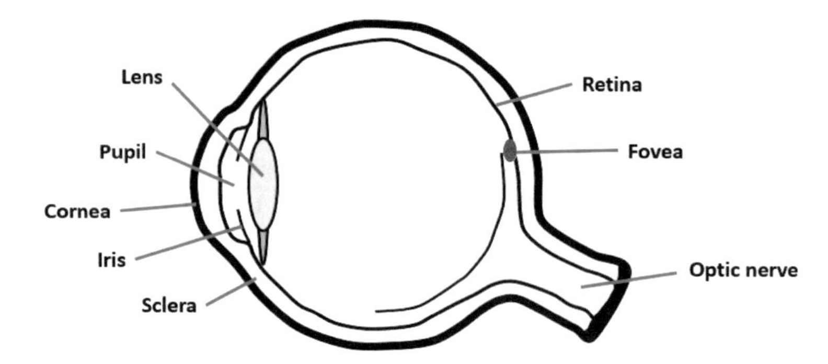

Figure 5.1 The human eye is a complex organ that is important for the visual system [196]. Moreover, it builds the major ingredient for all eye tracking studies.

principles from cognitive psychology into play as well as aspects related to the eye–mind hypothesis [268, 269].

5.1.1 Eye Anatomy

Figure 5.1 shows some of the important components in the human eye [196] (many others that are not relevant ones for eye tracking are not shown) that might be distinguished in the outer and inner ones. The outer ones typically occur as confounding variables in an eye tracking study, the inner ones are useful to measure and record eye movement reliably. For example, outer eye components that might lead to inaccurate and erroneous eye movement data or biases in the study can be the differences in the lengths of the eye lashes, even false eye lashes due to cosmetic reasons, or properties of the eyelids as well as diseases of the eyelids like eye twitching. These can cause problems for certain study participants and might have a negative effect on the dependent variables measured in an eye tracking study. Many more problems and diseases [336] related to the eyes can cause irregularities in an eye tracking study such as eyestrain, night blindness, lazy eye, cross eyes, color blindness, dry eyes, cataracts, or glaucoma, just to mention a few. Even artificial corrections of the eye sight like glasses or contact lenses might be outer eye components that have an impact on eye tracking studies. Hence, it is a wise decision to collect such information for each participant before starting an eye tracking experiment, to incorporate those issues later in the data analysis stage.

The most important inner components of the human eye from a large repertoire could be listed as the iris, cornea, pupil, lens, sclera, retina, fovea, rods, cones, and optic nerve (see Figure 5.1). These play crucial roles when setting up and conducting eye tracking studies. Actually, the human eyes have many similarities to digital cameras. The cornea which takes the role of the camera lens is hit by light. The iris imitates a diaphragm, similar to that of a camera which is responsible for increasing or decreasing the amount of light that falls into the eye, in particular, the light ending up at the backside of the eye, i.e. the retina. To reach this goal the pupil's size can be modified in order to regulate the amount of light falling in. The eye's lens plays the role of focusing visual objects just like the autofocus mechanism of a digital camera lens. The remaining light reaching the retina which is consisting of rods and cones, transforms this light into electronic signals, transmitted by the optic nerve to the visual cortex which stands for the region of the brain that manages the sense of sight [521]. From here, cognitive processes guide the actions and reactions of the human body functions, also the movement of the eyes.

The rods and cones placed on the retina take the role of photoreceptors. Several million of them [196] absorb light to transform it into nerve impulses which are further sent via the optic nerve to the corresponding brain region. One difference between rods and cones is their effectivity at different day and night times. Rods are important for vision at night since they are more sensitive to light than cones, hence if the light is very low the rods play the most crucial role for human vision. The cones instead, are useful during daytime when the light is very bright, i.e. there are many more photons than during nighttime. Rods and cones have an impact on color perception, but this impact is highest for the cones. Certain wavelengths, typically short, medium, and long ones, are considered, characterizing the cone photoreceptors into three classes. Research about these aspects describes the number of photoreceptors in the human retina as being approximately 6 million for the cones and 120 million for the rods, but these numbers vary from eye to eye. For eye tracking studies, it is important to use bright light in case the cones are more the focus of the results whereas in a comparative eye tracking study investigating how different light levels affect the eye movement behavior and further dependent variables we need profound knowledge about the functions of the photoreceptors [253].

5.1.2 Eye Movement and Smooth Pursuit

When the eye moves, the eye muscles initiate a movement in the desired direction. This happens very rapidly, i.e. the eye can be accelerated a lot and stopped very quickly to allow accurate visual attention to a certain point or object of interest. Extraocular muscles are responsible for these eye movements which are primarily under the humans' control, i.e. on a voluntary level. While looking around, the humans are free in their decisions where to look at and for how long. But in rare cases, we can detect some reflex eye movement actions that have a higher priority than the voluntary eye movements because they have to be done quickly, for example, due to some dangerous, unforeseen, or interesting situations. The voluntary eye movements are also typically without explicit awareness, to reduce the cognitive efforts and to ease the ways we pay visual attention to a visual stimulus, either to interesting static or dynamically moving objects or sometimes caused by body or head pose changes and movements. The goal of the rapid eye movements based on the actions of the extraocular muscles is to quickly navigate and direct the way the light falls on the retina, actually the small part called the fovea, to change the focus for the visual input to the visual cortex to make the information processible in the brain. A high degree of precision and rapidness of the oculomotor system is required to achieve the best visual attention scenario possible.

The gaze has to stay rather constant in a situation in which we have to watch a small visual object for a longer time, meaning the eye muscles must frequently adjust their position to keep the interesting visual part in the center of the fovea, in case we make small or larger head movements while at the same time fixating the same visual object with our eyes. Since such a scenario happens very often due to the fact that the head is typically not fixed and shaking around a bit, even more in a free walking eye tracking situation, we can see that the eye muscles take over a special and challenging job in keeping our vision as effortless as possible while solving given tasks. The responsible six extraocular muscles have different tasks, for example, rotation of the eye toward the nose, outward, upward, or downward eye movements. In normal eye health situations the muscles move both eyes simultaneously and synchronously, i.e. as conjugate movements, in contrast to disjunctive movements.

There are different ways to move the eyes that make a difference for the analysis and visualization of the eye movement data. Generally, the eyes never stay in the same position for more than a few milliseconds, making high frequent movements per second [11]. For eye movement studies it is of particular interest if the eyes move from spatial positions to other ones, staying at each for a while and make saccades, i.e. rapid eye movements in between. In contrast to this scenario the eyes might make continual movements, following a moving object smoothly, an effect for which the term smooth pursuit [177] was coined. For static stimuli the first scenario is typically detectable in the recorded eye movement data while for dynamic stimuli with animated visual features, the continuous following of a visual object, for example, a ball or a car, can cause additional data recording and analysis issues. The smooth pursuit effect actually allows us to follow moving visual objects while most people are unaware of doing a smooth pursuit and they are even not able to perform one without a visible moving object. If the visual object is moving too fast, like more than a velocity of 30 degrees per second, people tend to do so-called catch-up saccades, i.e. quick movements of the eye from one position to another one. For eye tracking studies containing smooth pursuit tasks it has to be noted that there is a general difference in the direction of smooth pursuits, i.e. humans perform better for horizontal and downward movements than vertical and upward ones.

5.1.3 Disorders and Diseases Influencing Eye Tracking

We might classify the diseases having an impact on the value and reliability of eye tracking studies and their generated results into two major classes. The first class consists of diseases or disorders that directly influence the vision due to malfunctions related to the eye. Those diseases could be color blindness, cataracts, glaucoma, retinal disorders, strabismus as well as refractive errors such as myopia, hyperopia, astigmatism, and many more [434]. The second class contains those that are not directly eye-related diseases but have a similar impact on the conduction of eye tracking studies and the reliability of the results. Examples for this type are Parkinson's disease which typically affects a human's motor system, or autism that belongs to the class of developmental disorders causing problems for social interactions and communications. Diseases belonging to this second class are typically related to senso-motoric disorders or communication problems that hinder study participants to take part in an eye tracking study generating results of high accuracy rates and with acceptable response times compared to people who are not suffering from such problems. Moreover, for most scenarios under investigation, the eye movements should be recordable in an efficient and well-calibrated way with a high degree of expressiveness about the visual attention behavior, an aspect that unfortunately oftentimes leads to the exclusion of people with a certain kind of disorder.

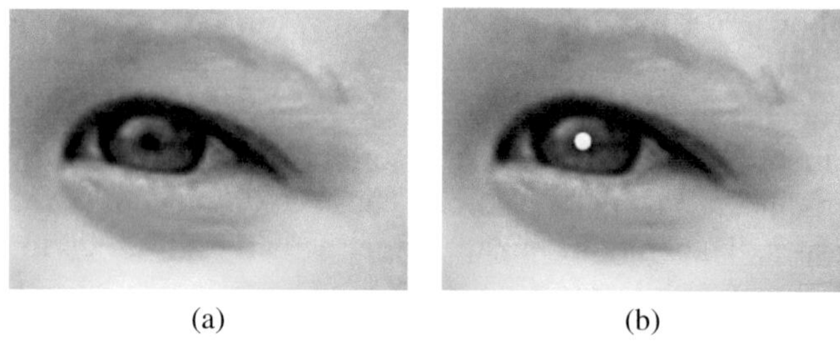

(a) (b)

Figure 5.2 Cataracts [392] affect the lens of the eye in some kind of degeneration process causing clouded and unclear vision: (a) clear vision; (b) an eye with cataract issues.

The causes of eye-related diseases are manifold, like infections, allergies, genetic issues, smoking habits, and even vitamin deficiencies. Moreover, diabetes, which has other causes, might be one of many reasons [437], having a negative impact on the eyes making them less powerful for vision. Diabetes causes a macular edema which destroys the sharp vision, leading to an effect of partial loss of vision or complete blindness. Hence, this issue can be problematic for eye tracking studies since many people do not know if they suffer from any form of diabetes [341] or not. Apart from diabetes there are various more disorders that have a negative impact on vision and sight. For example, visual impairments might also be caused by cataracts (see Figure 5.2) which affect the lens by a degeneration issue making it becoming opaque and causing some kind of clouded unclear vision effects. Cataracts [392] are mostly age-related problems but can even be caused by diabetes or traumatic issues.

Eye tracking could even be useful to identify a disease of the eyes [301] or even one that is not directly related to the eyes like Alzheimer's [25]. This could be done in a controlled lab environment or even in an online session in which the doctor inspects the eye remotely. Such a remote session might produce less exact and less expressive results than the lab examination but still some useful knowledge about a person's health status might be generated. Based on the outcome of the remote check-up, the online doctor might invite persons with strong symptoms indicating possible diseases for further more detailed examinations into the lab or other further steps might be taken. The eyes might tell more about a person's health status than we would expect. Eye tracking could even be useful to check for states like tiredness which could have serious impacts on tasks like driving a car or a truck. In the case of an eye tracking study in visual analytics, tiredness might cause a low task accuracy or high response times, a situation that might cause biases in a user

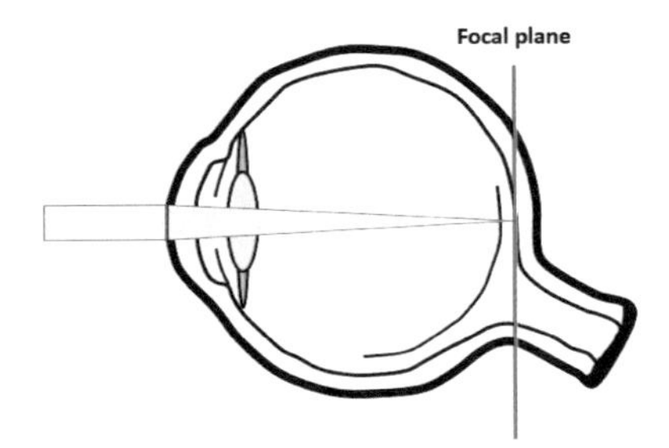

Figure 5.3 No corrective lenses are needed for normal vision.

study. In particular, for long-duration tasks, which might also occur in visual analytics eye tracking studies, this effect can have negative consequences on the reliability of the results. Finally, with eye tracking we might find out the emotional states of people [491].

For standard eye tracking studies it would be too expensive or even too frightening for people to take part if they have to visit a doctor or eye specialist first to check for diseases, hence the number of participants would be reduced tremendously for studies investigating issues in visual analytics. The best way is, consequently, to just invite the people for the study and ask them if they would like to fill out a corresponding form while explaining them the ethics and privacy issues related to this information providing process. In any case, it is good advice to check participants beforehand, i.e. they should fill out some kind of form that includes information about personal details to which diseases also belong. However, such private information should not be abused and hence it should be explained to the participants if they decide to provide such information. Moreover, the recorded personal data should be anonymized in a way that it cannot be recovered later on.

5.1.4 Corrected-to-Normal Vision

From the five senses sight, hearing, taste, smell, and touch, sight is reported to be the most important one for our daily life. Hence, it is important to find ways to correct the vision in cases certain visual acuity or color blindness issues occur, leading to effects of blurred images that might even affect the eyes for long-duration tasks in the sense of making them more tired than they would get in situations with normal vision (see Figure 5.3). Effects that are caused by eye strain, such as headaches, dry eyes, or even squinting might occur if the incorrect vision is not medicated. However, there is hope that those issues

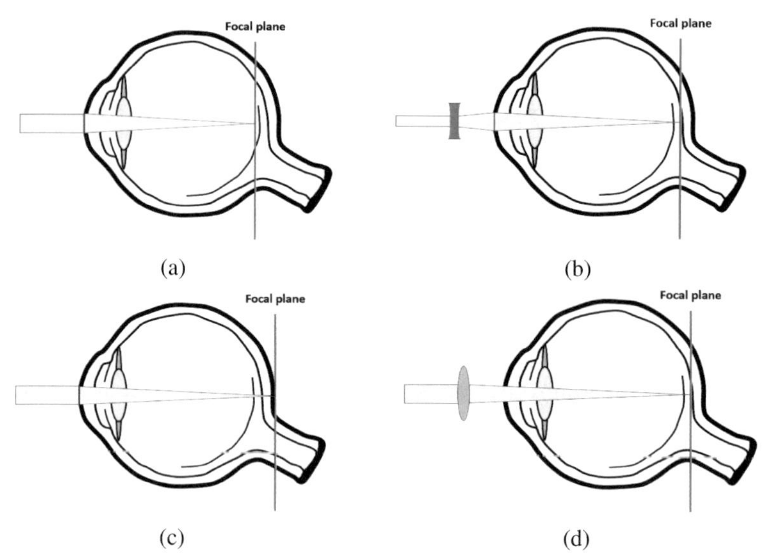

Figure 5.4 Refractive errors: (a) nearsightedness (myopia) and (c) farsightedness (hyperopia) can be corrected by special lenses (b), (d).

might disappear or at least be reduced when wearing glasses or contact lenses regularly. Actually, in many scenarios the vision can be corrected artificially by such non-invasive instruments like glasses or contact lenses. Typically, glasses are worn at a short distance in front of the eyes while contact lenses are worn directly on the eye surface. Both instruments have the purpose of correcting the vision, i.e. they function as corrective lenses (see Figure 5.4). To reach this goal they work as light benders when it enters the eye and hence they correct refractive errors. Most common sight issues are nearsightedness called myopia, farsightedness called hyperopia, or astigmatism.

The reasons for these effects are typically caused by aging, but might even have different causes lying in the genes of your family, i.e. they can be inherited from ancestors. The older people get, the less flexible their eyes are to certain situations, i.e. the negative effect is likely to get worse and can also vary a lot with increasing age. Moreover, both eyes do not suffer equally from these problems but they can have varying extents of refractive errors. However, even this inhomogeneity can be detected by an optometrist or ophthalmologist and corrected by differently shaped lenses, one for each eye, taking into account the extent of refractive error individually. For hyperopia, a convergent lens is used while for myopia, a divergent lens corrects the error. Also color blindness correction glasses are researched, offering a way to put the three primary colors in a certain balance in case there is a disorder or deficiency in one of the three primary colors.

For eye tracking studies these vision correction instruments can cause problems which typically have to do with the correct calibration of the eye tracking device or even with wearing a head-mounted eye tracker in combination with glasses for example. It is a good advice to collect information about the wearing of glasses or contact lenses, but also the age of the study participants might give a hint about certain sight problems, even if the participant is unaware of it and does not wear any vision correction instruments. The experimenter can at least conduct a visual acuity test, for example, by means of a Snellen chart [470], but whatever result comes out of this test, it should not be reported to the participant since the experimenter of a visual analytics eye tracking study is typically not privileged and educated enough to give such a result. But at least the study participant performing badly in the visual acuity test might be excluded from the data analysis part later on; however, the rules for this exclusion process should be defined and determined before starting the study to avoid biased performance data.

5.2 Eye Tracking History

Eye contact, eye movement, as well as gaze direction have played and still play a crucial role in silent human-to-human communication, for example, as a means to express emotions, interest, disinterest, or just as a way for an efficient form of social interaction. Typically, we use joint attention to a certain visual object to build this form of eye communication, i.e. some kind of visual object takes the role of the interface between human and human if words are missing. The need for understanding the complex visual and cognitive processes incorporated in joint attention might be some kind of starting point for research in eye tracking [518]. In the 19th century, researchers started to observe eye movement patterns, like Javal [250] who tried to describe eye movements during reading tasks. But an accurate recording was impossible due to the missing technical equipment in these times. The invention of the film camera in 1905 brought a big step from the observational description to a more accurate after-recording analysis, but still the development of the technology was not finished.

Actually, Yarbus [539] was one of the pioneers of eye tracking research with an already high accuracy, but long ago before the invention of powerful computers being able to process big data, asking himself which role the task plays in visual attention patterns in a given static scene. The major conclusion from such experiments was that, depending on the task, the spectators perform different scanpaths, i.e. their eye movements can vary a lot. However, although the recording of the eye movement data was quite accurate in these days, the quality of the data cannot be compared with that generated by eye tracking devices having undergone many stages of technological progress as we have them today. The devices to record the eye movements in these old days were mostly based on so-called suction caps [492], similar to contact lenses as we know them today for correcting refractive errors in the eyes.

The progress in the hardware technology brought various novelties into the field. With those novel inventions and more and more accurate eye tracking devices, many more challenges occurred, mostly focusing on the analysis concepts needed for a proper and insightful evaluation of the recorded eye movement data combined with additional data sources like physiological and verbal data [44]. Moreover, visual analytics was detected as a field to further uncover data patterns by including the human user with interactive tools containing visual components as well as algorithmic ones. But still, the field is moving forward since many open challenges still remain such as the link between visual attention, cognitive processes, and psychological issues that might guide the visual task solution strategies [305]. With steady progress we have reached a level at which even real-time eye movement data can be recorded and analyzed. Moreover, eye tracking can be used as an interaction modality, normally known as gaze-assisted interaction, demanding for very accurate eye trackers, depending on the interactive visual components provided by a user interface, for example, a complex visual analytics system.

5.2.1 The Early Days

Eye movement was recognized many years ago, but tracking it in a very accurate way was impossible due to missing instruments and devices in these early days. However, Wells and Erasmus Darwin studied afterimages as a means to examine eye behavior and movement patterns. Researchers around 1879 also tried to listen to eye muscle movements by means of a kymograph to get some insight into the corresponding eye movements. The actual roots of the technology lie in the 19th century in which researchers became interested in reading tasks, but professional equipment as we know it today was not available. Even a simple device could not be used, just human-to-human eye movement observations were possible, for analyzing

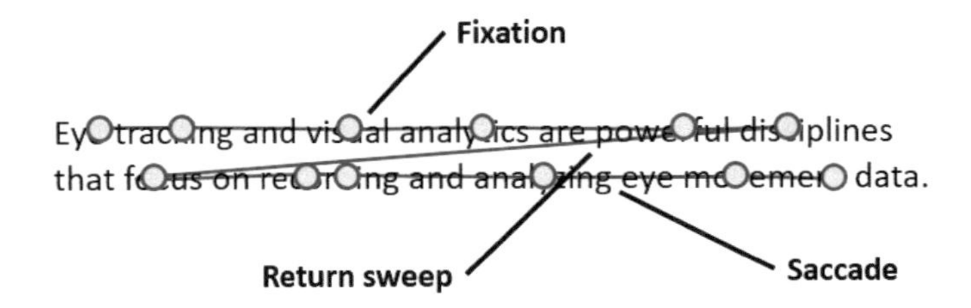

Figure 5.5 Eye movements during a reading task consist of short stops (fixations) and rapid eye movements (saccades). This insight was found by Hering, Lamare, and Javal around 1879 [263].

reading behavior [263]. For example, the experimenter used mirrors so as to not distract the reader from the actual task too much. Although this concept was quite novel and provided some insights into viewing behavior, it was far from being accurate since it relied on the observation performance of the experimenter and not on powerful, accurate, and fast measurement devices, cameras, sensors, and computers as we have them today, just a microphone was used and a mechanical device for better counting the eye movements. One of the first experimenters in these times were Hering, Lamare, and Javal [519] who studied the movement of the eyes during reading, which is reported to have happened around 1879. One major outcome of this first eye tracking study was the fact that the eyes do not move smoothly over the words while reading but instead, they make short stops and even jump back and forth with rapid eye movements in between those stops (see Figure 5.5). These effects of stop-and-go brought into play the terms that we refer to as fixations and saccades today.

The biggest challenge in these early days, however, was the eye tracking invasiveness that could cause pain or even eye damage in some cases. Around 1898, Huey, for example, used some kind of contact lens with an opening for the iris while the lens was connected to an aluminum indicator to uncover the eye movements. Delabarre also used an invasive way to research eye movements by attaching a gypsum cap to the eye, but with his approach he could only detect and draw horizontal eye movements, a clear limitation of the early eye tracking techniques. Some sort of breakthrough in eye tracking research might have been the non-invasive eye trackers that made use of reflected light and hence were often called optical trackers. Dodge and Cline studied the corneal reflection method but still had to record the eye movements based on this concept on a photosensitive photographic plate, later replaced by photographic tape. Also in their method it was only possible to record horizontal eye movements. In 1905, Judd, McAlister, and Steel were the first who presented a device that was able to record eye movements in both directions, i.e. horizontal as well as vertical with the drawback that the study participants had to sit or stand still during the experimentation. The most popular outcomes of these days, although eye tracking was in its infant ages, were that humans do not obtain information during rapid eye movements (saccades), that the eyes need some time to first initiate before they can start obtaining information, and the fact that humans have a limited visual field that only allows focusing on the visual objects in the center of the view sharply while the objects at the edge of the visual field are more observed in a blurred fashion [521].

5.2.2 Progress in the Field

The progress in the development of the film camera after 1900 brought novel ideas into play to enhance the methods and techniques in the field of eye tracking, for example in the research by Fitts in 1947. Instead of being invasive in the sense of injuring the study participants' eyes, the camera-based concepts were non-invasive, allowing to take part in eye tracking experiments without affecting the eye. Such non-contact devices were developed by Buswell around 1935 by reflected light on the eye that was recorded on film. Further more semi-invasive concepts focused on electric potential differences in the eye and used electrodes to measure those differences during eye movements, typically called electro-oculography [535]. A big problem was the dependency on the head movements, i.e. in typical eye tracking experiments in these times, the participants were not able to move and their head had to be fixed somehow. Some methods made use of chin rests, head rests, or bite bars. For traditional visual analytics studies, head movements might not be needed to make use of the full potential of the visual analytics system. But for more complex immersive analytics studies [347], as we find many of them today, studies applying eye tracking technologies have to be conducted in a way that people can freely move around to reach the entire visual space, for example to explore the represented data. Before the invention of visual analytics as an interdisciplinary field, Hartridge and Thomson researched the recording of eye movement by allowing a high degree of head movement, although their technology was not considered comfortable for the study participant.

Mobile eye tracking devices, like head-mounted or wearable trackers as we know them today, were a great success since they made eye tracking more comfortable and less stressful for the study participants. The invention of the computer and the steady progress in hardware and software technologies had a simultaneous positive impact on the development of the eye tracking research field. With the rubber suction cap invention applied by Yarbus [539], the field made progress again in these days. With his method he created a means to record and analyze eye movement at a comparably high accuracy. With his ideas, he found out that the task plays a crucial role for the differences in eye movement patterns. Also organizations like the NASA and the US Air Force were in focus when researchers started to develop eye tracking devices, or oculometers, typically for pilot experiments in aviation [43]. But although many studies were conducted with much money involved, eye tracking devices were still more or less created for the industry or military, but not for the everyday user, which is still a problem today due to the immense costs of the more accurate devices. Also researchers in the field of marketing detected eye tracking for their benefits and tried to figure out how advertisements are inspected to derive insights from the eye movement data to improve advertisement strategies.

We might say that the computational power of computers with their efficient algorithmic approaches are a key to success for fast and accurate recording, in particular the detection and extraction of important information from the eye to draw conclusions from the eye movements, but also for the analysis of the bigger and bigger getting spatio-temporal eye movement data. This technological improvement, based on computers and their powerful algorithms, have brought us closer to the eye tracking devices and software as we find them today. Those might be classifiable into remote and head-mounted or wearable ones, as well as based on electro-oculography, infrared-oculography, video-oculography, or scleral search coils. More and more application fields are applying eye tracking techniques, also due to cheaper and cheaper devices. Also eye tracking research and studies conducted at universities have increased over the years, leading to a vast amount of scientific publications ranging over all sub-disciplines related to eye tracking, even bringing international conferences into play, like the popular symposium on eye tracking research and applications (ETRA), as a meeting point for researchers from academia and the industry from all over the world.

Figure 5.6 An example of an eye tracking device as we know it today, known as the Tobii Pro Glasses 3. Image provided by Lina Perdius (Tobii AB).

5.2.3 Eye Tracking Today

No matter how efficient a technology currently is, we have to take into account the value of the recorded eye movement data which is typically based on the so-called eye-mind hypothesis [268, 269]. This hypothesis states that people tend to cognitively process, i.e. think about, the visual objects that they fixate as long as they pay attention to them. The problem here is that cognitive psychology and eye tracking typically run as two separate research fields, hence many aspects remain unresearched and unexplained, while more joint research might lead to many synergy effects from which both fields might benefit. Overt and covert attention come into play here that explain the effects of fixation and cognitive processing. Covert attention is denoted as the attention that we pay to some object we are not looking at while overt attention is the opposite of that. Consequently, covert attention makes the reliability of eye tracking data a bit questionable due to the missing links to the cognitive processing stages. Regardless of the eye–mind hypothesis, we can find different types of eye tracking technologies today that could be categorized by means of some major criteria, apart from niche concepts that we will not explain here in much detail (see Figure 5.6 for a modern eye tracker example known as the Tobii Pro Glasses 3). Those classifications might be based on invasiveness, flexibility, or data acquisition.

- **Invasive/intrusive vs. non-invasive/non-intrusive.** As described by Duchowski [161], for example, systems that are invasive or intrusive typically touch or have a contact with either the eye or the skin around the eye. This contact makes such studies uncomfortable for the study participants and in rare cases, at least in early times, people might have suffered from some eye injuries or at least their eyes had to relax some time after having taken part in such an eye tracking experiment. On the other hand, systems that rely on technical equipment that actually avoid these contacts try to make the study procedure more comfortable for the subjects, at the cost of finding a good way to get accurate and reliable eye tracking measurements, for example, if head movements in interactive applications must be allowed [360].

- **Remote vs. mobile/head-mounted/wearable/portable.** Looking at where the eye tracker is placed and how much flexibility it provides brings us to another categorization consisting of major classes. Those contain devices that are used in remote settings, placed away from a study participant without direct eye or body contact allowing contactless measurements, typically integrated into a computer monitor. The eye tracking devices could even be mobile eye trackers in the sense of being applicable everywhere, for example, in field studies, in which a high degree of flexibility is required. Moreover, such systems can be head-mounted, wearable, or portable which is achieved by better and better hardware technologies. Some mobile systems are so advanced that they can even track the eye movements in combination with head movements to reliably compute the points of eye fixations.
- **Electro-, infrared-, video-oculography, and scleral search coil.** Another classification is based on how the eye movement data is acquired. For example, certain sensors might be placed around the eye, measuring the skin potential during eye movements based on electric fields called electro-oculography. Also infrared light can be thrown on the eye to better record the effects of eye movements on the pupil positions. The amount of reflected light and, in particular the amount changes play a key role in this approach called infrared-oculography. Video-oculography, on the other hand, takes into account the video-recorded images based on single- or multiple-camera eye trackers. The corneal reflection technique makes it possible to record the position of the pupil given by the additionally reflected light. Finally, the search coil method uses wires in some kind of contact lens placed in a magnetic field causing voltages based on Faraday's law which give a hint about the eye position.

One of the biggest challenges today is the analysis and visualization of recorded eye tracking data. Moreover, if the data has to be analyzed in real-time, for example, for gaze-assisted interaction, we have to take into account the most powerful and most flexible technologies we can develop to guarantee eye tracking in free walking scenes, from a data recording perspective as well as from the algorithmic analysis power in terms of runtime efficiency. These aspects bring into play the criteria mentioned above like invasiveness, flexibility, or the method of data acquisition, which are related to aspects like cost, accuracy, or sensitivity, standing in a trade-off situation. Moreover, the eye tracking device in use also depends heavily on the application field as well as which environment the eye movement data has to be recorded. For traditional visual analytics systems this is definitely different than fields like marketing in which the study participants have a high degree of flexibility, typically walking around to find products meeting their needs.

5.2.4 Companies, Technologies, and Devices

In this section we present a list of popular eye tracking companies with the important eye tracking devices and technologies they have developed during the years of their existence. Moreover, we take a look into several criteria and aspects related to the designed technologies which we found on the corresponding web pages provided by the companies. We do not focus on listing all companies since there are many companies with varying backgrounds and scope of their research due to the fact that the whole field is progressing a lot, involving many international researchers from varying application domains. So that we do not prioritize one company over another by the order of their mentions in the list, we use a lexicographic order of the company names (see Table 5.1).

5.2.5 Application Fields

Due to the progress of hardware and software technologies the whole field of eye tracking steadily improves. Moreover, many companies (see Section 5.2.4) have been founded over the years with similar but also varying intentions to develop further concepts, devices, techniques, and technologies or to just provide support to certain fields of research. These fields also come in a variety of forms focusing on different goals demanding for a mixture of efficient interdisciplinary approaches with eye tracking among them. In the following we will briefly describe a list of interesting application fields for eye tracking without promising to be complete.

Table 5.1 Eye tracking companies with respect to hardware and software developments as well as focused applications, described by major buzz words

Company	Developments/Technologies	Applications/Focus
Argus Science	Binocular, 3D mobile, real-time	Marketing, sports
Blickshift	Visual analytics, data analytics	Usability studies
Ergoneers	D-Lab, Dikablis glasses, portable	Automotive, marketing
EyeSee	Online, facial coding, webcam	Virtual shopping
EyeTech	USB-connected, low-powered	Assistive, disabled
EyeVido	Browser data, data analytics	UX, usability studies
EyeWare	3D software, depth camera	Real-world interaction
GazeHawk	Webcam, crowd sourcing	Usability, comparisons
Gaze Intelligence	MRI, mobile, remote	Behavioral studies
Gazepoint	Biometrics, GP3 HD	UX design, usability
iMotions	Biometric sensors, real-time	Human behavior, UX
ISCAN	Real-time, head-mounted	Pilots, military
LC Technologies	Eyegaze, tablet communication	Assistive, disabled
Mirametrix	USB, attention sensing	UX, HCI
Pupil Labs	Open-source, wearable headset	UX design, marketing
Smart Eye	Head tracking, AI-powered	Automotive, aviation
SMI	Glasses, VR, RED500	Research, neuroscience
SR Research	EyeLink 2, portable	Academic research
The Eye Tribe	Tracker Pro, smart phone	Gaming, web usability
Tobii	VR headsets, wearable	Usability, VR, cars

- **Human–computer interaction/gaze-assisted interaction.** Using gaze as a means to interact with a user interface can be a great support to the standard interaction modalities like mouse, speech, touch, gesture, etc. [137]. However, when interacting by using gaze, so-called gaze-assisted interaction, we are confronted with extra challenging problems, one of which is the Midas touch problem [516]. This states that everything we look at is immediately interacted with, which can be a nasty feature. Hence, a certain time threshold, i.e. some kind of dwell time, should be decided when a visual component is interacted with, but it is not clear at what this threshold should be set in the default case, and in addition, it is user-dependent. A better solution would be to combine gaze-assisted interaction with some extra modalities [391] to better guide the interaction process. Moreover, the interactions could be linked to a database of all former users' interactions to somehow predict the way how we interact, maybe in a real-time setting which requires fast algorithmic solutions.

- **Physical disablement.** People who have a certain kind of disablement might be limited in the way they can interact with a visual stimulus. Eye tracking, in particular gaze-assisted interaction, can be of great support [517]; however, this kind of interaction suffers from the aforementioned problems that gaze-assisted interactions mostly have, the need to combine it with extra modalities such as speech, for example, based on voice recognition techniques. Moreover, in many cases the user interface provided for the physically disabled contains special visual components, for example, enlarged buttons, sliders, or menus, to allow better navigation with the eyes, otherwise controlling such a user interface using the eyes might end up as a frustrating experience.
- **Visual analytics/visualization/interaction.** Concepts from the field of visual analytics, for example major ingredients like visualization and interaction, are often examined in usability studies based on eye tracking [306]. Moreover, gaze-assisted interaction, apart from normal non-gaze-assisted interaction, provides a means to directly interact with the eyes to navigate in the system. Visual analytics is not only evaluated by eye tracking technologies, it is also a powerful concept to analyze the recorded eye tracking data, no matter from which kind of application field [14].
- **Medicine.** From a medical point of view, eye tracking can give a lot of positive advancements to the field. For example, eye movements can give hints about certain types of diseases or disorders like autism or Parkinson which are worth examining further [271]. Moreover, based on eye movements, a certain therapy or medication could be modified, adapted, or enhanced. Using web-based technologies, a doctor might examine the eyes of a person in an online meeting and decide what further treatment is needed, for example, inviting the person into the doctor's office [217]. Also in the medical environment we find many applications for gaze-assisted interaction, for example in a complicated surgery in which the doctor needs the hands while at the same time needing to interactively navigate an eye-controllable large monitor, hence eye tracking might be useful in a surgical training setting [340].
- **Marketing/product design/web usability.** How advertisements like print ads, online ads, or commercials are observed by possible customers and where, when, and for how long they pay visual attention can help to find out whether the design and the advertisement story of the product are adequate or not. An effective advertisement can attract more customers and hence, can bring a lot of money to the company that wants

to sell a product. Eye movements can at least give a hint about the visual attention strategies; however, they do not tell us what the customers are thinking, i.e. cognitively processing. There is a lot of research in this domain based on eye tracking with the goal of understanding the customers viewing strategy and, based on that, improve a product design or even the placement of products in a department store or on a web page [23].

- **Immersive analytics/VR/AR/MR.** Virtual, augmented, and mixed reality environments stand for some novel technologies in the field of data analytics, for example in the field of immersive analytics [347]. Combining them with eye tracking technologies can lead to powerful tools in the domain of data analysis while the human users are even more integrated in the data analysis process by taking into account their eye movements [140], even as a way to interact with the system by using the eyes as in gaze-assisted interaction. For example, recent technologies like the HTC Vive Pro Eye or the Microsoft Hololens 2 integrate eye tracking to improve the user experience. However, to allow a combination of eye tracking data with the rest of the system, real-time eye movement data analysis is required, supported by efficient algorithms running in the background.
- **Gaming/entertainment/sports.** The fields of gaming and entertainment benefit from gaze-assisted interaction in a way that the user has an additional means to interact, which makes the gaming experience much more realistic, for example if target-based shooting or foveated rendering is based on the gaze input. However, on the negative side, it is also more difficult to learn how to eye-control the games, even how to combine different interaction modalities. Moreover, analyzing and visualizing the recorded eye movement data [83], for example to explore visual attention strategies of the players for patterns and anomalies, is a challenging task due to the time-varying, interactive visual stimulus, even in a collaborative way as in multiple player systems. In the field of sports we might use eye tracking technologies to examine how active players perceive their environments during a match, for example to improve their winning strategy based on the detection of inefficient eye movements or the fact that they have been unaware of a certain situation [252].
- **Education.** We could inspect the teaching process from two perspectives, from the teacher and the students' side, for example who pays attention to what [147]. Such eye movement data could give

hints about the quality of a teaching strategy and, in particular, if the students were able to follow the written course content enriched by various illustrative figures or if there is room for improvement. Moreover, eye tracking itself could be taught as a novel concept to the younger generation to educate students and to create an interest in the technology. Not only the technology itself but also linked concepts like visual analytics should be learned to find insights in the recorded eye movement data [71].

- **Car driving/automotive.** The automotive industry is trying to investigate eye tracking technologies as assistive concepts during car driving [501]. This new kind of interaction inside a cockpit, while driving a car at the same time, can have benefits due to being less distracted to better focus on the main car driving task. Moreover, the recorded eye movement data from millions of car drivers in a long-duration task can give many insights about certain accident black spots or serious impacts on the driving ability after longer journeys. Understanding eye movements cannot solve the problems but it can give valuable insights into reducing the number of dangerous situations on the roads, i.e. eyes-off-road detection, for example, can be a useful strategy to create safer cars. Moreover, eye movements might be useful in an education setup in which the car driving teacher can watch the visual attention during a car driving lesson afterwards to identify possible negative issues.
- **Aviation/military.** Eye tracking could also be useful for analyzing pilot training sessions [430, 431], for example understanding their progress and to educate them while at the same time taking into account their eye movement patterns (see Figure 5.7). For example, a pilot might have had problems understanding the functionality of a dashboard or they might have missed a chunk of information that was needed to safely land a plane. Moreover, fatigue effects might be measured as well as possible skill levels which can guide the setup of the next training sessions. Also in the field of military, eye tracking devices have been used, for example, in a flight simulator for learning how to control fast military jets [267].

There is an endless list of applications using eye tracking as a means to either interact with or exploit the recorded data to find insights into the strategic visual attention patterns of the observers to enhance certain aspects based on the identified design flaws. Some extra application fields might be neuroscience, perception, cognition, psychology, communication, reading

Figure 5.7 Eye tracking technologies can be useful in the field of aviation, in particular, when training pilots to land a plane [430, 432]. Image provided by David Rudi (Copyright ETH Zurich).

research, activity detection, authentication, music score page turning [51], walking and hiking, and many more.

5.3 Eye Tracking Data Properties

It does not matter how advanced an eye tracking device is, the general benefit is that it produces data describing people's eye movement patterns, i.e. spatio-temporal effects varying from participant to participant, over the tasks at hand and shown stimuli (see Figure 5.8 for an illustration of a scanpath overplotted on a static stimulus). In the best case this data is measured at high tracking frequencies and the quality of the data is good enough to derive visual attention patterns that give hints about certain design flaws, for example, to improve a visual analytics system. However, apart from the quality of the spatio-temporal eye movement data, it can come in many facets, typically including information about the visual stimuli, the participants, fixations with their fixation duration, saccades, and additional physiological data as well as verbal feedback or extra personal data [44]. Although some eye tracking devices provide very fine-granular eye movement data, in space and time, for the analysis and visualization concepts, such data is mostly aggregated on a level that allows efficient methods targeted at insight detection as well as building, confirming, refining, or rejecting hypotheses based on certain user tasks.

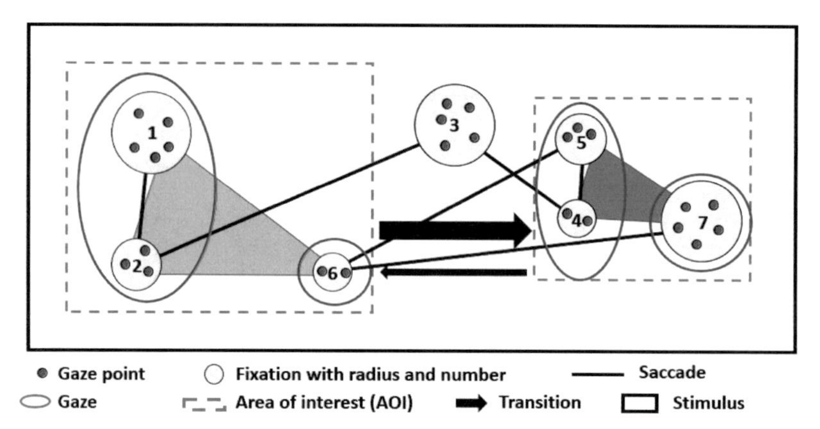

Figure 5.8 Eye movement data can be described as consisting of gaze points, which is the lowest level of granularity that is interesting for eye tracking in visual analytics. Those gaze points are spatially and temporally aggregated into fixations by modifiable value thresholds. The fixations with duration (encoded in the circle radius) contain saccades in-between, i.e. rapid eye movements. A scanpath is made from a sequence of fixations and saccades. Regions in a stimulus that are of particular interest are called areas of interest (AOIs). If we are only interested in fixations in a certain AOI we denote those by gazes. Between AOIs there can be a number of transitions indicated by the number of saccades between those AOIs [44].

The stimuli are a key ingredient in an eye tracking study since they carry semantic information that has to be understood and combined in a way to successfully solve the given tasks. These stimuli can have several properties and make a difference for the data storage, linking, as well as later data analytics in the form of algorithmic as well as visual concepts combined with interaction techniques, such as visual analytics. The stimuli can be just static or even dynamic, i.e. changing over time as in a video. Moreover, if the content is changing on user demand, like in interactive user interfaces, we can have many different static stimulus states that might form some kind of stimulus graph [68] instead of a linear sequence of animated frames or snapshots from a video in which each frame is typically watched only for a fraction of a second. In general, one challenge will be the linking of the stimulus data with the eye movement patterns before starting an analysis. Moreover, combining several different stimuli with similar characteristics attached by eye movement data is a challenging synchronization task as well as the storage of the data if it comes to long-duration eye tracking tasks in the wild where video sequences have to be recorded together with the eye movement data.

In this section we will have a look into the different stimuli in eye tracking studies. As a next step we describe fixations and saccades that model the visual scanning behavior of several people, for example when taking part in an eye tracking study. Areas of interest (AOIs) are useful to explore the visual attention patterns based on aggregated regions in a stimulus. Static AOIs are easier to handle than dynamic ones, for which some kind of matching function is required. Apart from the pure eye tracking data we can easily enrich the data by extra physiological measures such as EEG, galvanic skin response, or pupil dilations, to mention a few. Finally, additional metrics can be derived from the original data to get more insights from a different perspective, typically on a more aggregated view or in a form that links different data aspects to build a new kind of measure, for example the ratio of fixation durations and saccade lengths, or the time to first fixation to areas of interest. Actually, there is no limitation to what types of data can be used as extra input, it is just a question of what is meaningful for solving the tasks at hand from a data analysis perspective. Moreover, the runtime complexities for certain algorithms might need to transform the data in the desired format first which could have an impact on a possible interaction in a visual analytics system for analyzing eye tracking data.

5.3.1 Visual Stimuli

A visual stimulus in an eye tracking study is what we see, like a static or dynamic diagram or the real world, and it builds a basis for solving given tasks, either artificially made tasks in a user experiment or real-world everyday tasks like shopping, car driving, or watching football matches on TV. Each stimulus contains some kind of semantics, i.e. visual aspects and objects that can be compared and linked, and from which information can be deduced to apply it to another stimulus or another spatial region in the same one. In this process we add extra information stored in our brains' long-term memories that we have gained over the years which builds our level of experience. All of these aspects in combination support the finding, identification, or detection of a task solution, or in the worst case we might fail to find a solution, but maybe we refine our knowledge at least. A visual stimulus guides us somehow based on the powerful vision that has been adapted over the millions of years of mankind evolution. The information that we see in the stimulus is transformed to make it processible in the brain and a certain kind of reaction guides and controls the eyes to the next chunk of visual information that is required to come closer to a task solution. These

Figure 5.9 A car driving task generates a dynamic stimulus, in this case we see four snapshots at different time points (T1) to (T4) of a longitudinal eye tracking experiment with indicated points of visual attention [44].

eye movements are interesting since they describe some kind of path that is taken in the visual stimulus which hints at facts like how we solved the task, where we stopped for longer and shorter times, and if certain visual objects have been inspected several times in a row. However, the eye movements do not describe what is cognitively processed in the brain, they just give hints about the visual attention patterns.

Visual stimuli in an eye tracking study can come in several forms. Figure 5.9 shows an example of a dynamic stimulus from a car driving experiment.

- **Static visual stimulus.** A visual stimulus could be a static diagram, picture, poster, or any visual object that is not equipped with interactions and that cannot be rotated or for which there is no opportunity to walk around to inspect it from different perspectives. Typical examples are standard "old" diagrams before the invention of the computer or those for which interaction is turned off (see Figure 5.8) like in a technique-driven user experiment. Moreover, text could fall into this category for which researchers try to investigate reading tasks and the corresponding eye movements, also static scenes like pictures or paintings for which a task is asked, like in the work by Yarbus [539]. To keep the setup of an eye tracking study easy, as well as the data analysis later on, in most cases a remote eye tracking system is used, for example integrating the eye tracker in a computer monitor on which the static visual stimulus is shown.
- **Dynamic visual stimulus.** A dynamic stimulus is one that changes its content from time to time. This does not mean that the stimulus itself has to change necessarily but it could be possible that an observer has the possibility to watch the stimulus from many different perspectives, for example, walking around in a museum and inspecting a "static" sculpture. This visual object is always the same but the viewer's field of view changes dynamically while walking closer and farther away, also

with varying gaze depth [163, 166]. In a shopping task we have a similar situation, but here the viewer typically explores a larger dynamic scene while other customers cross the way and we might start conversations and communications. In a car driving task we see the dynamic stimulus passing by while at the same time navigating the car and also interacting with several electronic devices like the radio, telephone, or navigation system. Similar aspects hold for virtual environments in which people freely walk around or react to certain situations with body movements or gestures. Moreover, a video or animation belongs to the class of dynamic stimuli although the viewers cannot change their content, maybe just pause and play back. It is more like an autonomously changing linear sequence of static pictures. In visual analytics or many other application fields we come across interactive user interfaces which are by definition dynamic. For such a dynamic visual stimulus we require more advanced *head-mounted, portable, or wearable* eye tracking devices. However, analyzing such data is much more challenging because the dynamics of the stimulus has to be stored by a video and this demands for algorithmic concepts like dynamic AOI detection, matching and linking of the stimuli over time, or just identifying the group behavior and visual attention paid over space and time.

A visual stimulus can be mathematically modeled as a sequence of static pictures carrying a time point given as a natural-numbered index, no matter if the stimulus is static or dynamic. In a static case the sequence consists of the same stimulus all the time. This means the visual stimulus can be modeled as

$$\mathcal{S} := (S_1, \ldots, S_n)$$

if $n \in \mathbb{N}$ describes the total amount of time the stimulus was visually attended. In the static case we consequently have $S_i := S, \forall 1 \leq i \leq n$ where S describes the static visual stimulus. It may be noted that we do not use this model to store the sequence efficiently, it is just a mathematical illustration. To avoid storage issues, only the stimuli differences over time might be stored. Moreover, for an interactive stimulus we might additionally store the relations between the individual static stimuli from the sequence since they form some kind of interaction graph [68], i.e. the stimulus changes from one state to another one with the option to go back and forth between the same stimulus states. The semantics contained in the stimuli as well as the experience levels of the stimuli observers lead to evolving knowledge states that can also be mathematically modeled as

$$\mathcal{K} := (K_1, \ldots, K_n)$$

for $n \in \mathbb{N}$ describing again the total amount of time the stimulus was visually attended. Those knowledge states might be updated from time to time, but we cannot really describe how they are updated; however, they might have an impact on the visual attention strategy [269].

5.3.2 Gaze Points, Fixations, Saccades, and Scanpaths

An eye tracker registers gaze points at high tracking rates, called sampling frequencies. The frequency describes how often an eye tracker registers a gaze point per second. It can vary from tracker to tracker and depends on the purpose of the recorded eye movement data, i.e. for some scenarios a lower sampling frequency has benefits compared to higher frequencies. For example, if a scanpath has to be followed at very fine intervals, a higher frequency makes sense. However, the high frequency eye trackers come at the cost of being much more expensive due to the fact that the technical equipment is more advanced, for example, the cameras and the artificial illuminations must be more powerful. Moreover, a higher sampling frequency means more data to be handled as well as more unneeded data points which require additional algorithmic preprocesses. Typical eye trackers useful for visual analytics have a frequency of 60–120 Hz, but those with high sampling frequencies of 600–1200 Hz also exist. But again, the question is whether such a high tracking frequency really has any benefits. For example, in a case in which we are interested in so-called microsaccades or post-saccadic oscillations [377], we need eye tracking devices with high sampling frequencies.

If an eye tracker has a sampling frequency of n Hz, it means that n gaze points per second are registered. These gaze points are too fine-granular for data analysis, hence they are typically aggregated beforehand into so-called fixations. For this we need two parameters, a temporal distance and a fixation radius meaning gaze points that are next to each other over time and space are aggregated into a fixation, like a gaze point cluster for which we need a representative element, i.e. the fixation point in space. These two parameters are set by default in an eye tracking system but can even be modified on user demand, for example for generating more or fewer fixations from the recorded gaze points. The coordinates of such a fixation point might be derived directly as the center of the circle spanned by the fixation radius or the average point of all gaze points in a gaze point cluster. There are even more advanced methods [463] which are not required for eye tracking visual analytics. The fixation radius is typically so small that it is irrelevant where exactly inside the gaze point cluster the representative fixation point is located.

The total amount of time a fixation lasts is called fixation duration which can be interpreted as the longer a visual object is fixated, the more interesting it is and it attracts our attention; but on the other hand, it might also be confusing and some time is wasted in understanding its meaning. The eye movement from one fixation to the next one is called a saccade while the sequence of all fixations and saccades forms a scanpath (see Figure 5.8). However, there are also smooth pursuits that describe eye movements constantly following a moving visual object, i.e. saccades are not occurring very often due to the fact that all gaze points are close together over time and space by a small constantly shifting offset. In rare cases there might occur so-called catch-up saccades which are effects caused by visual objects that move too fast for keeping an eye on them or a distracting visual object that also attracts the viewer's attention at the same time. A scanpath describes a trajectory over space and time making it a challenging type of data to analyze if several people's scanpaths are involved in finding a common visual scanning strategy in the spatio-temporal data.

Mathematically, we can model a sequence of gaze points as

$$\mathcal{G} := (g_1, \ldots, g_k)$$

with $g_i \in X \times Y$ being a 2D spatial point if X and Y model the width and height of the stimulus in pixels for example. If we have an eye tracker of sampling frequency 60 Hz we get 60 gaze points in a second, i.e. $k = 60$ in the sequence above which means a gaze point is measured after every 16.67 ms approximately. If the gaze point sequence is mapped into a fixation sequence

$$\mathcal{F} := (p_1, \ldots, p_m),$$

we get much fewer points, i.e. $m < k$, depending on the given parameters. Each p_i, $1 \leq i \leq m$ describes a point in 2D as $p_i \in X \times Y$, similarly to the gaze points. However, each p_i is based on a group of g_i and, hence, some kind of aggregated or representative value based on g_i. Moreover, each p_i is attached by two time points t_{e_i} and t_{l_i}, expressing the time points the eye enters the fixation space and when it leaves it again. Consequently, each fixation has a fixation duration given by $t_{\mathrm{d}_i} := t_{\mathrm{l}_i} - t_{\mathrm{e}_i}$ in addition to the spatial position on the stimulus. Between each pair of subsequent fixations we have a saccade that describes the eye rapidly moving from one fixation to the next one in the sequence. All fixations in a fixation sequence and the saccades in between together form a scanpath. If several people are taking part in an eye tracking study we have many of those scanpaths, typically all with varying properties incorporating the visual attention behavior of many people. The sum of all fixation durations

$$t_{\text{response}} := \sum_{i=1}^{m} t_{\text{d}_i}$$

in a scanpath of one participant results in the response time for a specific task. By contrast, this means that each response time, for example in a traditional study without eye tracking, is actually a temporal aggregation of many response times of subtasks which might be identifiable when observing a scanpath.

5.3.3 Areas of Interest (AOIs) and Transitions

An area of interest is, as the name suggests, a certain spatial region in a 2D or 3D stimulus that might be of particular value for the data analyst. By defining areas of interest we can reduce the number of data points since the AOI definition is some kind of spatial aggregation technique. The fixations contained inside an AOI are denoted by gazes (see Figure 5.8). Even more metrics can be derived and values computed, just for those specific regions in a stimulus. For a static stimulus, the AOI definitions are typically also static, meaning each AOI just corresponds to a spatially connected subregion of a stimulus, i.e.

$$A_{\text{AOI}} \subseteq X \times Y$$

if X and Y are the width and the height of a stimulus in pixels for example. Although the entire stimulus can be defined as an AOI it is questionable whether this makes sense for a later analysis, but maybe if we want to separate on-screen and off-screen fixations from each other it might be a useful AOI selection. Such an AOI region can be a rectangular or any arbitrarily complex shape, but typically it is given by a closed region to better identify what is inside. Moreover, if the stimulus is not static but dynamically changing over time we need some kind of algorithm that identifies a selected AOI in each of the frames in a video, for example [264]. The challenging problem for this algorithm is that the AOI can move from one position to another one, for example, a moving car in a video or a pedestrian. This movement

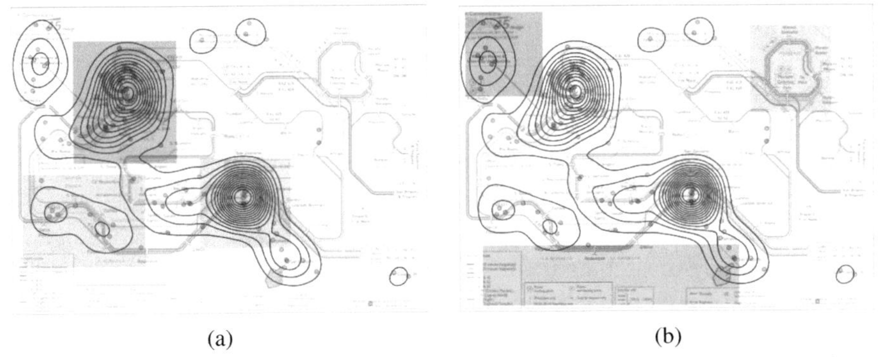

(a) (b)

Figure 5.10 Selecting areas of interest in a static stimulus can reduce the amount of eye movement data and can impact the eye movement data analysis since each AOI is some kind of spatial aggregation: (a) AOI selection based on hot spots of the visual attention behavior; (b) AOI selection based on the semantics in the stimulus [100].

can happen smoothly, but also abruptly, jumping from one position to a completely different one. Moreover, the visual object to be located inside an AOI can change its shape dynamically, it can rotate, get bigger or smaller, be partially occluded for a certain amount of time, and so on. All of these effects make an automatic detection of an AOI a difficult and sometimes error-prone process. However, defining dynamic AOIs manually is possible, but it is a very time-consuming process.

There are several ways to define areas of interest that are independent of the fact whether a stimulus is static or dynamically changing over time (see Figure 5.10 for an example of such an AOI definition based on user input).

- **User input.** Manually selecting areas of interest is a possible solution. For static stimuli this sounds doable, but for dynamically changing AOIs this can become a time-consuming process. However, a manual user-specified AOI definition and selection can be more exact than an automatic one due to the perceptual strengths of the human visual system and pattern recognition abilities [521]. Moreover, the selection can be based on rectangular, polygonal, or arbitrary shapes.
- **Automatic spatial subdivision.** A stimulus, static or dynamic, can be split into several subregions based on equally sized grid cells while the cell sizes are modifiable and adaptable by the user. Although this is a naive, semantics- and visual attention-ignoring process, it works quickly and might give some hints about the eye movement data. However, a more fine-tuned AOI definition should follow after this naive idea is applied.

- **Visual attention-based.** Conducting an eye tracking experiment, at least partially with a few participants, provides the opportunity to define areas of interest based on the already recorded visual attention paid to a stimulus. The relevant visual attention-based regions might serve as AOIs, for example using the hot spots of visual attention. Those hot spots can be identified by the human observer but also by advanced density-identifying algorithms. AOIs can be based on the density shapes but also on Voronoi regions taking the hot spot regions with their highest density points as Voronoi cell centroids into account.
- **Semantics-based.** Incorporating the semantic meaning of a stimulus' components is a valuable approach to exploring eye movement data. The question comes into play: which visual features in a stimulus are really carrying meaningful semantic information to solve a given task? For this AOI definition to work reliably, the human users are required with their experience and domain knowledge. The human does this faster and more accurately than a computer.
- **Hybrid approach.** A combination of several AOI definition and selection concepts is always possible. For example, the computer might identify hot spot regions in a stimulus while the human user can refine them, remove or add some based on the semantic understanding of the stimulus. Moreover, formerly defined AOIs should always be modifiable after a certain data analysis or visual representation process has been applied to the eye movement data with the AOIs in use.

The eye movements between pairs of AOIs are called transitions, typically given as a weighted transition matrix that describes the transition frequencies between all pairs of AOIs, either as a time-aggregated measure or as a time-dependent AOI transition sequence [80].

5.3.4 Physiological and Additional Measures

Apart from the eye movement data we could be interested in additional data sources, either recorded during the eye tracking experiment, maybe synchronized with the eye movement data, or additional data that is unrelated to the eye movements and maybe recorded before or after the eye tracking experiment. There could be any kind of additional data source but the biggest challenge is to link those extra data inputs with the eye movement data to get benefits for the insight detection process focusing on visual attention patterns and visual task solution strategies.

Additional data that might be recorded during each eye tracking experiment could be related to the physiology, i.e. functions and mechanisms in the human body. Typical measures are galvanic skin response, blood volume, blood pressure, respiration activity, EEG, ECG, EMG, skin conductance, or pupil dilation [289], among many others, data which could be acquired by additional sensors, in case those sensors produce time-dependent data of an acceptable quality that do not lead to misinterpretations later on. Moreover, body movements or gestures, eye blinking, vocal tone, emotions, and many more could enrich the repertoire of possible physiological data sources. Some care must be taken when measuring such additional data since most sensors are some kind of semi-invasive technologies, meaning they are partially connected to the human body and hence may cause some negative issues in an eye tracking experiment, in the worst case having an impact on the eye movements in the form of confounding variables and leading to a certain bias in the study results.

Apart from physiological data other study-related data can also be measured during the running experiment which is not directly related to the physiology, for example mouse clicks and movements, keystrokes, or touch interactions if those are allowed and supported in an eye tracking study. If we are not directly in a laboratory environment, but for example, in a car driving study, additional instruments might be used, like human–car interactions for which other properties might also be of interest such as the velocity of the car or the steering wheel angle, the volume of the radio, the communications and conversations with other car passengers, and so on. Moreover, verbal feedback during an experiment like think-aloud or talk-aloud are additional data sources that are worth considering in a later data analysis. To avoid noting down all the feedback, audio or video might be used. However, such textual feedback is typically difficult to analyze since it might require advanced text processing algorithms, no matter whether audio or video recordings are available, in particular, if the semantics of a text has to be integrated. The problem of transforming verbal feedback into a valuable and linkable data source is not a problem for studies with a few participants, for which it can be done manually, but in a future scenario we might have eye tracking experiments with thousands of participants, and then at least we need some kind of automated process that brings the massive verbal data into a suitable form for a later data analysis step.

A third type of additional measurements could come in the form of qualitative feedback after the study. Moreover, any kind of personal details about a study participant, such as gender, age, level of experience, visual deficiencies, and so on, as well as confounding variables might be worth including as an important data part in the whole study data base. Physiological data is typically measured during the running experiment, but it might even be a good concept to measure, at least some, additional physiological data before or after the experiment, for example to compare it with the data during the experiment to understand if the participants had some varying performance states that might cause changes in the data interpretation step. If tasks are asked to be solved in the study, the task accuracy (or error rate) might be of particular interest as well as the response time that can be directly derived from the fixation durations. However, no matter how much data and which types are recorded, they might be useful for a later analysis. Actually, we can measure as much as possible, the storage of the data is often not the problem; for a later data analysis step we can still decide which data sources are important and worth integrating into a visual analytics system focusing on finding insights in eye movement data as well as the combined extra data sources.

5.3.5 Derived Metrics

Not only the recorded and measured data might be of interest but also additional data that can be derived from the originally given data, be it combined from eye movement data, extra physiological, or even further data sources such as human interaction data or verbal feedback for example. In this context we often find the term derived metrics that focus on combining two or several of the data sources [299] to come up with a completely new one, but still containing the input from the given ones [452]. Examples from this concept are fixation duration for which we might be interested in the average fixation duration, either for each scanpath of the participants to compare the participants [300] or for all scanpaths of one stimulus to compare the stimuli. The same idea can be found for saccade lengths or saccade orientations and directions. Also the mean, median, standard deviation, variance, and even further statistical measures might give further insights into the properties and distributions of the recorded data. In most cases, the statistical data only serves the purpose of providing a quick way to compare different data dimensions like the participants or stimuli.

Further metrics might take into account the length of a scanpath in terms of number of fixations, or the total length in terms of distances that the eye has moved over a given stimulus. Moreover, the scanpath-enclosed maximal stimulus region might be of interest which could give a hint as to how much of the input information has been covered by the visual attention, which can also be estimated by the enclosed region of the fixation points in a scanpath, i.e. given by a minimal polygonal shape that encloses all fixation points. Also areas of interest provide a way to form some kind of derived metric since they all contain additional insights for specific stimulus regions like the numbers of fixations to certain areas which allow comparisons between stimulus regions, even for comparing participant groups based on their visual scanning behavior. Moreover, metrics like time to first fixation, time spent in an area of interest, number of AOI revisits, number of transitions between two AOIs, and many more provide interesting additional values.

If several metrics are combined we can further extend the repertoire of possible metrics; however, since this repertoire is very large, we might better consider which ones make sense for a certain data analysis task beforehand and then select the ones that are interesting for these specific tasks. For example, a combination often seen is the ratio between fixation duration and saccade length, or the saccade orientation (in degrees) divided by the saccade length, average saccade length in a certain area of interest, average distance between all pairwise fixation points, and many more. All of these derived metrics serve their own purpose and can be analyzed for further statistical properties and also for correlations between two or more of them, for example, asking oneself if the fixation duration and length of subsequent saccades are positively or negatively correlated [299].

5.4 Examples of Eye Tracking Studies

There is a huge body of eye tracking studies due to the fact that this technology is applicable to many research fields. In this book we will specifically focus on eye tracking studies in the context of visual analytics, including visualizations, diagrams, plots, and charts, and also interaction techniques, reading tasks like texts, labels, source code, maybe for understanding algorithms, user interface design and usability, and the like. Finally, we explain eye tracking studies that focus on visual analytics as a whole, i.e. in which all of the aforementioned concepts are linked to some degree with the goal of finding insights in datasets, and how the recorded eye movement data and additional data sources are analyzed and visualized will be described in more detail in Chapter 6. To allow the human user to interact with the system we also need user interfaces and their design, hence we will also focus on this aspect and the usability issue investigated by eye tracking studies.

Not all of the aforementioned concepts can be evaluated by eye tracking in the same way since they contain varying and also differently complex scenarios making use of a variety of ingredients to achieve a maximum of usability for a visual analytics system. For example, visualization techniques are often studied if they can be understood and compared to other visualization techniques, like in a technique-driven evaluation, showing the same dataset. From the perspective of eye tracking it is interesting to find out where and when people make errors when inspecting such a stimulus and, in particular, we are interested in the question why they made them, to reduce the number of design flaws in a visualization, diagram, chart, or plot. For interaction techniques this situation is a bit more challenging since the stimulus becomes dynamic, i.e. changes over time [68]. However, the recorded data can still give insights about which interaction is useful or which one causes delays and misunderstandings. Here we can distinguish between two scenarios, interactions supporting gaze and those that do not support gaze. We often have to read labels or additional textual information, for example, as a details-on-demand feature or as error messages, further textual instructions, or source code, for example in visual analytics systems for software developers [453]. For user interfaces it is even advice based on the eight golden rules for interface design [461] to add textual components. The most difficult eye tracking setup in this context comes into play if all of those issues are studied in combination, like in a fully fledged visual analytics system.

5.4.1 Eye Tracking for Static Visualizations

Visualization techniques are a way to depict data graphically. In many cases there are alternative visualization techniques for the same dataset that might be better or worse in terms of user performance for a given task. Standard user studies measure error rates and response times when people try to solve the given task with several visualization techniques. The researchers expect to understand the differences between the visualization techniques by taking into account the measured performance in a comparative way. Although this idea is already quite useful and has been applied many times, for example in graph and network visualization [76], it does not provide any details about the visual attention behavior over space and time. Error rates and response times are some kind of aggregated measures, but eye tracking can give detailed insights into the visual scanning process over space and time which is lost if only error rates and completion times are analyzed. However,

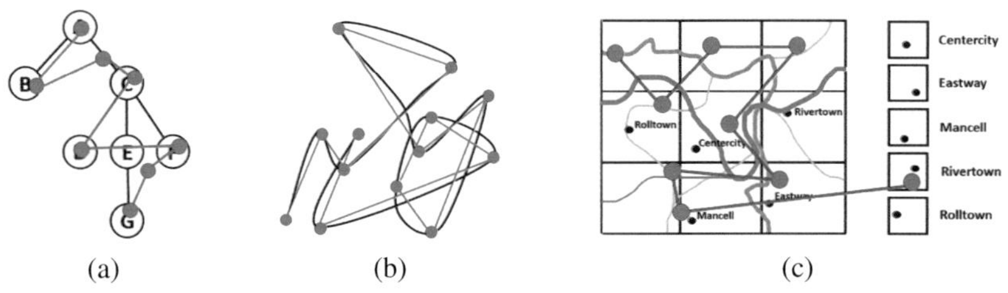

(a) (b) (c)

Figure 5.11 Visualization techniques have been explored a lot by applying eye tracking. (a) Node-link tree visualizations [78]. (b) Trajectory visualizations for bird movements [369]. (c) Visual search support in geographic maps [371].

the recorded data is much more challenging to analyze, in particular if we are interested in differences over space, time, study participants, stimuli, or even further derived metrics (Section 5.3.5). These differences might even be used to classify the data dimensions into groups, classes, or time points and intervals based on the eye movement behavior, for example groups of study participants, groups of stimuli, groups of time periods, or spatial subdivisions of the shown stimulus. Moreover, a combination of all of those aspects is imaginable building a several level subdivision of the data dimensions.

To investigate the usefulness of node-link diagrams for hierarchical data, some of the existing layouts and link representation styles have been evaluated in an eye tracking study [78]. Traditional, orthogonal, and radial tree diagrams served as independent variables and the eye movement data of 40 participants was recorded who tried to find the least common ancestor in a corresponding tree diagram of highlighted leaf nodes. As further independent variables, the number of highlighted leaf nodes, the tree orientation, and the sizes of the hierarchy datasets were varied. The Tobii T60 XL eye tracking device, integrated into a monitor, was used to record the eye movements of the study participants when paying visual attention to the static diagrams. As a major goal it was found that people tend to make fewer errors and had lower response times when using the top-to-bottom traditional tree diagram for this specific task. The eye movement data additionally showed that the reason for the longer response time for the radial diagram type was a cross checking behavior until the solution node was confirmed by a mouse click [72]. An example stimulus from the tree eye tracking study in an orthogonal layout overlaid with a gaze plot can be seen in Figure 5.11(a).

Another eye tracking study was based on static diagrams for trajectory visualizations for bird movement behavior [369], recorded by making use of GPS. Several edge representation styles like arrow-based, equidistant arrows, equidistant comets, and a tapered edge representation style (see Figure 5.11(b) for a stimulus showing a trajectory) were compared. The rendering order was also investigated in the study. Twenty-five participants were recruited to answer tasks focusing on long, dense, complex, and piecewise linear spatial trajectories visually encoded in each of the edge representations for the trajectory visualization. The given tasks asked for tracing paths, identifying longest links, and estimating densities of trajectory clusters. As an eye tracking device the Tobii T60 XL was chosen, partly due to the fact that the researchers had some experience with it in earlier eye tracking studies and that it was a suitable eye tracking technology for evaluating static diagrams visually depicted on a computer monitor. Tapered edge representation styles [238] have been identified as the best option for a trajectory visualization in this eye tracking study, also by evaluating the recorded eye movement data statistically and by presenting the results in the form of simple statistical diagrams.

General geographic maps contain lots of labels to provide a textual overview for the viewer to better orientate oneself in the map. However, locating a given label in a map in which we do not know where the label is actually located is a tedious task and requires some kind of time-consuming search strategy. If the search is supported by additional hints like within-image, grid reference, directional, or miniature annotation [371] we can typically reduce the map into relevant regions in which the label is to be found and hence the search time is tremendously reduced (see Figure 5.11(c) for artificially generated labels for a grid reference annotation overplotted with a scanpath). In an eye tracking study we can additionally find out how the observers pay visual attention if certain search support in a map is given. Moreover, based on a comparison of all the tested search support techniques we can identify which one is actually the best one for the task of finding a label in a geographic map. Thirty participants were recruited to solve such a task with several techniques by showing them artificially generated static geographic maps to avoid learning effects. A Tobii T60 XL eye tracking device integrated into a monitor was used to record the eye movements. The major result of this study showed that all annotations outperformed the within-image annotation while the eye movement data showed that the visual search patterns differ from annotation to annotation with the miniature annotation producing the lowest response times.

Public transport maps [103] are designed to support travelers in the task of finding a route from a start station to a destination station, for example when planning a journey through an unknown city. The map is some kind of

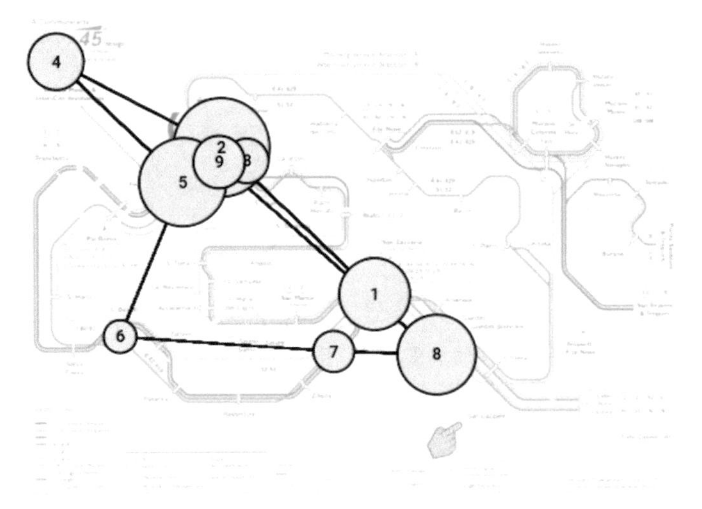

Figure 5.12 Public transport maps for different cities in the world (in this case Venice in Italy) [372].

visual stimulus used to efficiently solve such route finding tasks. By using eye tracking we might find out if the map contains certain design flaws [372], if its design is better or worse than another one for the same city, if the map complexity in terms of number of stations and metro lines plays a role for the visual attention strategies [85], or if visual enhancements in the form of color coding, legends, or sights [65] have any impact. Twenty-four public transport maps from cities all over the world were used as stimuli by changing between colored and gray-scale stimuli to investigate if color has any impact on the user performances as well as visual task solution strategies [372]. Route finding tasks were asked in this study by highlighting the start and destination stations to reduce cognitive effort when finding those relevant points for answering the task (see Figure 5.12 for a public transport map overlaid with a scanpath). Forty participants were recruited and a Tobii T60 XL eye tracking device was used. The major result showed that the original task of finding a route between two highlighted stations was actually subdivided into many subtasks [63] containing cross checking behavior, i.e. the found route was not confirmed immediately but after a careful check of the found route. Moreover, color coding was found to be very important for lowering the response times, while the saccades became much shorter for gray-scale maps due to the fact that the observers had to more carefully follow a gray-scale metro line which is obviously easier if color is supported.

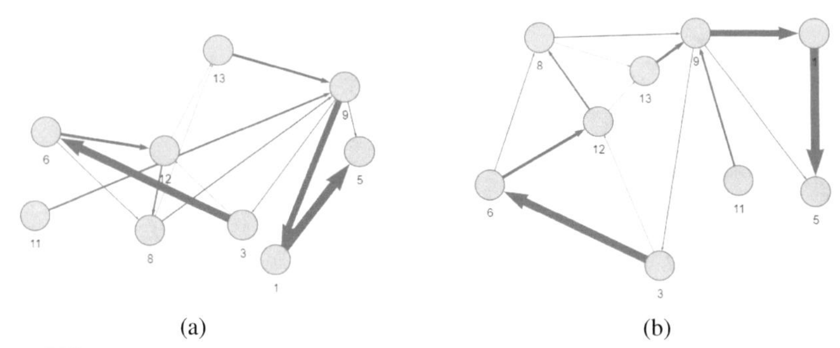

Figure 5.13 Graph layouts with different kinds of link crossings, crossing angles, and the effects of geodesic-path tendency can have varying impacts on eye movements [245].

Node-link diagrams for graphs and networks (see Figure 5.13 for examples illustrating the ideas in the graph layout eye tracking study) can be visualized in several layouts, all following certain aesthetic graph drawing criteria [29, 405] with benefits for a certain well-defined graph task [323]. The eye movement data [245] can give additional insights into the task solution strategies applied by typical graph visualization users, for example, which impact link crossings, crossing angles, or the geodesic-path tendencies might have on task performance and why they could be problematic. With an iViewX head-mounted eye tracking device created by SensoMotoric Instruments (SMI) and by recruiting 16 participants such effects of the graph layout were investigated. As a major result it was found out that small crossing angles can lead to increased response times and more eye movements around the crossing areas for route finding tasks while node location tasks are not that affected by these issues. Moreover, the criterion of geodesic-path tendency shows that routes in a graph might be harder to follow by the eye.

There are many more eye tracking studies taking into account visualization techniques, for example focusing on visual variables in 3D visualization [335] by recruiting 36 participants and by using a Tobii T120 eye tracking device. An SMI RED 250 eye tracker was used to investigate 2D and 3D visualizations in cartography [402]. Visual exploration patterns in general information visualizations have been studied by using a Tobii X120 eye tracker by recruiting 23 participants [5]. Scatter plots [178] and scatter plot matrices [451] have been a focus of eye tracking. A remote Eye Tribe eye tracker was used in the scatter plot matrix approach in which 12 participants were recruited. Ontology visualization has also been explored by applying eye tracking [193]. Indented list and graph approaches were evaluated by recruiting 36 participants recording their eye movements with a Tobii 2150 eye tracker. Even Euler diagrams have built the basis for an eye tracking study [438] by asking 12 participants and by using a Tobii X2-60 eye tracker. Further topics under eye tracking investigation include 2D flow visualization [231], flow maps [157], program visualization [33], clutter effects [358], or radial diagrams [204].

5.4.2 Eye Tracking for Interaction Techniques

Interaction techniques are useful to bring a static diagram to life and allow the users to navigate in the visual depiction of data. However, evaluating the usefulness of an interaction technique is much more challenging than evaluating a static diagram for example. From an eye tracking perspective, the major difference is that the stimuli are dynamic due to the interactive feature that they provide. However, the recording of the eye movement data is not a challenging issue, it is more problematic to analyze the dynamics of the data later on, since the stimulus content varies over time, typically leading to many more visual components with various visual variables that have to be matched first to the eye movement data and additional data sources. In this section we describe some of the eye tracking studies that have been conducted in the field of interaction related to visualizations, but it may be noted that there is a much longer list focusing on research in this direction.

When including interaction in a visualization or user interface we have two major options. The first one incorporates standard interactions like mouse, keyboard, voice, touch, and so on to modify a given stimulus. These types of interactions have to be recorded by video and audio and have to be matched with the additional data from the eye tracking device, i.e. the dynamic data has to be synchronized to allow the identification of insights over space, time, and participants. A second option to incorporate interaction in a visualization or user interface is by means of gaze, i.e. the field of gaze-assisted interaction comes into play here. The major benefit of gaze-assisted interaction is that the eye movement data is already recorded and matched with the stimulus, otherwise it would be problematic to interact with a stimulus based on gaze. Moreover, standard interactions can be combined with gaze interaction [510], building some form of hybrid or multi-modal interaction.

The browsing behavior in web pages was investigated by recording eye movements to understand this kind of interaction [156]. Searching and scanning are two popular viewing strategies with the goal of finding relevant information, often resulting in an "F"-shape pattern if the web page contains mostly textual information, as found out by other eye tracking studies [464]. However, the visual hierarchy of a web page plays a crucial role if this "F"-shape pattern occurs or if it looks different, which also has an impact on the way we interact with the content. Forty-eight participants were recruited to explore the browsing strategy in web pages while the Tobii X120 eye tracker was placed in front of a monitor. This interaction is not an active but rather a passive interaction technique [476] in the way that the user communicates with a stimulus to get back important information, but the stimulus is not changed. However, it can give insights into where the user might find information and hence needs to use scroll bars, and so it is some kind of passive "explore" interaction [544]. The major result of this eye tracking study is that the top-down model based on the visual hierarchy is typically preferred, but the "F"-shape pattern of viewing also plays a significant role.

Another interaction task, namely navigation in web pages, showed interesting results based on the viewing behavior [543]. Eighteen participants had to work with typical web pages and had to respond to typical navigation tasks while they had to browse through the different linked pages to locate a task's result. During this scanning procedure their eye movements were recorded with a Tobii 1750 eye tracking device. The major result of the study is that participants tend to use reference and identification points to better track their locations during the navigation task which is some kind of mental map preservation to not be completely lost in such a complex task, hence people tend to reduce their cognitive load somehow to orientate themselves in the web page contents, in particular when the content is changing by self-initiated mouse clicks on web page links.

Interacting with large displays plays a larger and larger role these days, in particular in visual and immersive analytics applications. Two gaze-based interactions for an individual user have been developed and evaluated by an eye tracking study [283]. Walk-then-interact as well as walk-and-interact are two different setups in the system that are investigated in the eye tracking experiment. Twenty-six participants were recruited while the Tobii REX eye tracker was used to measure the gazes and let the people reliably interact with the system. The major result was that the system was in general accepted by the users and that the interaction kick-off time was decreased to only 3.5 s, making it pop out from the repertoire of existing solutions.

Eye tracking for interacting in a plane's cockpit or to understand how pilots interact in a cockpit are valuable research fields [339], in particular for training sessions, i.e. new pilots learning to do their job reliably in an airplane, for example, by educating oneself in a flight simulator. The intelligent cockpit supports pilots as much as possible with their tasks, and if eye movements are incorporated in this process we might get even more insights into the viewing behavior as well as interactions with the displays and components in a cockpit. Twelve pilots' eye movements were recorded using a Smart Eye system while performing manual landing tasks. The major result of this study showed that the pilots did not visually attend all of the instruments, displays, components, and environments equally well, possibly failing to see relevant information, and hence leading to false interactions in the cockpit. The eye movement data is quite complex since it includes long-duration tasks with many options of AOIs to inspect over space and time.

Click-down menus are very important components in user interfaces to allow the user to interactively select a certain option. Eye tracking studies can show how users search menu items and which impact they might have on the visual attention behavior [105]. Eleven participants' eye movements were recorded with an ISCAN RK726/RK520 eye tracker. One major result of the study showed that people's eye movements, in particular response times and fixation numbers, generally tend to increase with a larger number of menu items. Hence, it might be a wise idea to reduce the number of menu items in a user interface if efficient interactions are required to solve a certain task.

Also in the field of interaction we can find many more eye tracking studies as well as interaction support by means of gaze-assisted interaction techniques [508]. For example, focusing on interacting using different kinds of displays is a research topic on its own [355]. Large displays [549], stereoscopic displays [9, 489], as well as pervasive displays [270] have been and have become more and more the focus of eye tracking research, which are all useful for visual analytics. Searching strategies in different contexts also typically include interactions and are evaluated or supported by using eye tracking [7, 153, 215]. Game playing is a popular field of research including more and more eye tracking interaction techniques [222], with a new workshop on eye tracking in games and play (PLEY). Also aviation, in particular for training pilots, requires understanding of the usefulness of interaction techniques [499]. Another emerging field, also useful due to the many online meetings in the time of Covid-19, is remote eye tracking, for example as a way to interact remotely by using one's eyes [122, 548].

5.4.3 Eye Tracking for Text/Label/Code Reading

Although a picture is worth a thousand words, there are still textual enhancements in any visualization or diagram. In particular, in the field of visual analytics, there are textual descriptions, error messages, or hints, for example, in a details-on-demand output which requires reading tasks to be understood. In general, each diagram needs some kind of scale with indicated values, physical units at possible diagram axes, or legends, to just mention a few. At the most fine-grained level of the visual representation we might wish to get an insight into the raw textual data, for example in source code that has to be checked for the causes of bugs or performance issues [88]. All of those textual elements are important in a visual analytics system or in a visualization since they allow us to finally derive meaning by understanding the semantics or the context of visual information, but on the negative side, textual representations are much more time-consuming to read and to understand than a good diagram representing the same kind of data.

Eye tracking studies in the field of text reading have been conducted as one of the first ones in the field [418], with the ground-breaking result of the effect that the human eye does not smoothly move over the text but makes small jumps and regressions back to certain words. These differences in eye movements are also caused, in addition to other reasons, by the more or less complex meanings of the individual words and the time it takes to understand them, using peripheral and central vision. Moreover, in some situations we have to jump back to an already read word because we now have to understand it from a different context, for example. The general question with eye tracking is not only how text is read and how the eye behaves, but also whether eye tracking can be a useful technology to guide a reader or to give extra special information based on the word or sentence that is currently in focus. Such an eye tracking-based reading assistance technique can also be helpful in visual analytics, for example, only showing the most important textual information and then, based on the eye movements, add more and more of it until the user is confident. Such a scenario could be useful for text labels in a geographic map, in which the gaze tracking can be used as some kind of automatic semantic zoom function that adds more and more labels to a geographic map.

Label positions in online forms have been evaluated by using eye tracking [143]. The position of textual components and elements in a user interface or a visualization is as important as its content. In many cases users are already familiar with label positions due to some given design rules and hence they might expect to find the relevant textual information exactly at a certain position. Eleven participants were recruited with nearly no experience with eye tracking. The eye tracking device in use in this study was a Tobii 1720. The major results showed that left-aligned labels should be avoided. Moreover, columns are not a good choice and should be avoided, or if this is not possible, right-aligned labels should be used.

An eye tracking study in the medical domain investigated how medical information is actually read, and probably understood, by patients [208]. To find insights in these aspects, 50 participants were recruited while the EyeLink 1000 eye tracker was used to record eye movements during reading. Major results of the study showed that there are differences in the reading and understanding of the medical textual information compared to a simplified variant of these texts. Hence, the eye tracking study shows that we have to be careful with the information we provide for people who are no domain experts, for example, in a visual analytics system in which textual descriptions might be used in various forms.

Highlighting of text is a powerful tool to guide viewers' visual attention to specific relevant text fragments [121]. Such a feature is, in particular, useful if a lot of text has to be shown that might be separated in relevant and not that relevant information. However, the complete text has to be shown for those readers who are not that familiar with a certain aspect while the experienced user gets the most important parts in a highlighted fashion. Such a highlighting effect of textual content is only possible these days due to lots of possibilities for digital reading. An eye tracking study with six participants showed that highlighted text areas attract the attention of people. The highlights seem to pretend some kind of special importance. The used eye tracking device in the study was an SMI EyeLink I.

Musical scores are some kind of special textual information that has to be read by the musician to play successfully. The general problem with such music sheets is the page turning which might be controlled by gaze instead of using the musician's hand [51]. Ten participants were recruited, all of them with experience in music playing, in total eight pianists, one violinist, and one euphonium player. An SMI RED500 eye tracker was used to measure, record, and analyze the eye movements to make such a page turning functionality a successful tool. The major result of the evaluation of the system showed that the gaze-assisted page turning tool reduced the page navigation time by 47%, a value that the researchers obtained by comparing to existing music score reading systems.

Inspecting source code, in particular for identifying code defects that cause syntactical or semantical problems during the program execution, can consume a lot of time for the code developer or program maintainer. Hence, eye tracking might be an option to investigate how people strategically find such defects in source code [454]. Understanding the visual attention behavior might help to improve the search strategy to save some valuable time in the program development stage. Fifteen participants were recruited to let them find code defects in small program snippets. Their eye movements have been recorded by using a Tobii 1750 eye tracker. The major result of the source code reading experiment was that people typically scan through the entire code line by line and, finally, end-up in a critical code region that is then inspected in more detail, meaning the search space in the source code is first reduced step-by-step before one concentrates on the relevant smaller code pieces (Figure 5.14 shows a code snippet and an example visual scanning strategy).

The reading of textual information is a relevant part in a visual analytics system, but not that researched in combination to other visual analytics components. However, the pure reading task has been evaluated a lot with eye tracking techniques, also in various contexts. Those focus, for example, on attempts to improve the reading performance [116], to create a reading assistant [214], to analyze reading behavior [468], or to simplify texts [462]. Moreover, some other approaches investigate textual [457] and label [371] placements or look into health applications that are full of textual information [354]. Also source code as a form of textual information

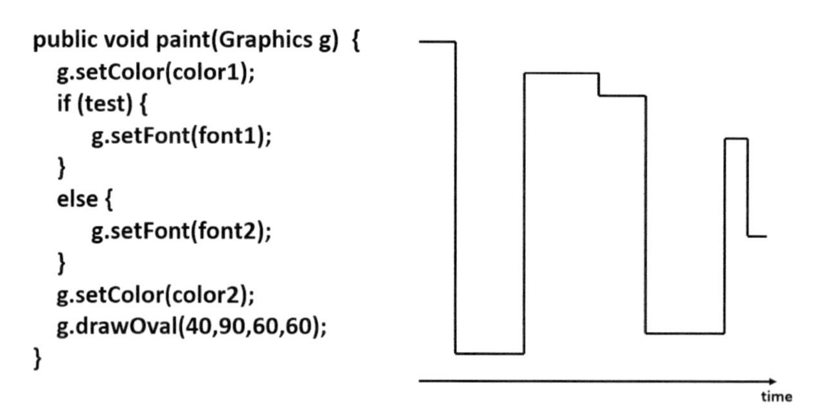

Figure 5.14 Finding a bug in a source code typically requires to scan the whole piece of code before one concentrates on specific parts of it [454].

might be of particular interest for eye tracking studies, focusing on code enhancements [212], reading skills [504], a dual space analysis [547], or educational purposes for novices in programming courses [542], to mention a few.

5.4.4 Eye Tracking for User Interfaces

While visualizations and diagrams might follow visual design criteria [503, 521], user interfaces should also be based on well-considered design principles, for example by following the eight golden rules for interface design [461]. A user interface builds the playground in which humans can communicate with machines and where they get feedback on which the humans base their decisions and further interactions. The user interface is some kind of general language that is understood by both sides and hence a crucial ingredient in any visual analytics system. However, from a design perspective it is expected that a user interface is as user-friendly as possible and that the human user has full control over it and consequently, over the machine. In some scenarios, it is expected that the machine or computer is capable of including automatically generated feedback that might support the human users with their decisions; however, it is not a wise idea to give the machine the full control. There are many examples for user interfaces in these modern days like ticket machines, GUIs of navigation systems, dashboards in a plane's cockpit, while visual analytics itself requires a powerful and effective user interface full of interlinked functionalities, visually depicted by interface components, visual elements, and diagrams, all equipped with interaction features.

Eye tracking plays the role of exploring how people visually communicate with a machine via a user interface [112, 403], typically a visual or graphical user interface, meaning either analyzing their eye movement behavior to understand design flaws or exploiting the gaze behavior as an additional modality to interact with the individual interface components. In this process, the layout and the sizes of the components as well as their linking are important design criteria, aspects that can be explored by recording and analyzing eye movement data. The visualizations in a visual analytics system can be as well-designed as possible, but if the user interface is not able to present them in a well-designed way, they are only half as powerful as they could be. Hence, the user interface, as a communication means between the humans and the machine, is also worth investigating in eye tracking studies. Although static snapshots of user interfaces can already serve as stimuli, it might be a wise decision to incorporate the dynamic version of a user interface in an eye tracking study with some relevant functionalities, but negatively, the recorded eye movement data based on such dynamic stimuli is also much more challenging to be analyzed later on, with the goal of detecting design flaws worth improving.

Knowing the semantics of a user interface can be a great benefit for interacting with a system but also to better understand the connections of certain interface components. An eye tracking study was conducted to investigate such semantic effects [353], in particular for controlling user interfaces which are typically designed in a way that they create a lot of information overload that actually clutters the interface and leads to a degradation of performance at some task [426]. Twenty participants were recruited and split into two groups using an SMI REDn and a Tobii EyeX eye tracking device. The major result of the study was that the developed web browser called GazeTheWeb let people perform search and browsing tasks much faster than a corresponding emulation approach.

Multi-modal user interfaces allow people to interact in several ways by using more than one standard interaction device. In an eye tracking study such multi-modal interactions have been evaluated [36] although such combinations have mostly been the focus of HCI research in recent years. Eleven participants were recruited to use two mice and speech as input in a user interface. The eye movements were recorded using a Tobii x5 eye tracker. The usability study showed that such interaction modes are well accepted by the users although it takes some learning time to completely adjust to the new interaction scenarios.

Item list interfaces are frequently occurring types of interfaces, in particular, in the world wide web and also in recommender systems. An eye tracking study was conducted [197] to analyze the user behavior when being confronted by the task of choosing a movie to watch while a textual and an image-based variant was used for the stimuli. Sixty-four participants were recruited and a Tobii X2-60 eye tracking device was used to record the eye movements. The major outcome of the study was that there are differences in the visual attention behavior depending on what type of variant was shown in the item list interface. The researchers argue that the visual appeal of the images had an impact on the user behavior and, in particular, on the visual attention strategies.

The product selection has a big impact for certain online shops, for example, to modify their selling strategies. However, the online web interface also plays a crucial role in attracting viewers' attention to certain products.

Recommender systems try to help customers with a reduced and well-prepared repertoire of products coming in effective layouts like list-based ones or grouped into categories. An eye tracking study [210] was conducted to investigate the user behavior in recommender interfaces by making use of the Tobii 1750 eye tracking device and by recruiting 21 participants. The major outcome of the study was the fact that the better organized interfaces, focusing on a good layout, seem to attract the users' attention to more products than the standard layouts.

A gesture-based user interface brings into play another way to interact with the interface components. This could be useful for large-scale displays, for example, high resolution powerwalls, combined with eye tracking technologies, this could lead to great synergy effects. Five participants were recruited to test such a gesture-based kinetic interaction [488] while a Tobii x2-60 eye tracker was used to measure the eye movements. The major result of the study was that there was a relation between the eye fixation and an active point which could be used in the future to design and implement a gaze-controlled interface.

Many more eye tracking studies in this special topic of user interfaces are related to graph displays [456, 550], to online learning behavior [216] or education in game environments [545], as well as to hand gesturing and multi-modalities [286]. Also tangible user interfaces have been under investigation [53], adaptive user interfaces [119, 128], and personalized focus-metaphor interfaces [319].

5.4.5 Eye Tracking for Visual Analytics

Visual analytics is an interdisciplinary field, at least combining concepts from visualization, interaction techniques, text reading, and user interface design as mentioned before. Moreover, there are many more fields worth mentioning but most of them do not provide a visual stimulus that might attract the visual attention of an observer. For example, an algorithm running in the background is an important ingredient in a visual analytics system, but actually, we can only see its output, maybe in a visual form, or the menu and the parameters that have been selected by a data analyst in a user interface. In particular, a visual analytics system might allow us to inspect the running algorithm from iteration to iteration, but still, a visual output is required to provide details about such internal processes. Visual analytics systems are complex and require long-duration tasks in case the entire system with most of its functionality has to be evaluated, i.e. some kind of real-world setting. We might argue that in a simple visualization technique the user tasks might be much easier and faster to solve, for example, in a comparative user study, but for visual analytics it is a wise decision to evaluate the system as a whole, meaning the tasks to be solved are more explorative tasks with lots of options to solve them and lots of system components to use.

Eye tracking is also useful for larger systems [307] compared to simple visualization techniques. However, we must say that the recorded eye movement data, be it for gaze-assisted interaction, to use it for detecting design flaws, or as a recommender system supporting users at task solving, is much more complex and is recorded over much longer time periods than in traditional eye tracking studies investigating visualization techniques, text reading tasks, or the effectiveness of layouts of user interfaces. A visual analytics system might even be installed on different kinds of displays and in varying environments; there can be a multitude of interaction techniques incorporated, even working in a multi-modal fashion combined with gaze interaction. Also augmented, virtual, or mixed reality techniques can be part of a visual analytics system, typically integrated in immersive analytics tools [347]. Moreover, additional data sources can come into play like physiological measures, body movements, verbal feedback like think-aloud or talk-aloud, and many more. All of these should be considered in an analysis of the eye tracking study data to make the best of it in order to detect the design issues and, consequently, to identify a way to improve the visual analytics system.

Eye tracking applied to visual analytics systems is a relatively novel concept and hence, there is not that much research focusing on the entire visual analytics system, rather on specific components that are under investigation from the perspective of visual attention. An eye tracking study was conducted [379] taking into account tasks to explore networks. These tasks were not given beforehand but the system was able to detect them based on eye movement behavior and suggested visual adaptations. Twelve participants were recruited while an SMI RED120 eye tracking device was used. As a major outcome of this line of research it was found that there seems to be some accuracy improvements for the network task of checking if two nodes are connected.

Using scatter plot matrices for depicting multivariate data with the goal of identifying correlations can be a challenging task, in particular if the user of such a system is not able to focus on the most important views and data aspects to solve given tasks. Such a problem was investigated by making use of eye tracking [451] to support the visual exploration of scatter plot matrices

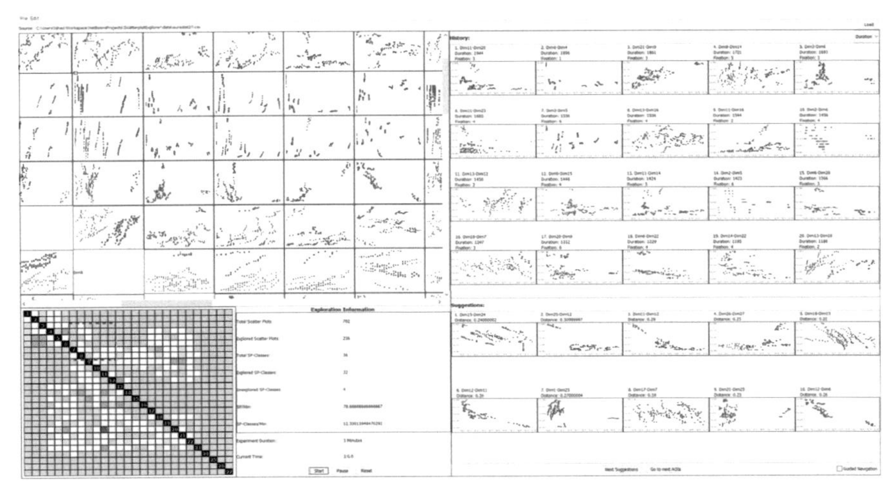

Figure 5.15 A recommender system for scatter plot matrices equipped with eye tracking technologies to support the data analysts [451]. Image provided by Lin Shao.

(see Figure 5.15) based on user input coming from eye movement behavior, similar to a recommender system. 12 participants were recruited while the Eye Tribe SDK was used to measure and transform the recorded eye tracking data. The benefits of this experiment showed that such a system can get higher pattern recall in comparison to a different interaction modality like mouse input for example.

Also in the field of time-series visualization it is of particular interest to identify temporal patterns, for example, to compare them with other patterns on different levels of temporal granularity. A visual analysis can be a tedious task if too many of those time-series patterns are displayed, hence some kind of recommender system might support the viewer at those tasks [467]. Thirty participants were involved in an eye tracking study in which an Eye Tribe eye tracking device was applied to incorporate the eye movement data into the data analysis and recommendation process. The evaluation of the system showed that it is possible for the observers to quickly identify time-series patterns that are of particular interest.

Visual analytics systems might even involve user-adaptive information visualizations [481] that allow the configuration of a system in a user-specified view with an adapted layout of the user interface, important views and visualizations, active interaction techniques, and so on. Eye tracking can help to give additional user-specific input for finding a suitable solution for each individual user. An eye tracking study with 35 participants was

Table 5.2 Examples of eye tracking studies focusing on aspects in visualization, interaction, text reading, user interface design, as well as visual analytics

Scenario	Participants	Eye tracker	Ref.
Hierarchy visualization	40	Tobii T60 XL	[78]
Trajectory visualization	25	Tobii T60 XL	[369]
Public transport map	40	Tobii T60 XL	[372]
Geographic map	30	Tobii T60 XL	[371]
Graph visualization	16	SMI iViewX	[245]
Webpage browsing	48	Tobii X120	[156]
Webpage navigation	18	Tobii 1750	[543]
Large display interaction	26	Tobii REX	[283]
Plane landing	12	Smart Eye	[339]
Menu selection	11	ISCAN RK726/RK520	[105]
Label positions	11	Tobii 1720	[143]
Health document reading	50	EyeLink 1000	[208]
Text highlighting	6	SMI EyeLink I	[121]
Music score page turning	10	SMI RED500	[51]
Source code reading	15	Tobii 1750	[454]
UI interaction	20	SMI REDn/Tobii EyeX	[353]
Multi-modal interfaces	11	Tobii x5	[36]
Item list interface	64	Tobii X2-60	[197]
Interface layout	21	Tobii 1750	[210]
Gesture-based interface	5	Tobii x2-60	[488]
Network exploration system	12	SMI RED120	[379]
SPLOM recommender system	12	Eye Tribe	[451]
Time-series patterns	30	Eye Tribe	[467]
User-adaptive system	35	Tobii T120	[481]
Problem solving	28	Tobii T120	[244]

conducted taking into account such challenging problems. A Tobii T120 eye tracker was applied to predict the best possible scenario for a user, typically focusing on the tasks to be solved by interpreting visualizations. The major finding of this research was that there are promising initial results indicating that the predictions made have some positive value for the system; however, there are still a lot of open future challenges to make it a real-time adaptable system.

Problem solving belongs to visual analytics which can occur in at least two ways, i.e. by using interactive visualizations or by applying algorithms for analytical problems, but in the best case, both concepts work in combination which is actually the power of visual analytics. However, understanding the problem-solving behavior of people could be of particular interest and eye tracking can play a crucial role in detecting patterns in the human users' visual attention behavior [244]. Twenty-eight participants were asked to focus on problem solving while a Tobii T120 eye tracker was used. The researchers identified several varying strategies concerning the information processing stages.

Although eye tracking applied to visual analytics systems is an emerging topic, only a few examples for this important topic exist so far. In particular, it could generate valuable insights for detecting design flaws in a system, for exploiting eye movement data as recommendations for the users, or for applying gaze-assisted interaction as an additional way to interact with the visual analytics interface components as well as the visualizations. Some of the approaches not discussed above also take into account general eye tracking support for visual analytics systems [466], a combination of eye tracking for evaluating visual analytics systems, but also for visual analytics of the eye tracking data [46], eye tracking in personal visual analytics [312], or comparisons of several interactions in two separate systems [427].

Table 5.2 summarizes some of the example eye tracking studies for each of the five separately discussed fields involved in visual analytics.

6

Eye Tracking Data Analytics

Each eye tracking study or each gaze-assisted interaction produces a lot of spatio-temporal data in form of scanpaths with fixations and saccades. The recorded data generates a valuable source of information, hence it could be stored in a database to make use of it later on, for example, to improve the accuracy of a recommender system based on the eye movements of various people. Although this concept opens new doors for powerful user interfaces and visual analytics systems it comes at a cost. Typically, such data has to be used in real-time to make fast recommendations or predictions. This effect actually requires that the data is somehow preprocessed and transformed in a suitable data format that allows fast access to it. Storing the raw eye tracking data is not a good solution, the data must be prepared in a clever and efficient way to make it usable for later purposes. On the other hand, if the data analysis is not required in real-time, for example in typical eye tracking studies, we can store the data in its raw format first and later on, if all of the scanpaths are recorded, we can start to think about how to process, transform, and algorithmically analyze the data with the goal of finding patterns, correlations, and insights in it.

Various approaches exist that make use of pure algorithmic concepts, of visualization techniques, or even of visual analytics systems that combine algorithms, interactive visualizations, and human users with their perceptual strengths. In a real-time eye tracking data analysis we typically rely on the pure algorithmic results since an algorithm can produce a fast and accurate solution to a given well-specified problem. If the goal of the eye tracking data analysis comes more in an explorative way for which more time can be wasted than for a real-time analysis, researchers mostly transform the data in statistical data and inspect the results in form of simple plots and diagrams. These diagrams help to quickly compare the data reduced to a certain well-defined aspect or data dimension. However, eye movement data consists of spatial, temporal, and participant data dimensions, hence reducing

Figure 6.1 Applying visual analytics as a combination of algorithmic analyses and interactive visualization to eye tracking data can provide useful insights into visual scanning behavior over space, time, and participants [309]. Image provided by Kuno Kurzhals.

all of those to simple statistical graphics can lead to wrong conclusions [15] or missing correlations between data dimensions that might have been visible in a case in which more complex visualization techniques are used. Negatively, such visualization techniques demand for learning and understanding the new concepts [71] which is actually difficult for non-experts in visualization. Consequently, to visually explore eye tracking data reliably we should have some profound knowledge in eye tracking and visualization at the same time which reduces the number of people actually using visual depictions of eye tracking data.

Positively, standard and well-known visualization techniques like visual attention maps [50] or gaze plots [203] which show space, time, and participant dimensions are already well established in the eye tracking community in a way that they are well understood and they can be found in many results sections in scientific research papers related to eye tracking data. Moreover, they have been so popular that they are typically built-in features in today's commercial eye tracking analysis software. However, from a perceptual and visual perspective such visualizations are not as powerful as expected. Visual attention maps aggregate over time and participants, while gaze plots lead to visual clutter if too many scanpaths have to be shown on a visual stimulus. These drawbacks are one major reason why many more advanced eye tracking data visualizations have been developed [47], all trying to show as many insights about the recorded eye tracking data as possible. Visual analytics (see one example of a visual result in form of gaze stripes in Figure 6.1) adds one powerful strategy to this existing repertoire of analysis techniques [14], but again, it requires knowledge from several domains, other than from eye tracking, to find knowledge and meaning in the eye tracking data which is definitely one of the biggest challenges in this whole domain combining eye tracking and visual analytics.

6.1 Data Preparation

Normally, the eye tracking data comes in a raw data format from the eye tracking device, before it gets processed into the form that is required for a more advanced data analysis or as input for a visualization or visual analytics technique. The major ingredients of the eye tracking data contain scanpaths, with aggregated fixations with fixation durations as well as saccades that can be derived from two consecutive fixation points. Each scanpath is typically produced by one person trying to answer a certain task. This requires an efficient data collection and acquisition process. Moreover, some of the recorded eye movement data elements must be better organized, to focus on the most relevant one first, in case quick real-time decisions have to be made, for example on a certain data dimension or in case of a heterogeneous data source on the most relevant datasets first. Eye movement data might need to be annotated or even anonymized. The eye tracking data also needs to be in a specific computer-understandable format which is due to the fact that not all eye tracking devices work with the same data format. Finally, if the eye tracking data comes in several data files, we need to find a way to link them together, in case more than one of those files is required for a later analysis.

The algorithmic concepts in a visual analytics system need a proper input data to reliably work. However, in some situations it is unclear how this can be achieved successfully, for example, it might not be known from the beginning which data dimension is the most important one; this might only be known during the runtime of the system. Consequently, it is difficult to organize and restructure the data in an efficient way without any extra knowledge about the users and the tasks to be solved, i.e. which data dimensions are mostly in focus. A similar aspect holds for the data annotation which typically happens manually, i.e. by the users themselves in a time wasting process. This means that eye movement data is typically enriched by extra information in a post process and not in real-time unless a clever algorithm is able to do that in a fast and automatic way. This definitely depends on the way in which the annotation has to be done, for example if complicated semantic information from a visual stimulus has to be included in a data analysis process, we need the human users to add this extra information to the data. If the semantic information is clear from the context, then the algorithm might do the data annotation step.

6.1.1 Data Collection and Acquisition

Acquiring the data is supported by the eye tracking device in each of the eye tracking studies, for example, if a visual analytics system is evaluated. In our modern days, we can record massive amounts of eye movement data in case long-durating tasks are asked, which is a typical scenario if more complex visual analytics approaches are evaluated compared to traditional static diagrams or simple visualization techniques. However, we must still rely on the accuracy of the recorded eye movement data, which could be problematic for certain persons who are not suitable for taking part in eye tracking studies since they suffer from certain disorders or other issues which make the tracking of the eyes error-prone. Moreover, the eye tracking device itself might be wrongly calibrated although today the integrated software and hardware is quite advanced to prevent such errors whenever possible. It is a good advice that a well-experienced experimenter guides the study participant to avoid such measurement errors as much as possible.

In a gaze-assisted visual analytics system, however, such guidance cannot be expected in the future since nobody wants to wait for long calibration times each time he or she starts interacting with a system. Hence, it is a good idea to develop eye tracking devices in such a way that they support quick and reliable data acquisition, no matter which scenario we are confronted with. But actually this is one of the challenging topics in eye tracking research, at least in gaze-assisted visual analytics systems. One solution for a future scenario would be to combine the collected eye movement data with some already recorded data to see potential problems and either clean the recorded data based on formerly measured data records or to annotate the eye movement data with certain events such as calibration errors, which is typically integrated as a feature in most of the modern eye tracking devices. Although the data acquisition step does not seem to be a part of the data analytics process, it might be one of the most important ones since this is the data we will base our evaluations on in a later data analysis step. For this reason it is a crucial idea to annotate, either manually or automatically, the data using reliability aspects like error-proneness or uncertainty. All of these aspects do not only hold for eye movement data but for any extra data sources that complement the eye movements with the goal of deriving further insights from user behavior.

6.1.2 Organization and Relevance

The recorded eye movement data from a user study in visual analytics or visualization has to be organized in a proper way to allow fast access for an analysis later on. This is in particular even more required if the data is heterogeneous, for example, consisting of several data sources all storing important or unimportant attributes and variables with values about the stimuli, the participants, as well as their behavior during the eye tracking experiment. Organizing those data aspects in several files might be a good idea to separate them and allow to quickly achieve the part of the data that is relevant for an analysis or a visualization technique. If such a data part selection process happens during the program execution, the interactivity of a visual analytics system or a visualization tool might suffer. However, in some situations it is unclear which tasks are crucial for a data analyst, hence it is important to have all the data at hand, but in an organized way, to quickly react to the users' wishes and to enrich the data already under investigation with additional data sources based on the users' requirements. An example would be to organize the stimuli separately from the scanpaths, since in some situations a visualization does not need the visual stimuli, but only an aggregated view on the scanpath data is shown, for example as a scarf plot or statistical graphic, in which the stimulus cannot be integrated and aligned directly. However, the user might wish to see the stimuli, and hence, an extra request for the stimuli could solve this problem.

Organizing data also means to filter out irrelevant data right from the beginning, but this requires knowing what the user is planning to explore with the eye movement data later on. This is actually a challenging problem, meaning that most eye trackers store and provide all of the data on the finest possible data granularity; however, in most situations the given granularity or data extras are not required for one or the other task. As long as the data size or disorganization does not lead to a degradation of performance at some task, this situation is acceptable. In cases in which we see such performance issues we might consider reducing or reorganizing the data sources for the most relevant parts for solving the data exploration tasks before running a visualization tool or visual analytics system. For example, if we realize that a scanpath visualization suffers from visual clutter due to the fact that the tracking frequency in the data is too high, although the higher frequency and denser points do not lead to many more insights, we might temporally aggregate the data beforehand to reduce the number of points to be drawn. One might argue that this could happen while working with an interactive scanpath visualization; however, if the drawing is the problem, then the visualization can suffer from performance issues and clutter effects which might both be mitigated by reorganizing the data and only focusing on the relevant time granularity for example. This is just one example from many others, but it still requires knowledge about the tasks that have to be solved with a data analysis tool. Reorganizing data and only focusing on relevant parts have to be taken with care anyhow.

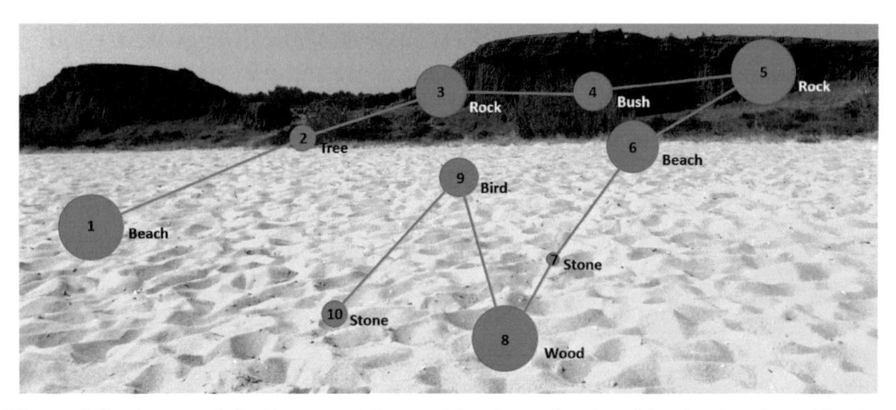

Figure 6.2 A manual fixation annotation tool has been developed to step-by-step add extra information to the fixations, for example based on the semantics of a stimulus [370].

6.1.3 Data Annotation and Anonymization

Annotating the recorded eye movement data with extra events or semantic information from the stimulus should be done as early as possible in the data analysis process. Such a data annotation supports data analysts in setting the found visual patterns in special context, for example, based on the semantics of the shown stimulus, in a case where the stimulus cannot be shown directly in the eye movement data visualization. Moreover, special events such as user behavior or verbal feedback might even be used to annotate the data to give quick feedback later on when the data is visualized. The annotation can be done algorithmically, but this requires that the annotation algorithm can be specified in a proper, well-defined way, otherwise a more manual annotation is needed [370], taking into account the perceptual strengths of the human users (see Figure 6.2 for a tool supporting the human annotator in such a time-consuming task). Data annotation might even be done after the data is already visualized, meaning visualization can be used as a special support to first detect patterns and then annotate the data with these patterns. However, in a visualization the pattern detection is typically done by the human observers due to their perceptual strengths and rapid pattern detection abilities, hence this visualization-based annotation step requires a manual procedure. In some situations, for example when an annotation algorithm cannot be clearly specified, a manual annotation is not avoidable, consequently, a time-consuming process including humans is required.

Some kind of opposite effect to the data annotation, in which data is added to the already existing data, might be data anonymization, a step in which data elements from the data source are removed, for example for data privacy reasons. This step typically removes all personal information from the data source, which means all data elements that might somehow be used to recover the person behind a certain scanpath for example. A typical strategy is to replace all person names with identifiers that are just random natural numbers. It should not be possible to link these identification numbers with the filled out study participant forms to definitely avoid recovering personal information like the person's name. A problem is definitely the fact that certain person's names might be recovered in the case that a certain attribute value only exists once and hence can be mapped uniquely to this person. An example would be if we just had one person who is wearing contact lenses, then this person can be identified easily later on in the recorded scanpath data by just looking up this special attribute. The safest anonymization effect in an eye tracking study is to completely avoid the recording of a person's name, but just in case the persons are paid for participation in an eye tracking experiment we normally need the names. However, this extra payment information should be separated from the recorded data and maybe should be destroyed as soon as possible to avoid anonymization problems later on.

6.1.4 Data Interpretation

Eye tracking data can come in many formats, typically depending on the eye tracking device in use. To reliably work, the data has to be interpreted in a way that the data elements it is composed of can be stored into variables in a data analysis or data visualization tool. To make this possible a certain kind of template is required that perfectly describes which data elements have to be expected and in which order those are stored. If this data interpretation process does not work properly, the data might be not parsable and readable at all or the data might be read while some or all data values might be attached to the wrong attributes or even mixed up completely. This has the negative effect that the data analysis leads to wrong conclusions. In most scenarios the human user has to keep an eye on the raw data first to understand its structure and the order of data elements to check if the underlying data template matches the one given in the data reading and parsing algorithm. This is in particular important if the eye tracking data comes from several eye tracking devices, all typically having different kinds of data outputs. One option to handle this problem is to write a data parser for each of the eye tracking data sources separately or to transform each data source into the same format based on the same template.

If several data sources are stored for a later data analysis, this problem might get even worse. For example, the scanpaths from an eye tracking study as well as additional verbal feedback, interview data, personal study participant data, physiological measures, and so on might be worth investigating. All of those data sources could be stored in different formats, for example, stemming from various eye tracking experiments conducted by different groups of researchers. Finding a data analysis tool that helps to identify data patterns and insights in those data sources is a difficult task since, in the worst case, each data scenario has to be adapted to the tool's data reading and parsing requirements. If this adaptation happens on the tool side it can be challenging for the tool developer to create a parsing function for each of the data formats. Maybe only the most important ones might be supported by a tool, the rest has to be brought in the correct template by the research group who generated a dataset. This is typically the best option since they are best in interpreting their own data and deciding which values in the dataset belong to which attributes for example.

6.1.5 Data Linking

We often face the situation that the eye movement data is not the only available data source, but many more describe the user behavior, for example, storing time-varying facial expressions or verbal feedback during a study. Moreover, many personal details are available that have been typically stored before running an eye tracking study. All of these data sources have to be linked in a certain meaningful way, otherwise they cannot be taken into account in combination later on in a data analysis or visualization process. Even the eye movement dataset might be stored in several data sources, for example, the scanpath data for each participant might be given on a text file while the corresponding stimuli, static or dynamic ones, are stored in other files, maybe in a stimulus directory. Hence, to remap the scanpaths from the eye tracking experiment with the stimuli that have been seen at a certain point in time, a well-defined key or identifier is required that allows finding the right stimulus file for each of the scanpaths. This is in particular problematic for dynamic stimuli that occur as video sequences which have to be matched with the corresponding scanpath, but also over time, meaning the dynamic stimulus actually consists of many static stimuli while the scanpath data describes which static stimulus from the video has been watched at a certain time. Hence, the linking does not only happen between several separate data sources, but also over time, in case we have to deal with dynamic visual stimuli, which is a typical scenario in visual analytics.

The general problem might occur if the data sources have to be linked during the runtime of a data analysis or visual analytics tool. For example, a user might decide to visually explore eye movement data by also taking into account an additional data source describing interview data that has been acquired by asking the study participants after each of the experiments. Then this extra data has to be linked with the primary eye movement data first and then additional views or visual output have to be incorporated into the original visualization for example. Although the visualization of the extra data might not be a problem at all, it might take some time to link the two data sources before the visualization can be modified and updated. Even if such a linking process just takes a few seconds it might even lead to a bad user experience. Consequently, the question comes up which data sources should be linked beforehand, i.e. before the analysis tool is running to avoid such waiting times during the data analysis process. Moreover, another question is how the data sources can be linked, meaning if there is some common key that can be used to do that successfully, for example, as in our scenario above, the study participant might be the key to link the data sources. In some situations it is not clear how the linking can be done reliably without asking the user of a data analysis tool which makes the data linking a quite challenging topic.

6.2 Data Storage, Adaptation, and Transformation

All the data from all data sources recorded before, during, and after an eye tracking experiment has to be stored in an efficient way, for example, in several, typically linked files, or in a database allowing fast access to it during a later analysis process. For example, if a real-time reaction on the eye movement data is required based on the previously recorded data elements, the data should be stored in a clever way to allow such fast interpretations to support the real-time experience. This means that the new data to be stored has to be modified in a certain way to bring it into the desired format to allow fast access to it. Data storage is not just putting the data into a file or database, the data is already adapted to a certain common situation. This is even more challenging if the eye tracking data comes from several eye tracking devices, having different data formats, but still following the same general goal. Hence, it might be a good idea to transform the data already in this early stage while it is stored or added to the general eye tracking database for later usage.

Important stages in this whole process include the checking of the data for inconsistencies or redundancies. Consequently, the data sources get validated, verified, and typically cleaned and freed from errors or missing data entries whenever this is possible. This builds some kind of data enhancement or data enrichment process in which certain data elements can be removed, added, or even annotated by special events or uncertainty values to indicate that they might otherwise lead to misinterpretations. Moreover, in this step, additional analysis-relevant data metrics can be derived and values for them can be generated. Computing those values beforehand can save a lot of computation time during the running data analysis system. Hence, this whole stage focuses on storing the data in an efficient way, on adapting it in a way that it can be accessed as quickly as possible later on, to save valuable computation time during a running data analysis or visual analytics system, and finally, on allowing data transformations that might modify a raw dataset into a computer-readable one, for example, based on bringing it into a given format that is understandable by a data analytics tool. This kind of data transformation is just responsible for adapting the format of the data, it does not draw any conclusions from the data nor does it derive any data patterns, like an explicit algorithmic data analysis process would do which will be discussed in one of the following sections. Generally spoken, when a visual analytics system for eye tracking data is started, the underlying data to be analyzed should be in the most appropriate data format as possible, for example, to avoid long runtimes while using the visual analytics tool and while interacting with it. Nobody wants to wait too long for the results of a data analysis, but in some situations it is unavoidable.

6.2.1 Data Storage

It sounds like the data storage process is not as important as all the others, but it may be noted that the wrong storing of eye movement data can have a tremendous impact on a data analysis or visual analytics stage later on. In addition, if a real-time analysis is required we need fast access to well-structured and already preprocessed data sources, otherwise there might be missing data chunks during the real-time analysis which would lead to a degradation of performance of the underlying data analysis system. However, in many situations, fast data access to all required information is not possible, hence a real-time reaction is not possible, meaning it does not matter how efficient the data is stored in these scenarios. For eye tracking data it also depends on which data dimension the data analysts are primarily interested in, like the spatial information from the visual stimulus, the temporal information, i.e. how the situation changed over time, for example, the stimulus and/or the eye movement behavior, or the eye tracking study participants, individually or clustered into participant groups. The primary data dimension typically decides which algorithms and visualizations are used later on and hence the data should be stored in a way that this primary aspect can be accessed as fast as possible while the secondary or tertiary data dimension plays a minor role.

For example, if a visual analytics system focuses on supporting an analysis of participant clusters, the individual participants are of primary interest and not the visual stimuli. However, the stimuli might be used as a later details-on-demand request to see where certain participant clusters paid visual attention, but actually for identifying the participant cluster this data is not as important as the participants themselves, maybe with their personal information. Things are not that easy in most situations. To decide on the order of relevance of the data dimensions and how to structure and organize them is a difficult problem. On the one hand it definitely depends on the data analysis tool or visual analytics system that works with this kind of data, but on the other hand it also strongly depends on the user tasks. From the scenario above, it might be efficiently stored if the major focus is on participant clusters based on personal information, but if the users decide to switch to a more scanpath- and stimulus-related grouping of the participants instead of the personal information, the performance of the system might suffer from it.

Today, with only small eye movement datasets, we might argue that this is actually not a big challenge, but in future scenarios, for example, with millions of scanpaths and dynamic stimuli of hours of lengths, we might get into serious performance issues if this problem is not treated well enough. One scenario could be the tracking of the eyes of millions of car drivers over longer driving distances. On the one hand, we would like to analyze the scanpaths after the driving tasks have been finished which actually gives us enough time to process the data, i.e. making the storage problem actually not that relevant. However, if we are facing a scenario in which we already have millions of scanpaths from car drivers and now a new car driver is eye tracked, we might want to get real-time feedback based on the eye movements of the one car driver while at the same time taking into account the existing scanpaths, typically stored in a database, prepared for such purposes. There are many of such future scenarios in which we might face this challenging problem, the larger the eye tracking data sources get, the more insights we can extract from them, but at the cost of thinking about efficiently storing and managing the data.

6.2.2 Validation, Verification, and Cleaning

After the data is stored and successfully brought into a format that allows fast access depending on the tasks solvable by a data analysis or visual analytics system we should incorporate another follow-up step to the storage process. In the best case this validation, verification, and data cleaning process might happen simultaneously to the data storage process, but in some situations the data sources must be inspected as a whole to validate and verify them. For example, checking if a certain data element is missing in the data source can only be done if the whole data source is available. In this stage it is important that incorrect data is removed or at least annotated as being incorrect or inaccurate, for example caused by calibration errors or even by linking it to additional physiological data that provides further insights about eye tracking study participants' performance, telling us if there is a chance that the eye movement data might be error-prone and not worth including in a later data analysis step.

Moreover, the data might be redundant or inconsistent, as well as incomplete. The incompleteness might be treated differently, depending on how crucial this aspect is in a data analysis step. For example, if missing data elements are located in a dataset they can be treated as being missing, meaning they can just be ignored in the data analysis. Moreover, they can be interpolated by taking into account the neighbored existing data elements or a certain derived pattern might be used that looks similarly as the data points with the missing data element, helping to close the data gaps. This similarity is typically based on similarity values that are generated by algorithmically comparing two patterns. If the similarity values are above a certain user-defined threshold it is decided to apply the data completion process based on such a found similarity pattern. However, no matter which data correction or completion strategy is applied there is always the chance that this procedure does not produce the right data points. Hence, the best option in such a scenario might be to not modify the data and just indicate in a data analysis or data visualization that there are certain missing data elements. However, if too many of those data points do not exist, it remains questionable how reliable a data analysis will be in the end. Cleaning eye movement data is a challenging process for low-level eye tracking data [444]. Positively, additional data sources might be a good option to more efficiently and reliably clean the data since they allow us to take into account further data perspectives compared to just one individual data source alone.

6.2.3 Data Enhancement and Enrichment

The originally recorded data is typically not sufficient enough for a data analysis or a visualization. In many situations we have to first enhance and enrich the data to get it in a suitable form to detect insights later on. The data enhancement and enrichment mean that the data can be attached by additional attributes but even superfluous and redundant data elements might be removed in this step. Actually, everything that adds more value to the data itself could be regarded as a data enhancement and enrichment. For example, in an eye tracking study we often have the situation that data might contain scanpaths with a high uncertainty, maybe due to calibration problems or other participant-related issues that lead to such data problems. There are several options to enhance the data; in the best case we let the study participants with badly recorded data do the experiment again, but this might cause additional costs and the participant has some kind of learning effect from the original experiment, hence this should be taken with care. Another option to enhance the data is by closing the gaps algorthmically, if there are just a few, for example by interpolation or by deriving a similar data pattern from other parts of the scanpath or other participants' scanpaths. However, this always happens at the cost of not being correct since it is never data from the real experiment. If there are too many data gaps we should think about replacing the eye tracking device or modifying the study setup in a way that makes the data more reliable.

Also metadata, i.e. data about data, plays a crucial role in this process. Such data describes, for example, error reports, performance feedback, quality of the results, trustworthiness of the data, uncertainty values, provenance and lineage of the data, to provide an information about the context in which the data was recorded. Each data element, for example, a scanpath related to an individual participant, might even have privacy information attached to it. Moreover, metadata can even be based on additional computations applied to the data which lead to data enrichment in the best case. Also physical units are important metadata that describe at what scale the data was measured, for example. In particular for eye tracking studies, the tracking frequency given in Hz could be important to understand the temporal granularity of the data. This could be very relevant information if the recorded data stems from different eye tracking devices with varying properties. Consequently, the type of the eye tracker should also be stored as well as additional meta information about the experiment.

6.2.4 Data Transformation

Transformations of data are typically those that bring the data into another format. Such a new format can be based on different aspects, mostly depending on the users' input and tasks. For example, the data might be aggregated over space and time or the scanpaths, and additional data of individual participants might even be aggregated into groups or clusters of participants. However, the aggregation strategy might cause some problems due to the fact that it is sometimes not clear how the aggregated values have to be computed to show a representative data element for the aggregated time, space, or participant group. For the time dimension the scale is typically changed from the most fine-granular temporal dimension given by the tracking frequency of the eye tracker to a more coarse-grained temporal scale, for example, to inspect the data on a per second basis instead of on a per millisecond one. Such an aggregation can come with several challenging problems. For example, for a dynamic stimulus shown in an eye tracking study it is pretty unclear how to also aggregate this stimulus over time. For a static stimulus it is no problem at all since the content of the stimulus is always the same over time, but a dynamic one like a video might change its content from time to time. In such a case it might be a good idea to not just aggregate the temporal dimension based on the scanpath data, but to take into account the dynamic stimulus information and maybe base the time periods on the sequences of the dynamic stimulus that carry the same information, i.e. where the dynamic stimulus shows a static content. Moreover, if the dynamic stimulus shows a static content for a longer time, this time period might again be split into sub-periods and then all of them might be aggregated while the representative stimulus content can easily be derived from the static content of those subsequences, i.e. it is guaranteed that it is always a static stimulus.

One goal of a data transformation is always to reduce the complexity and size of a dataset, but at the cost of losing information. Hence, data transformation should be taken with care and it should be clear right from the beginning if the transformation process is still acceptable so as not to remove relevant information that cannot be regained later on. Another concept is to work with the raw data in a data analysis or visual analytics system, but deciding during the runtime of the system which parts of the data to be transformed might lead to a performance degradation and, consequently, it is a wise decision to preprocess the data and transform everything that can be transformed while taking into account the fact that after the transformation the granularity of the data is changed. This means if we aggregate or even normalize the given recorded raw data beforehand, for example, changing from a lower to a higher scale, we no longer see the lower scale in the data analysis or visual analytics tool. One solution could be to still keep the raw data, just in case the tool user makes a request to it which might lead to runtime performance issues, but in the case that the data is used on the higher scale, the tool might perform quite well.

6.3 Algorithmic Analyses

Algorithmic analyses focus more on the computer-supported generation of numbers, patterns, correlations, rules, and insights from the given data, typically applied to individual data entries but also in comparison between several data entries over space, time, and participants if we speak about eye tracking data. Moreover, algorithmic analyses should take into account any further available data sources with the goal to effectively and efficiently generate new perspectives on the given data. Algorithmic solutions can be distinguished from visual solutions by the fact that algorithms must be clearly specified to run properly while a visual solution shows visual patterns that have to be interpreted by the human users based on their perceptual abilities to rapidly recognize patterns. However, just recognizing patterns is not the key to the solution. The solution based on visualization comes from the linking of the visual patterns to the data patterns which is a benefit of pure algorithmic solutions that come up with the generated data patterns directly, but negatively, those data patterns might be hidden in a flood of generated solutions. Interesting and popular algorithmic operations in this context are finding better structures by ordering and clustering, reducing dataset sizes by summarizing, classing, classifying, aggregating, and projecting data, or by allowing comparisons, for example, by normalizing the data or by applying multiple sequence alignment methods to find a consensus matrix for a list of scanpaths [84].

There are various algorithmic approaches that also vary in their runtime complexities. Each algorithm is typically based on a certain task to be solved by extracting the relevant information from a given dataset or several of them in combination. Actually, there are two options to apply an algorithm which is, on the one hand, an offline approach that starts computing after the data has been recorded completely, i.e. as some kind of post-processing. The other alternative might be denoted by the term online approach which indicates that the data has to be algorithmically analyzed during the recording, i.e. in real-time. The second option is typically more complicated to implement since it requires the algorithm to react quickly, i.e. in real-time, on a given input. This could be interesting for gaze-assisted interaction for which the data of many eye tracked people might be taken into consideration to generate a quick recommendation while interacting. Moreover, in any scenario in which the user dynamically interacts or inspects a visual stimulus, an algorithm might generate real-time solutions and hints during this dynamic process. The offline approach, on the other hand, is typically useful when we have enough time to analyze the data, for example, in an eye tracking study for which we have recorded all the data beforehand. As a post-process, i.e. a data evaluation and analysis, we might apply a visual analytics system with various algorithms [14], to find design flaws in the given stimuli during the eye tracking study.

6.3.1 Ordering and Sorting

There are several ways to bring a certain structure or organization into a dataset based on ordering and sorting algorithms. Also for eye tracking data there are some ways to order or sort them, typically depending on a well-defined criterion on which an order can be computed. The easiest way to do this is to reduce the scanpaths or any other data source to a certain quantity like the lengths of the scanpaths, the completion times for a task, the average fixation duration, and many more. Based on such a quantitative value we can derive a one-dimensional order of the data, for example, for the list of study participants, maybe to see who produced the longest fixation sequence and who the shortest one. If the data is more complex and cannot be easily reduced to quantities such an approach cannot be applied in the same way. One example would be the list of visual stimuli that should be brought into a certain order. There is no unique property to apply an ordering strategy for the list of stimuli, however we might create one that meets our needs. For example, we might compute the distribution of the individual colors in each visual stimulus [309] and then use a priority list among the individual colors to derive a unique one-dimensional order for the stimuli. This procedure could be helpful to create an order based on a certain primary color that has to be investigated for its impact on visual scanning strategies. No matter which data aspect we use to apply an ordering or sorting strategy it typically focuses on a certain well-defined user task, for example exploring the impact of color coding of the visual stimuli on the scanning strategies as described above.

Some of the data dimensions bring their own order, for example the time dimension. It is clear which the first fixation is and which one the last in a scanpath and it is a bad idea to reorder the temporal aspect, in particular in a case in which we are interested in temporal patterns. In a different scenario in which we might be primarily interested in the visual stimulus, the temporal order might be less interesting, for example for reasons of visual clutter reduction. The temporal order, on the negative side, restricts the order of another data dimension. This is a general problem for any kind of ordering strategy. If we base our order on a certain attribute or data aspect we have to resolve this issue for all the other data aspects, meaning the order typically focuses on a primary aspect for which we have lots of opportunities in eye tracking data, especially if we take into account additional data like physiological measures or personal information as well as verbal feedback and so on. For a visual analytics system it could be important to have already computed the orderings for the most important data attributes, to avoid long computation times during the runtime of the system. However, ordering for all aspects as a pre-process would be a challenging task, just because there are so many data aspects that might be taken into account for ordering or sorting. Moreover, there could be combinations of data attributes, for example, resulting in a matrix-like scheme expressing quantitative distances or similarities between pairs of data elements. For example, we might compute the scanpath similarity for all pairs of eye tracking study participants which would lead to a matrix filled with real-valued numbers. Without any order such a matrix would not help to identify group patterns among the participants, hence matrix re-ordering techniques [34] are of special interest here to better organize the data for a later visualization, in particular group eye movement data based on similar scanpaths [300]. For ordering more attributes we need more advanced techniques [299], however in such a multivariate dataset it is challenging to indicate the order for all attributes simultaneously.

6.3.2 Data Clustering

Clustering is a popular algorithmic approach to bring structure to a dataset. The idea is based on putting data elements or objects together that share similar properties while non-similar ones are not in the same cluster. Hence, by this strategy we can obtain a separation of all data elements based on

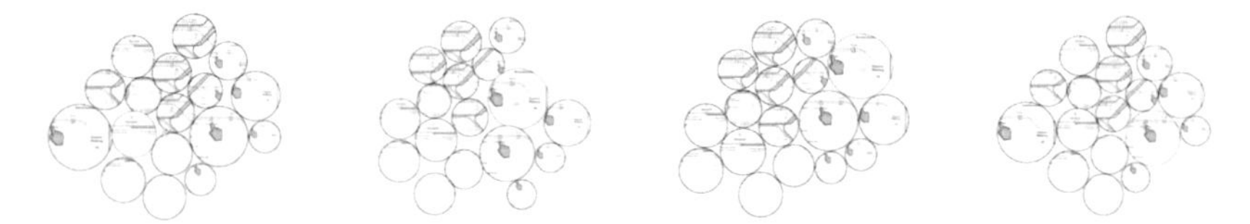

Figure 6.3 The Antwerp public transport map was visually explored in an eye tracking study. The visual attention hot spots were used to split the static stimulus into sub-images which are then grouped by a force-directed layout taking into account the transition frequencies between the individual sub-images [98]. Different parameters can be modified such as cropping sizes, cluster radius, or number of sub-images displayed, for example.

certain well-defined properties, even a fuzzy clustering could be generated, allowing a data element to belong to several clusters at the same time, but to each only to a certain probability. There are various techniques for computing a clustering [180] applicable to data from a multitude of application fields. A general idea is the fact that between each pair of data elements there exists a distance or similarity value that is used to generate good separation into clusters, for example, based on finding a hierarchical organization among the elements as in a hierarchical clustering [273]. A k-means clustering [82], on the other hand, actually tries to find a separation of the space, attaching the data elements to spatial regions belonging to certain sub-spaces and hence some kind of spatial separation is computed. In eye movement analysis, such an approach could be used, for example, for detecting the eye [411], i.e. before it comes to the actual eye movement data recording. Moreover, the visual stimulus might even be split based on formerly recorded visual attention hot spots and then the resulting list of images from the stimulus can be reorganized, maybe by using some kind of force-directed layout (see Figure 6.3) that takes the eye movement transitions between the split images into account to let those images attract or repel each other [98].

In the field of eye tracking, a hierarchical clustering also makes sense for a list of scanpaths that might be split into hierarchically organized groups of scanpaths, each group reflecting a certain similar visual scanning behavior [309]. Such a clustering is then useful to subdivide the group of participants, based on their viewing behavior, into several groups, while each group of participants might be inspected further by taking into account additional data. The clustering, on the one hand, brings a structure into the eye movement data on a scanpath basis [175] while the computed structure is further investigated for reasons causing these different behaviors, hence, if it is attached to additionally recorded extra data we might get hints about the different viewing behaviors that cause the changes in the scanpath data. It may be noted that the longer the scanpaths are under investigation, then the probability that they share common data patterns is normally lower. Moreover, there are always small variations in the fixation pattern for each scanpath, even if they are very similar. This problem might be mitigated by either using spatio-temporal thresholds for each fixation that still consider fixation points as being similar if they are located within the same radius, for example. Another idea to manually or automatically annotate or manipulate each scanpath is based on the semantics of a stimulus, hence identifying the visual object a fixation belongs to in a stimulus. However, this can be a time-consuming process, in particular, if it is done manually [370].

6.3.3 Summarization, Classing, and Classification

Eye tracking data might get quite big, in particular in future scenarios [44] in which eye tracking devices might be integrated in certain devices of daily life such as cars or mobile phones. In such data situations it is of great help to reduce the complexity and size of the data in a clever way while still preserving most of the original data patterns. This means, no matter how much we reduce the data we should still be able to identify the overall data patterns, which seems to be a challenging problem. However, concepts like classing or classification are suitable summarization approaches to reduce the dataset size in a way that they compute representative elements for each of the time periods, spatial regions, or participant groups to mention the most important data dimensions in an eye tracking study. For example, classes could be identified, even beforehand, and each time period could be checked for the class it might fall into. Such classes might be slow, medium, and fast, for example, to express the space inspected per time unit by each participant which is similar to a velocity measurement. By using such a subdivision into velocity classes we could aggregate or summarize each scanpath into a sequence of classes instead of a sequence of fixations, hence, the size of the temporal aspect in the data is reduced tremendously. Negatively, this happens at the cost of losing fine granular information. Moreover, each scanpath might be reduced to a sequence of areas of interest (AOIs) that have been visually attended over time. However, this requires defining the AOIs based on the visual stimulus and its content, also taking into account semantic information.

If we go one step further we might classify each participant based on a certain well-defined property, for example, the dynamic pattern of slow, medium, and fast movements as described above, or the sequence of areas of interest they create during the visual scanning strategy. Hence, such a classing and classification might be a useful concept to guide a clustering, for example a hierarchical one, taking into account the participants and their scanning behavior reduced to certain classes based on well-defined properties. On the one hand, we lose information by the classing, while on the other hand, the follow-up clustering algorithms might produce faster results since the input data is no longer that accurate, but it is condensed to the most relevant aspects that are still detailed enough to generate a suitable and expressive clustering, for example. The classing actually works for any data dimension, we only have to decide how the classes are created and how many we plan to get in the end to base further data organization, structuring, or clustering on. Moreover, if another scanpath is recorded it is faster and easier to classify to which category it belongs than taking into account the whole scanpath with all its detailed fixations and fixation duration. In particular, in the field of fatigue detection of car drivers based on eye tracking [285] we can find many approaches making use of classification concepts, an approach that typically requires real-time computations to provide fast results.

6.3.4 Normalization and Aggregation

A general problem with eye movement data comes from the fact that not all eye tracking participants are equally fast and hence, to better compare the data on a temporal basis it makes sense to normalize each of the scanpaths, i.e. stretching all of them to the same length. This stretching might help identify which regions in a stimulus or areas of interest are covered by the participants and which particular sequential order they follow in the visual attention process. The comparison might even work without a temporal normalization but it requires more effort to reliably do the comparisons. A temporal stretching somehow leads to some kind of temporal alignment, but it is not guaranteed that this approach works for any eye tracking study. In some situations the visual attention sequences are completely different and even a normalization will not help. Another challenging problem might occur if the data values vary a lot, for example, completion times or average fixation duration, hence really large ones overwhelm the small and tiny values and, consequently, the small values are hard to compare due to the fact that the big ones will reduce them in size a lot if those are visualized later on, for example as a bar chart. One way to allow fair comparisons of such quantitative data can be achieved by applying a logarithm function to all of the values or even more advanced concepts that transform the quantitative data into another format that allows better comparisons [230]. Choosing the logarithm idea actually transforms the quantities to exponents, allowing the small values to be compared visually. However, the users of such a logarithmic scale should be informed about this data transformation to avoid misinterpretations.

Aggregation, as mentioned earlier, is also a useful concept to reduce the dataset size and to provide a better overview of the data, in particular, if the data reaches sizes that no longer allow it to be visually depicted without needing the user to scroll a lot. In some scenarios we can find a combination of these concepts, i.e. the eye movement data is aggregated, normalized, and finally, transformed to a different scale. All of this follows the goal to get a scalable overview of the eye movement data for as many data dimensions as possible to allow fast comparisons. This first overview strategy is a fruitful concept to support a data analyst with a starting point for further exploration processes, algorithmically as well as visually. The aggregation strategy can be manifold and could be based on several approaches, including simple and more complex statistical ones. For example, if we plan to inspect the fixation durations in a scanpath and plan to temporally aggregate those, we may ask the question what the result of such an aggregation strategy might be. We could generate the sum of the fixation durations but without normalizing the sum by taking into account the number of fixations in a time interval, this approach might be misleading, hence the average or mean value might be of special interest here. Also the minimum valley or maximum peak might be interesting aggregation measures for each time period. More statistical values such as the median or the standard deviation could give even more detail. Also a combination of several of those aggregation measures could be useful, in particular if those are visually depicted later on, for example in box plots.

6.3.5 Projection and Dimensionality Reduction

Eye tracking data can even be regarded as multivariate or high-dimensional data consisting of a multitude of attributes [299], making it hard to identify patterns, in particular correlations among the attributes. Consequently, it might be a good idea to project the high-dimensional data to a lower dimension, typically 2D or 3D, for example, visualized by means of a scatter plot. The general concept behind the dimensionality reduction is the preservation of the original data structures as much as possible, meaning similar data points from the original high-dimensional data should be located close to each other in the projected lower-dimensional data. Moreover, high-dimensional data points that are not similar should be placed far apart in the projected lower-dimensional data. Although this is a powerful concept, the property of preserving the data structure in the projected data cannot be reached in all data situations due to the fact that a projection to a lower dimension does not leave as many options to place the data as a high-dimensional structure would offer. A pure algorithmic analysis of the high-dimensional data would also be an option, however the large number of attributes as well as the observations providing values for each attribute make an algorithmic approach sometimes computationally intractable [195]. Hence, a projection might be the last chance to provide an overview about certain data patterns, even at the cost of losing some of the important information from the originally non-projected data.

We can find many dimensionality reduction techniques in the literature [176], which are divided into two major classes called linear and non-linear techniques. Popular candidates for the linear class are the principal component analysis (PCA) [241] or multi-dimensional scaling (MDS) [497] while non-linear techniques are the t-distributed stochastic neighbor embedding (t-SNE) [511] or the uniform manifold approximation and projection (UMAP) [351] to mention a few from a large repertoire of existing techniques. For eye tracking data it might be of interest to interpret the scanpaths as feature vectors and to investigate whether a dimensionality reduction method can uncover similar and dissimilar scanpaths after they have been projected to a lower-dimensional space. However, interpreting the entire scanpaths as vectors will not lead to a good solution since they contain too many variations and build too long feature vectors that are hardly similar. A better option is to first transform, aggregate, and normalize the scanpaths, to first reduce the variability of the input data for the dimensionality reduction. As a final stage, the transformed scanpaths can be projected to 2D and be visualized as scatterplots [66, 79]. As another add-on in the visual output of the projection the user might wish to interact with the projection tool and modify typical data parameters like the length of the scanpaths to be explored or even other threshold parameters that reduce the data complexity, i.e. feature vector lengths.

6.3.6 Correlation and Trend Analysis

If several attributes exist in eye tracking data, for example fixation durations, saccade lengths, maximum area covered in a stimulus, and so on, it might be of particular interest to analyze whether these attributes stand in a certain correlation behavior. This means, for example, that an increase in value for one attribute also increases the value of another or even more other attributes. In contrast, the increase could also lead to a decrease for other attributes. The first observation is called a positive correlation while the second kind is denoted negative correlation. Not only the static correlation behavior of attribute values might be of interest but also the dynamic ones, i.e. it could be of particular value to identify the correlation behavior over time between two or even more attributes. For example, the average fixation duration in a certain time period might correlate in a specific way with the average saccade length in the same time period. If this time period is moved just like a sliding time window over the entire scanpath we could analyze if the dynamic correlation pattern changes or stays the same. The scanpath with the attributes fixation duration and saccade length is just a simple example for such a correlation analysis but there is no limitation to extend it to any kind of quantitative attribute, static or dynamic. The quantitative eye movement metrics could even be set in correlation to other metrics not directly related to the eye [540].

If the dynamics in the eye tracking data plays a crucial role for further investigations we might consider trend analyses [175]. These algorithmic approaches take into account a time-varying dataset and compute a trend in it based on certain attributes and data properties. For example, for eye movement data it might be of interest to analyze how the fixation duration changes over time since there is some kind of evidence that the fixation duration can give hints about certain task solution strategies and how much effort a viewer is putting into a certain task. A similar aspect holds for the saccade lengths, i.e. analyzing the saccade lengths can also give insights into the modification of a certain viewing behavior or scanning strategy. Trend analysis can help to uncover increasing or decreasing effects in a dynamic dataset, but also constant behavior, oscillating or alternating effects, as well as outliers and anomalies. Moreover, considering correlations we might even combine trend analysis with correlation analysis to identify countertrends in a dynamic dataset, for example, an increasing trend pattern might hold for one attribute, but compared to another one which shows a decreasing trend pattern, the combination of both patterns would uncover a countertrend, i.e. one dynamic attribute shows an opposite effect compared to another one or even many more. The trend analysis might even be carried out for an individual attribute, like the saccade length over time, and then if it is applied to all scanpaths of all eye tracking study participants, those trend detection results might even be usable for grouping participants. However, the grouping strongly depends on the attribute under investigation and also how long the corresponding scanpaths are, for example.

6.3.7 Pairwise or Multiple Sequence Alignment

Eye movements, i.e. scanpaths, can be interpreted as sequences of characters stemming from a common alphabet. This means we can transform each scanpath into a finite string consisting of a series of characters while there should be a limited number of those characters to reduce the variability in those computed sequences. This has the benefit that scanpaths can be compared by applying either a pairwise or multiple sequence alignment method [58], typically known from bioinformatics in comparing DNA or RNA sequences. The general idea behind this concept is to compute some kind of consensus matrix [84] which gives an impression of similar and dissimilar regions in the list of eye movement sequences, i.e. scanpaths (see Figure 6.4). The transformation of a scanpath into a sequence of characters can be based on several parameters, for example each fixation might be translated into a character based on a subdivision of a stimulus into areas of interest while each area of interest is modeled by a unique character. The more AOIs are present, the more characters will be encoded into a corresponding scanpath, hence making the chance quite low of finding a good consensus among many of those scanpaths. For this reason, it is a good idea to let a user interactively adapt the separation of the visual stimulus into areas of interest to see the impact of the sub-division on the output of the consensus matrix. However, such an approach is quite time-consuming, in particular if the scanpaths are long and consist of many characters. Positively, there is a lot of research in the field of bioinformatics that supports quite fast solutions to this algorithmic problem.

There are string-based sequence alignment methods which have also been used for comparing eye movement data, for example, the Levenshtein distance [54]. The alignment algorithms have been applied, for example, in combination with clustering approaches to find a good grouping of eye tracking study participants based on their scanpath patterns or to find specific areas of interest [162, 404]. Also the Needleman–Wunsch algorithm goes in a similar direction [367] and has also been adapted to work for eye movement data [525]. Tools like SubsMatch [297] or MultiMatch [148] are interesting offspring from this line of research, apart from many others. However, if it comes to dynamically changing scanpaths, generated by various people in a long-duration task, those alignment methods soon reach performance issues if they are applied to the raw scanpath data. In this case, some kind of aggregation or filtering algorithm has to reduce the size and complexity of the data before it comes to an efficient alignment. This is actually the

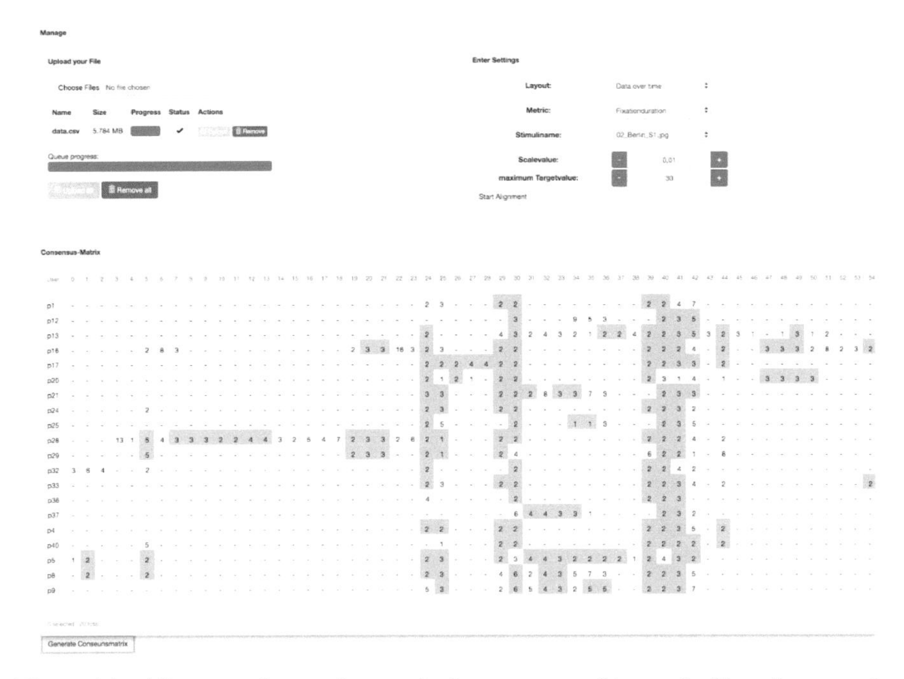

Figure 6.4 Alignment of a set of scanpaths from an eye tracking study. First, the scanpaths are transformed into character sequences based on user input, before they are aligned [84].

challenging bottleneck of this algorithmic technique to compare the eye movement patterns by aligning them to identify dynamic pattern groups that can be used to classify scanpaths, for example.

6.3.8 Artificial Intelligence-Related Approaches

Powerful concepts for data analysis have been developed in the field of artificial intelligence (AI) [326], and more and more have also found their way to the field of eye tracking [455]. The idea behind artificial intelligence is to mimic human intelligence as much as possible to include the power of the computer as a way of faster detecting solutions to challenging problems that the human alone could not find that quickly and the computer not that precisely. AI also includes fields like machine learning [10] and deep learning [390, 527], describing strategies to make the computer learn from given datasets to apply generated models from a training phase on new data elements, for example, to classify them or to predict future scenarios. The machine gets more and more experienced and is finally able to apply the computed rules to new situations. The learning can happen in different ways like supervised, semi-supervised, unsupervised, or as reinforcement. Multi-layered neural networks [384] are often used to train, as a mechanism to perform complex tasks in larger and larger datasets to which eye tracking data also belongs. Although artificial intelligence has generated various fast, efficient, and quite accurate methods, the whole discipline is just about to start to take into account eye tracking data. The major reason is that the available eye tracking data today might still be too small to make reliable predictions based on artificial intelligence, using machine and deep learning approaches.

However, a few problems have been tackled in the field of eye tracking by making use of AI-related concepts. For example, convolutional neural networks have been used for analyzing real-time eye tracking data with focus on interactive applications [77]. In most scenarios the research focuses on eye images, for example, to train a machine learning algorithm based on a multitude of such images to classify or predict newly seen images. Such an approach is, in particular, useful for detecting negative performance issues of car or truck drivers, for example, if their eye movements indicate fatigue effects that might cause accidents [115]. Such image-centric tasks are hard to solve by standard algorithms due to the vast amount of data and features to be explored, in particular, if real-time analyses are required. Machine learning, on the other hand, can be used to make fast predictions and classifications, however a certain large amount of training data is required to generate accurate and meaningful results. For example, a model for predicting where people look in images [302], predicting gaze fixations [133], or saliency in context to predict visual attention [249] are typical research areas.

6.4 Visualization Techniques and Visual Analytics

In most cases the output of the algorithmic computations is too complex and too large to be inspected by just reading the generated textual information. Typically, algorithms take a dataset or several of them as input and produce a new processed dataset which would still require a time-consuming exploration process to understand and to find patterns. Consequently, a visual encoding of such datasets is a powerful idea since the visual output is typically easier and more rapidly understandable by the human user than the textual counterpart. However, it should be guaranteed that the visual encoding is based on appropriate visual variables that help to quickly identify visual patterns that can be remapped to data patterns in order to analyze the data. In particular, in the field of eye tracking, the data can be based on at least three major data dimensions which come in the form of space, time, and participants in an eye tracking study. Depending on the tasks [304] the users of a visualization technique or visual analytics system plan to solve, the visual encoding can vary a lot as well as the interaction techniques that are integrated into the provided visual depictions of the data. Moreover, the way in which the data actually exists or is being transformed plays a crucial role for the visual metaphor and the visual variables in use. For example, it makes a difference for the visualization technique whether we are interested in the raw fixations to a stimulus or to spatially aggregated areas of interest, while the dynamics of the data plays a role for the visual depiction, static data is definitely easier to visualize than its time-varying variant, i.e. several instances of the static data.

Visualization techniques for eye tracking data exist in various forms [47], either focusing on individual aspects in the data, or incorporating more and more data dimensions and derived values, typically attached to one or more of the provided visualization techniques focusing on the primary aspect in the data based on the primary task or tasks a user wishes to solve with a visualization, or at least get a hint about a certain visual pattern that initiates further exploration and analysis processes. Visual analytics goes one step further than traditional visualization techniques since it is an interdisciplinary field that combines concepts from algorithmics, statistics, human–computer interaction, visualization, perception, cognitive processing, and many more. Hence, with visual analytics we can actually get power from both sides, the machine and the human side, to build models and hypotheses for our eye tracking data guided by interactive visual depictions of the interesting pieces of the data, finally leading to insights and knowledge from those large and heterogeneous eye tracking data sources, in particular in future scenarios when eye tracking data grows and grows [44] with many more extra data sources about human behavior and additional personal feedback. Visualization and visual analytics are not built to solve the problems in eye tracking data, but in cases where the data is visually encoded in a perceptually and visually effective way, we can recognize visual patterns that can be remapped to data patterns in the best case, leading to the formulation of new hypotheses and also to the confirmation, rejection, or refinement of already existing hypotheses. Visualization plays the role of guide through our large eye tracking data since it allows us to navigate, scroll, filter, and finally, explore the data.

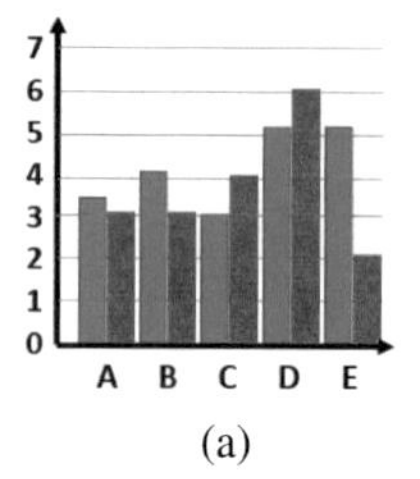

(a)

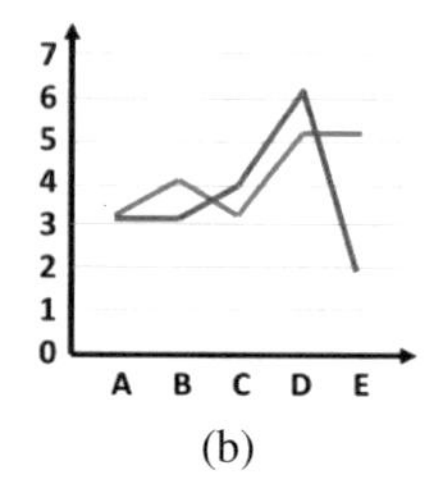

(b)

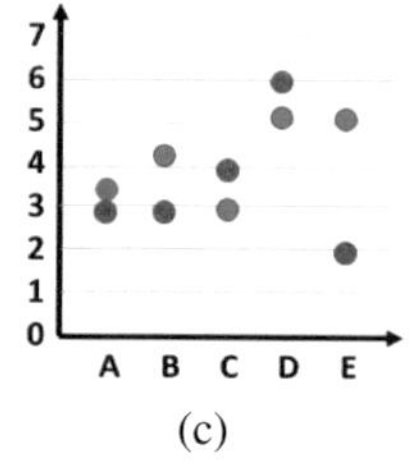

(c)

Figure 6.5 Statistical plots can be useful to get an overview of the quantitative values in an eye tracking dataset: (a) a bar chart; (b) a line graph; (c) a scatter plot.

6.4.1 Statistical Plots

The complexity and size of eye tracking data might be reduced to only a few quantitative numbers expressing properties about the data. Although the general data dimensions like space, time, and participants are not explicitly derivable from such quantities, they might be worth visualizing due to the fact that they can give an overview for comparing eye tracking data in several dimensions. Famous examples of such statistical graphics are box plots for the value distribution of one attribute, histograms and bar charts (see Figure 6.5(a)) for visually exploring weighted distributions of values of one attribute, line charts (see Figure 6.5(b)) for time-dependent attributes and correlations between several attributes, or scatter plots (see Figure 6.5(c)) for identifying correlations among two attributes, just to mention a few. Such statistical plots have already been used for eye tracking data, for example, for visualizing children's eye movement behavior when watching TV as a line chart [213], primarily to identify eye movements and saccades over time. Also fixations can be displayed in a line chart, for example, for different tasks to allow temporal comparisons [20]. Several more derived metrics like mean values for saccades, fixation duration, and many more might be worth investigating by using line charts [469].

In addition, bar charts and histograms have been used a lot to depict distributions of attribute values, for example, for showing fixation duration [129] or eye position accuracy [155], even extended to 3D bar charts for visual attention distributions in 2D TV screens [55]. Bar charts can show the weighted distribution, for example, the number of quantities that fall into a certain bin illustrated by a bar or line in a histogram. Box plots, on the other hand, do not encode the number of quantities that fall into a certain bin, but they can show the statistical distribution of all quantities over the value range while typically indicating the middle fifty percent of all the values as a bar separated by the median line that reflects the value exactly in the middle of the distribution. Box plots can be useful to show fixation deviations with the goal of analyzing the deterioration of the eye tracking device calibration [240] or as a summarized comparison of study participants and their normalized scanpath saliency scores [158]. Analyzing pairwise correlations can be done with scatterplots while each attribute is mapped to one of the axes. Also, in the field of eye tracking, scatter plots have been used, for example, to plot correlations between response latency and angular disparity [268]. Moreover, scatter plots can be used for comparing different species' eye movements like humans and monkeys, for example, by plotting amplitudes and velocities of the recorded saccades [35]. To explore more than two attributes we might use star plots, for example, to visualize scanpath properties [203] or fixations [365]. Apart from star plots we can use parallel coordinate plots to show correlations between several attributes like derived eye tracking metrics [299].

6.4.2 Point-based Visualization Techniques

Some visualization techniques do not focus on statistically derived data from original eye tracking data. Those visualization techniques more or less take into account the fixation data without further aggregating it and try to visually encode those fixations over space and time, for each participant individually or even aggregated for groups of participants. The fixation data is typically given as x- and y-coordinates with a fixation duration and a time stamp. Due to the fact that the fixation data is not further spatially aggregated, for example, into areas of interest, we denote a visualization of it by the term point-based visualization. One major visual focus is the fixation point evolution over time, i.e. in the x- and y-dimensions in space while the spatial dimension stems from the displayed static or dynamic stimulus in an eye tracking study. To reach this goal of exploring the temporal aspect of the eye tracking data we can use a timeline visualization which uses one axis for the time dimension while on the other one more typical time-varying property or attribute from the eye tracking data are represented. To show such time-dependent scanpath data on a point-based perspective we might split the x- and y-coordinates and plot both separately (see Figure 6.6) in two timeline plots [203, 209]. Moreover, a 3D variant would also be possible like a space-time cube [293] in which the x–y-plane is used for the coordinates of the fixations while the z-axis shows the evolution over time. However, 3D charts have to be taken with care due to the fact that they generate occlusion

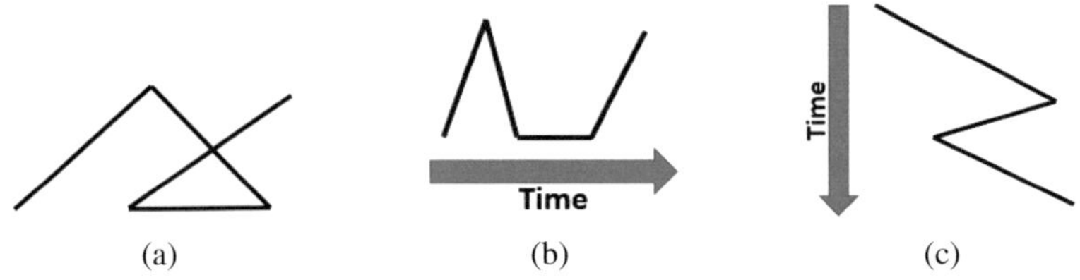

Figure 6.6 Splitting the fixations from a scanpath into their x- and y-coordinates: (a) the original scanpath; (b) a timeline for the y-coordinates; (c) a timeline for the x-coordinates.

effects and are difficult to interpret because of missing reference points to the axes. As a negative consequence of splitting the x- and y-coordinates of a scanpath to show them separately in timeline plots we cannot easily identify the temporal changes in the spatial dimension, i.e. in the visual stimulus. To see this effect we can use the popular gaze plots [448], however, if the scanpaths are quite long or many participants' scanpaths have to be shown at the same time, we reach a problem denoted by visual clutter [426].

Instead of showing the fixation points over time we might be interested in inspecting the fixation data with an additional view on the shown stimulus, i.e. the spatial dimension. This could be done using a transparent overlay on the stimulus, however, the dynamics of the fixation data can also be shown, for example, by an animated diagram known as a bee swarm visualization [1], also for a dynamic stimulus [345]. However, animation is typically considered to be problematic for comparisons over time [505] due to the fact that a viewer has to remember lots of visual patterns in the short-term memory to reliably do the comparisons. A static side-by-side visualization might be the better option for such data although the display space is a limitation of the static representation. If the fixation data is temporally aggregated as well as over groups of participants we denote such a visualization as visual attention map, fixation map, or heat map [49, 50, 473]. This aggregated representation of the fixation data does not show the time-varying behavior, but serves as a great overview of visual attention hot spots—regions in the visual stimulus that attracted much attention. Such hot spot regions are typically used for defining areas of interest (see Section 6.4.3). The visual depiction of these hot spots can be based on several criteria like fixation count, fixation duration, relative fixation duration, or participant percentage, to mention a few. Moreover, the visual appearance of the attention maps can be based on several visual variables (see Figure 6.7 for visual attention maps enhanced by contour lines); typically color coding [50, 167] is used to visually depict

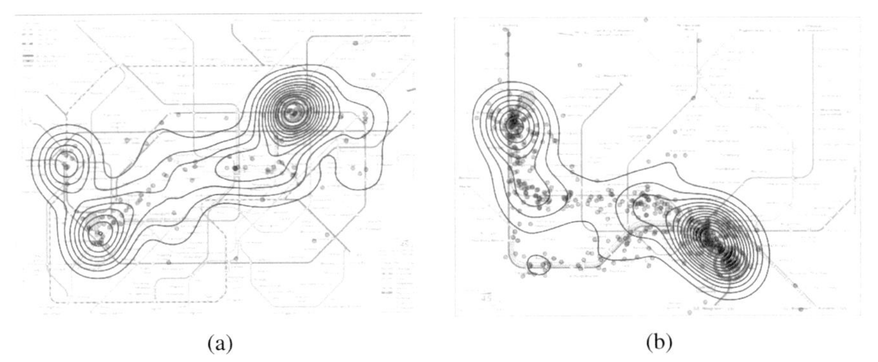

(a) (b)

Figure 6.7 Two different visual attention maps from a public transport map eye tracking study. In this case the hot spots of visual attention are indicated by contour lines [100]. Route finding tasks in the maps of: (a) Tokyo, Japan; (b) Hamburg, Germany.

the visual attention at a certain point, but also luminance [515], contour lines [100, 158, 206], or even 3D effects [321, 530]. However, although attention maps seem to be powerful concepts they also have to be taken with care [235] to avoid misinterpretations of the data. For a dynamically changing stimulus it is challenging to generate a visual attention map. However, synchronous attention of several participants can be computed and then this can be visualized over time [357]. Also motion-compensated attention maps based on optical flow concepts for moving objects have been explored [311]. On the other hand, for 3D visual stimuli, the 3D visual attention is typically directly incorporated in the stimulus [394], maybe also by 2D projections or by coloring the entire 3D visual object by using the visual attention map color [483].

If we are interested in seeing the connected fixations, i.e. the whole scanpath, over space and time we can use scanpath visualizations [375], meaning a sequence of fixations and saccades [235]. A standard visualization for a scanpath is a composition of circles of different size with the circle center where the fixation was done in the visual stimulus, encoding the fixation duration in the circle size while the saccades are represented as straight lines connecting subsequent fixations [448]. The scanpath visualization is typically overplotting the visual stimulus for contextual reasons (see Figure 6.8 for examples of scanpath visualizations). Also the velocity of fixation data can be encoded [318] while in the early days the fixations were not visually indicated, just the saccades [539]. Scanpaths can even be used to derive other visual effects, for example the convex hull

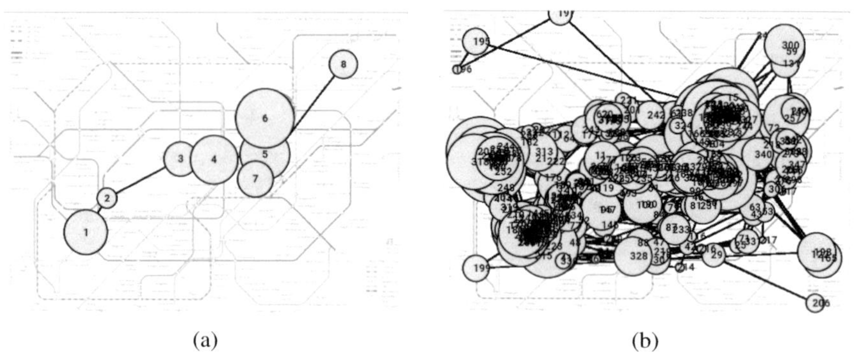

(a) (b)

Figure 6.8 Scanpath visualizations for (a) one participant and (b) 40 participants [100]. The scanpath visualization in (b) can hardly by used for data exploration.

or surrounding area of a scanpath [205] which might be an indication of how much area was covered during visually scanning a stimulus. Although scanpath visualizations show the spatio-temporal behavior of the participants' scanning strategies, they mostly suffer from visual clutter [426] making them only useful for shorter scanpaths for one or two participants. Bundling scanpaths can be an option [118, 225, 251] but at the cost of modifying the straight lines and hence leading to misinterpretations of the scanpath data. Moreover, the data-to-ink ratio might be reduced in the scanpath visualization [203] or the vertical and horizontal movement directions might be shown separately at the surrounding borders of a visual stimulus [94]. To avoid long scanpaths being shown we might only show the parts of the scanpaths that are located in a sliding time window that is animated over the entire time axis [523]. Finally, 3D scanpaths can be shown by overplotting the original visual stimulus [165, 394, 484] or by using a modified 2D representation of the 3D stimulus and by mapping the scanpath to the new 2D projection [415].

If the visual stimulus and the temporal information are of interest, a space-time cube visualization can provide insights [293]; however, occlusion, distortion, or irritating depth perception effects might occur. The benefit of space-time cubes is the fact that they can be used for static [330] as well as dynamic [164, 311] visual stimuli (see Figure 6.9). Space-time cubes require various interaction techniques to rotate, navigate, and scroll in the dynamic data, with the goal to get an overview about space, time, and participants.

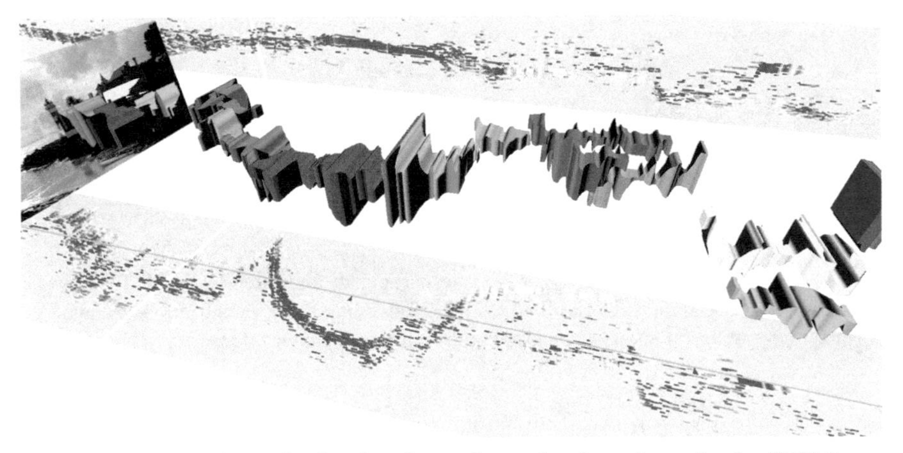

Figure 6.9 A space-time cube showing clustered gaze data for a given stimulus [311]. Image provided by Kuno Kurzhals.

6.4.3 AOI-based Visualization Techniques

A spatially aggregated view on the fixation data provides insights into certain connected regions in a visual stimulus in combination with the visual attention paid to the scene, even over time. Moreover, those areas of interest (AOIs) or regions of interest (ROIs) can be set in relation to each other, for example, by computing the number of pairwise transitions between them, generating a static [202, 205, 329, 393] or dynamic AOI transition graph [73] or a dynamic Sankey diagram called AOI rivers [80]. Those visualizations can be enhanced by additional eye tracking metrics [169, 465], even a node-link diagram [82] instead of a transition matrix can be shown. AOIs can be defined in several ways, typically taking into account the hot spots of visual attention, the semantics of a stimulus with its visual objects, or a naive grid-based sub-division of the stimulus space. The hot spot-based approach can by done algorithmically [404, 435] or manually, by defining rectangular or arbitrarily shaped borders around hot spots or by using a clustering algorithm, for example k-means producing a Voronoi-like sub-division of the space. One benefit of AOIs is that additional metrics can be derived based on the formerly defined spatial sub-division [261, 401], for example, the time to first fixation, the frequency of visits of an AOI, or the order of AOI visits, to mention a few. If we are interested in the temporal aspect of AOI visits we have several options, for example, using an animation or a static side-by-side or stacked display as well as some form of rapid serial visual presentation [477]. A static representation for the AOI visits for one or several participants is shown in

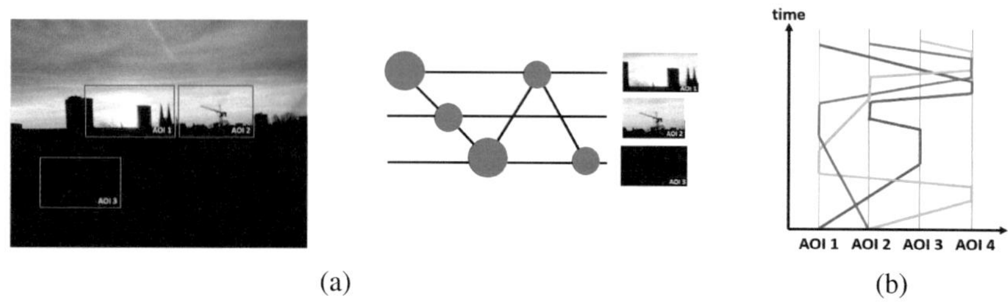

(a) (b)

Figure 6.10 AOI visits over time, either for one participant and three AOIs [414] (a) or three participants and four AOIs in parallel [417] (b). Extending these visualizations to many participants, many AOIs, and long scanpaths can lead to visual clutter effects.

Figure 6.10. The fixation duration in an AOI can be shown by additional visual variables [138], for example, by the circle size [414]. However, visual clutter can be a problem if too many participants are shown [417]. But for one participant, observations like reading tasks where each word can be defined as an AOI, a visual approach like this might be quite useful [40, 474].

An attention map can also be used to visually encode the fixation numbers and additional derived metrics in formerly defined areas of interest, statically but also evolving over time [126]. Moreover, if the changes over time are of particular interest, a scarf plot [100, 308, 421] is recommended which encodes the participants on parallel timelines and indicates the duration of AOI visits by differently long color coded rectangles [235, 523] while the color coding is used for the correspondence [423] between the AOIs shown in the scarf plot and those annotated in the stimulus (see Figure 6.11). The AOIs can even be visually enhanced by thumbnail images to show the contextual information from the visual stimulus [502]. Also 3D visual stimuli can be encoded in a scarf plot [484]. As a negative issue we cannot see the dynamic AOI transitions between AOIs in a scarf plot, just the AOI visits over time, which are visually encoded as well in an AOI river [80] by using merging and splitting of sub-rivers following the visual metaphor of a Sankey diagram (see Figure 6.12). One problem with the graph-based transition visualizations, matrix or node-link-based ones, is the fact that the contextual information to the shown stimulus is lost [48, 184, 498]; however, due to this visual independence from the stimulus, the visualization technique has all the freedom to place the visual objects for all AOIs wherever they fit in the display. Finally, apart from the transitions that form a graph we can generate a hierarchy of AOIs to indicate the major AOI branches that the participants follow during a visual scanning strategy [423, 502, 525]. A

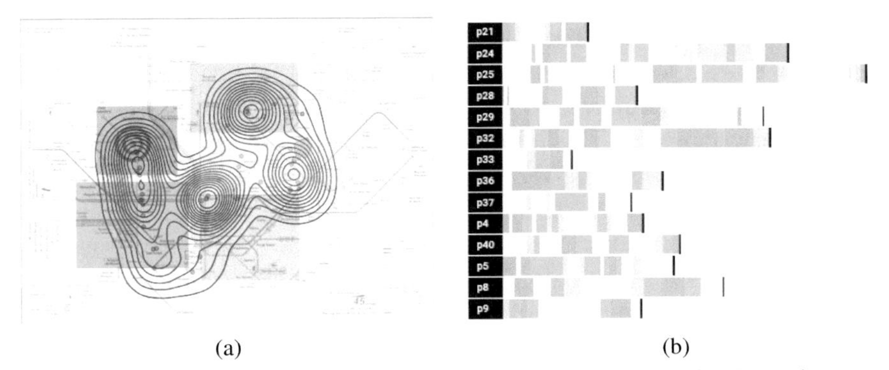

(a) (b)

Figure 6.11 Annotating a visual stimulus, overplotted with a contour visual attention map, with color coded AOIs (a); the AOI visits over time can be seen in a corresponding scarf plot (b) that uses the same color coding as in the annotation view.

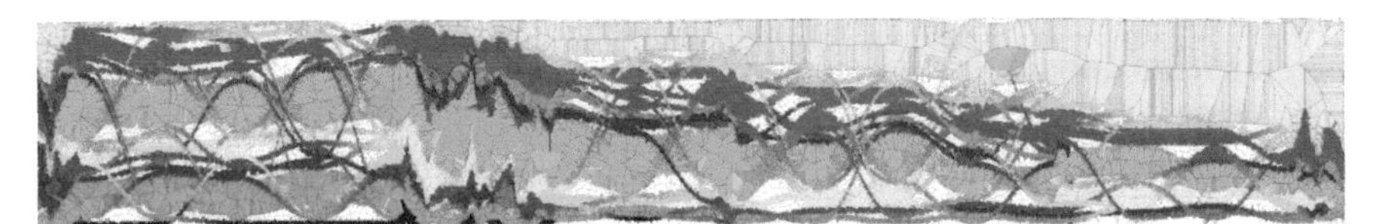

Figure 6.12 The dynamic AOI transitions can be shown in an AOI river visualization [80] with an enhancement by Voronoi cells.

combination of graph and hierarchical aspects among the AOIs can be used to compute a hierarchical graph layout for the AOIs together with their transition frequencies [81].

6.4.4 Eye Tracking Visual Analytics

Since eye tracking gets more and more interesting for various application fields, the number of available datasets increases day by day. Moreover, the size and complexity of those datasets also changes, from simple scanpaths as in early times to various other complementing measures, metrics, and additional data sources. This progress in eye tracking technology and the need for analyzing the data from various application fields requires new powerful concepts to which visual analytics also belongs [14]. Eye tracking data consists of so many different aspects generating a multitude of user tasks that require solutions or at least hints where to look further to find answers to hypotheses. Visual analytics does not directly give answers but it gives users the control to interact and navigate in the visual depictions of the data. Moreover, it supports algorithmic approaches, also allowing building of models and visualizing them in a way that users can step-by-step explore their eye tracking data. To reach this goal, visual analytics applies techniques from many application fields [277] due to the fact that it is an interdisciplinary approach. However, visual analytics cannot solve all eye tracking exploration problems [412], but at least it provides ideas that show how to come closer to solutions, in cases where the data problem seems to be algorithmically intractable or the required visualization produces a non-scalable representation of the data. For example, real-time eye tracking data analysis [482] is a challenging problem that needs powerful concepts from various disciplines to keep up with the flood of data that is recorded, even from several eye tracking devices at the same time, maybe generated in scenarios involving VR/AR/MR or in particular, immersive analytics.

Gaze stripes [309] or ISeeCube [308] focus on visual analytics of video data by providing visual and algorithmic concepts in combination. Both visual analytics concepts can also be modified to make them applicable to static visual stimuli. ISeeColor [390] combines interactive visualizations and automatic recognition of independent objects by applying deep learning approaches focusing on semantic segmentation. Also the identification of reading patterns is in focus of eye tracking visual analytics [538]. Making a visual analytics tool for eye tracking accessible via the internet (see Figure 6.13) is a good idea to reach many users and to get feedback [22].

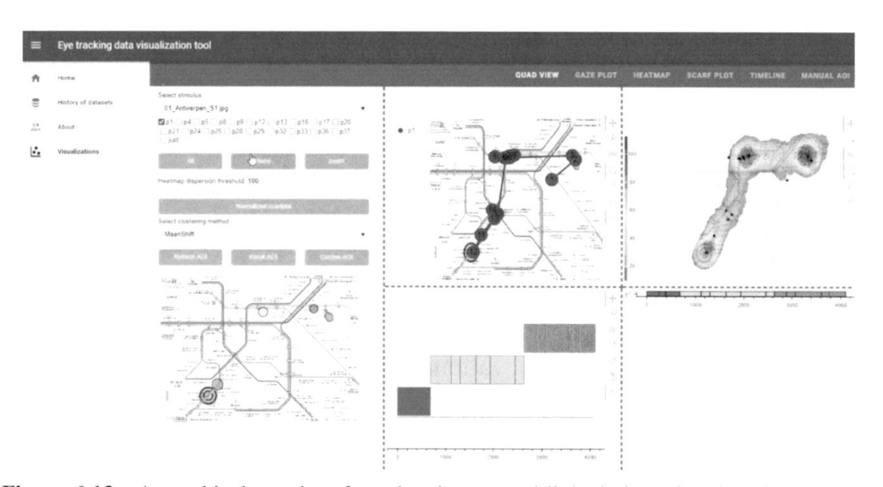

Figure 6.13 A graphical user interface showing several linked views for visually exploring eye movement data: a clustered fixation-based visual attention map, a timeline view on the visually attended AOIs, a scanpath visualization, a visual attention map with color coded hot spots, and a scarf plot for an overview about the inspected AOIs [22].

Moreover, researchers can upload and share their eye tracking data with others, as well as the found insights. ETGraph is a system for eye tracking visual analytics based on graphs [211]. Also in the medical domain there are tools making use of eye tracking data and trying to analyze them with visual analytics concepts [472].

7

Open Challenges, Problems, and Difficulties

In the previous chapters we read a lot about visualization, visual analytics, user studies, and eye tracking, but although these research disciplines provide various concepts and techniques to tackle problems and challenges at the intersection of eye tracking and visual analytics [528] we are aware of the fact that there are numerous open questions, problems, and difficulties. Some of them might be solved in the future, some of them are quite hard to solve with currently available technologies. In the following sections we discuss some of those future problems without explicitly stating that the discussions will take into account a complete list of all those challenges. On a top level we can define a sub-division of the problems into two major categories focusing on eye tracking as a technology incorporating user study issues as well as eye tracking device problems, for example, concerning the recorded data. Secondly, visual analytics with its interdisciplinary character builds another category for future challenges concerning aspects related to data analytics and interactive visualization, but also the human users with their perceptual and visual abilities. In general, it seems as if the data and the human users somehow stand in the center of the involved topics, building some kind of interface between eye tracking and visual analytics. The visual analytics systems might be evaluated by human users generating eye tracking data, while visual analytics is used again to analyze this eye tracking data but based on the human users with their tasks and hypotheses in mind.

7.1 Eye Tracking Challenges

Apart from technological problems related to the eye tracking devices, general user study problems can occur that lead to erroneous or inaccurate study data. In particular, in the field of eye tracking, people might suffer from visual and perceptual disorders that do not allow reliably recording eye movement data from those participants. Moreover, there might be calibration problems that lead to negative aspects concerning the recorded data. The eye movement data can be inaccurate in many ways, making it hard to use or to make predictions based on the users' visual scanning behavior. Hence, it is a good advice to check the reliability of the recorded data before it is analyzed and visualized to avoid wrong conclusions drawn from it. No matter how accurate the data is, the eye-mind hypothesis leads to the fact that the data is regarded as useful or not for describing what people are cognitively processing while they are visually inspecting a static or dynamic stimulus which also makes the interpretation of eye tracking data a challenging field of research. The eye–mind hypothesis can have a negative impact for the whole eye tracking data analysis and visualization community, depending on whether we believe in it or not.

Another eye tracking challenge is caused by the costs that come with each eye tracking device [44]. Although those costs have been decreasing over the years due to the progress in hardware technology like faster processor speed or improved digital video processing and the fact that the market for selling those devices is growing a lot, it can still be quite expensive, but this typically depends on the application domain and parameters like accuracy and tracking rate, i.e. all the required technologies involved in building a suitable eye tracking device. Moreover, nowadays there is some kind of competition between eye tracking companies (see Section 5.2.4) trying to design the best solution for any kind of application scenario, a fact that can also lead to cheaper devices, but mostly for general applications using the standard devices. One such emerging field of eye tracking research focuses on web-based or online eye tracking studies that are particularly useful in pandemic times such as those we are facing at the moment. However, online studies are typically uncontrolled. For eye tracking studies, this aspect also means that each participant must be equipped with an eye tracker or the eye movement data has to be recorded by other novel devices, for example, by using a webcam which, on the negative and challenging side, does not allow producing quite accurate eye movement data over space and time. This is also one of the challenging difficulties for eye tracking integrated in smartphones, either for analyzing the design of an app or for using it in the mode of gaze-assisted interaction to modify something in a user interface or to navigate in an app.

Allowing eye tracking to be used in smartphones might make this technology applicable in a collaborative manner, i.e. making use of the scanpaths of several people in different places in the world, but this powerful idea causes even further challenges related to data privacy issues and ethical aspects in particular, if the recorded eye movement data is made publicly available online. However, some future tasks, for example, by using a visual analytics system can only be solved in a collaborative manner due to the size and complexity of the involved data and knowing where each individual participant was looking at over space and time is very important to find design flaws in the visual interface, the visualization, the interaction techniques, but also in the collaboration and communication between the participants. Hence, it might be important to show the recorded eye movement data to other participants as well which contradicts the data privacy rules. In particular, gaze-assisted interaction can be a problem, either for the individual participant or even a group solving a collaborative task based on eye tracking. The Midas touch problem is well-known when it comes to interacting with visual objects, typically solved by allowing multi-modality interactions including voice, mouse, or gestures [391], or taking into account the dwell time or including eye blinks.

Leaving the laboratory for the real world, i.e. changing the eye tracking study setting from a more controlled environment to an uncontrolled field environment, causes problems for the tracking devices, the reliability and accuracy of the data, but also brings in new challenges for mapping and matching the data with the recorded dynamic visual stimuli. In addition, if eye tracking is used in virtual or augmented reality scenarios we might face additional difficulties that cause data analysis problems later on which are mostly also due to the unclear matching of the seen stimuli and the recorded eye movement data, also including aspects from cognitive psychology, perception, attention, memory, and many more related fields. Building a link between what we see and what we think or cognitively process is one of the key problematic issues we are facing today [305]; however, both fields, eye tracking and cognitive psychology might benefit from such research results, helping us to build better user interfaces, visualizations, interaction techniques, visual analytics systems, and the like.

There are many more challenges in the application area of eye tracking which make the whole field worth researching and which leads to an increase in the number of people involved in the community. Moreover, eye tracking can be found in many other application fields, making it a valuable scientific discipline.

7.2 Eye Tracking Visual Analytics Challenges

Data analysis, visualization, or visual analytics have been successfully applied to eye tracking data, but due to the fact that the data itself gets bigger and bigger, stemming from several heterogeneous data sources creates more and more challenges for all of the data analytics and visualization fields. Moreover, real-time analyses require the most powerful concepts to keep up with the pace of the growing eye tracking datasets. For example, having recorded various scanpaths beforehand and trying to analyze a new incoming scanpath in the light of the existing data might be a great idea, but to achieve fast solutions, i.e. in real-time, we need advanced algorithmic approaches that can efficiently tackle such data scenarios. Fields like artificial intelligence, machine learning, deep learning, data mining, and the like play more and more important roles in the data analysis, however visualization and visual analytics are suitable concepts to involve the human users with their perceptual and visual strengths. But negatively, the human users are typically not able to solve real-time data analysis problems, they can more or less guide the analysis process on a visual exploration basis. For example, human users can decide which kinds of algorithms to apply to a certain data problem or they can include additional information in the data analysis process, maybe the semantics given in a visual stimulus which is something that an algorithm can hardly involve automatically in the analysis, unless it is not trained with various stimuli beforehand.

For a visual analytics system it might be a challenge to reliably connect itself with the recorded eye movement data. This is in particular problematic if several eye tracking devices are used or a new one that has not been used before comes into play which requires first adapting to the new data format. Moreover, it is unclear if the visual analytics system is able to keep up with the growing dataset sizes, in particular if dynamic stimuli like videos are included. It is also unclear if such a system can handle both types of stimuli, static and dynamic ones, also those with actively changeable content like interactive user interfaces. The scalability issues might arise if eye tracking is integrated in smartphones one day, producing vast amounts of data worth analyzing, even in real-time which might cause further problems for storing the data, transforming it, and finally, accessing it again to make predictions or recommendations based on the formerly recorded data. Also the display has a crucial role for a visual analytics system, meaning small-, medium-, or large-sized displays, all having their benefits and drawbacks. Those are typical problems that come with growing dataset sizes including research from the field of big data [44], involving various disciplines to analyze the data with the goal to detect rules, correlations, patterns, and finally, insights and knowledge, even in real-time. Another aspect from the perspective of visual analytics systems could be the idea of letting users interact with the system while at the same time their eye movements are recorded. This data can be analyzed while users further work with the system and, based on the outcomes of such an analysis, the visual analytics system might be adapted to some degree, making the whole concept some kind of dynamic visual analytics system based on user behavior like visual scanning strategies.

Many application fields could benefit from eye tracking as well as visual analytics in combination. For example, the field of medicine might apply eye tracking to later analyze and understand how doctors behaved during surgery, i.e. where they looked over time. Moreover, a similar scenario holds for aircraft pilots who first control a plane, for example, in a landing maneuver. Recording the eye movements and analyzing and visualizing the data later on can help to identify which visual elements the pilot has missed during the maneuver. This again might help to improve the landing strategy for later training phases. Such insights could be helpful to better train young doctors or pilots to get more practice for future surgeries or landing maneuvers. Education in eye tracking as well as visual analytics [71] is a deciding factor to train young researchers and to make them aware of the challenges in both fields, but also the benefits and synergy effects that might come with such a combination. On the negative side, it is quite difficult to educate young students since the fields of eye tracking and visual analytics both involve that many concepts that it is impossible to teach these topics in a short time period, hence only the tip of the iceberg can be the focus of education.

References

[1] TOBII TECHNOLOGY AB. Tobii studio 1.x user manual, 2008.

[2] Moataz Abdelaal, Marcel Hlawatsch, Michael Burch, and Daniel Weiskopf. Clustering for stacked edge splatting. In Fabian Beck, Carsten Dachsbacher, and Filip Sadlo, editors, *Proceedings of 23rd International Symposium on Vision, Modeling, and Visualization, VMV*, pages 127–134. Eurographics Association, 2018.

[3] Athos Agapiou. Remote sensing heritage in a petabyte-scale: satellite data and heritage earth engine© applications. *International Journal of Digital Earth*, 10(1):85–102, 2017.

[4] Wolfgang Aigner, Silvia Miksch, Heidrun Schumann, and Christian Tominski. *Visualization of Time-Oriented Data*. Human-Computer Interaction Series. Springer, 2011.

[5] Jumana Almahmoud, Saleh Albeaik, Tarfah Alrashed, and Almaha Almalki. Visual exploration patterns in information visualizations: Insights from eye tracking. In Gabriele Meiselwitz, editor, *Proceedings of International Conference on Social Computing and Social Media. Applications and Analytics*, volume 10283 of *Lecture Notes in Computer Science*, pages 357–366. Springer, 2017.

[6] Basak Alper, Benjamin Bach, Nathalie Henry Riche, Tobias Isenberg, and Jean-Daniel Fekete. Weighted graph comparison techniques for brain connectivity analysis. In *Proceedings of ACM SIGCHI Conference on Human Factors in Computing Systems*, pages 483–492, 2013.

[7] Mohammad Alsaffar, Lyn Pemberton, Karina Rodriguez-Echavarria, and Mithileysh Sathiyanarayanan. Visual behaviour in searching information: A preliminary eye tracking study. In Saïd Assar, Oscar Pastor, and Haralambos Mouratidis, editors, *Proceedings of 11th International Conference on Research Challenges in Information Science, RCIS*, pages 365–370. IEEE, 2017.

[8] Bilal Alsallakh, Luana Micallef, Wolfgang Aigner, Helwig Hauser, Silvia Miksch, and Peter J. Rodgers. The state-of-the-art of set visualization. *Computer Graphics Forum*, 35(1):234–260, 2016.

[9] Florian Alt, Stefan Schneegass, Jonas Auda, Rufat Rzayev, and Nora Broy. Using eye-tracking to support interaction with layered 3d interfaces on stereoscopic displays. In Tsvi Kuflik, Oliviero Stock, Joyce Yue Chai, and Antonio Krüger, editors, *Proceedings of 19th International Conference on Intelligent User Interfaces, IUI*, pages 267–272. ACM, 2014.

[10] Massih-Reza Amini. *Machine Learning, 2nd Edition*. Eyrolles, 2020.

[11] Richard Andersson, Marcus Nyströ, and Kenneth Holmqvist. Sampling frequency and eye-tracking measures: how speed affects durations, latencies, and more. *Journal of Eye Movement Research*, 3(3):1–12, 2010.

[12] Keith Andrews. Evaluation comes in many guises. In *Proceedings of CHI workshop on BEyond time and errors: novel evaLuation methods for Information Visualization (BELIV)*, pages 7–8, 2008.

[13] Keith Andrews and Janka Kasanicka. A comparative study of four hierarchy browsers using the hierarchical visualisation testing environment (HVTE). In *Proceedings of 11th International Conference on Information Visualisation, IV*, pages 81–86. IEEE Computer Society, 2007.

[14] Gennady L. Andrienko, Natalia V. Andrienko, Michael Burch, and Daniel Weiskopf. Visual analytics methodology for eye movement studies. *IEEE Transactions on Visualization and Computer Graphics*, 18(12):2889–2898, 2012.

[15] Francis John Anscombe. Graphs in statistical analysis. *American Statistician*, 27(1), 1973.

[16] Daniel Archambault and Helen C. Purchase. The "map" in the mental map: Experimental results in dynamic graph drawing. *International Journal on Human-Computer Studies*, 71(11):1044–1055, 2013.

[17] Daniel W. Archambault and Helen C. Purchase. The mental map and memorability in dynamic graphs. In Helwig Hauser, Stephen G. Kobourov, and Huamin Qu, editors, *Proceedings of the IEEE Pacific Visualization Symposium, PacificVis*, pages 89–96. IEEE Computer Society, 2012.

[18] Daniel W. Archambault, Helen C. Purchase, and Bruno Pinaud. Animation, small multiples, and the effect of mental map preservation

in dynamic graphs. *IEEE Transactions on Visualization and Computer Graphics*, 17(4):539–552, 2011.

[19] Richard Arias-Hernández, John Dill, Brian D. Fisher, and Tera Marie Green. Visual analytics and human-computer interaction. *Interactions*, 18(1):51–55, 2010.

[20] Stella Atkins, Xianta Jiang, Geoffrey Tien, and Bin Zheng. Saccadic delays on targets while watching videos. In Carlos Hitoshi Morimoto, Howell O. Istance, Stephen N. Spencer, Jeffrey B. Mulligan, and Pernilla Qvarfordt, editors, *Proceedings of the Symposium on Eye-Tracking Research and Applications, ETRA*, pages 405–408. ACM, 2012.

[21] Ivan Bacher, Brian Mac Namee, and John D. Kelleher. Scoped: Evaluating A composite visualisation of the scope chain hierarchy within source code. In *Proceedings of IEEE Working Conference on Software Visualization, VISSOFT*, pages 117–121. IEEE, 2018.

[22] Hristo Bakardzhiev, Marloes van der Burgt, Eduardo Martins, Bart van den Dool, Chyara Jansen, David van Scheppingen, Günter Wallner, and Michael Burch. A web-based eye tracking data visualization tool. In Alberto Del Bimbo, Rita Cucchiara, Stan Sclaroff, Giovanni Maria Farinella, Tao Mei, Marco Bertini, Hugo Jair Escalante, and Roberto Vezzani, editors, *Proceedings of International Workshops and Challenges in Pattern Recognition, ICPR*, volume 12663 of *Lecture Notes in Computer Science*, pages 405–419. Springer, 2020.

[23] Michel Ballings and Dirk Van den Poel. Using eye-tracking data of advertisement viewing behavior to predict customer churn. In Wei Ding, Takashi Washio, Hui Xiong, George Karypis, Bhavani M. Thuraisingham, Diane J. Cook, and Xindong Wu, editors, *Proceedings of 13th IEEE International Conference on Data Mining Workshops, ICDM*, pages 201–205. IEEE Computer Society, 2013.

[24] Moshe Bar and Maital Neta. Humans prefer curved visual objects. *Psychological Science*, 17(8):645–648, 2006.

[25] Oswald Barral, Hyeju Jang, Sally Newton-Mason, Sheetal Shajan, Thomas Soroski, Giuseppe Carenini, Cristina Conati, and Thalia Shoshana Field. Non-invasive classification of alzheimer's disease using eye tracking and language. In Finale Doshi-Velez, Jim Fackler, Ken Jung, David C. Kale, Rajesh Ranganath, Byron C. Wallace, and Jenna Wiens, editors, *Proceedings of the Machine Learning for Healthcare Conference, MLHC*, volume 126 of

Proceedings of Machine Learning Research, pages 813–841. PMLR, 2020.

[26] Scott Bateman, Regan L. Mandryk, Carl Gutwin, Aaron Genest, David McDine, and Christopher A. Brooks. Useful junk?: the effects of visual embellishment on comprehension and memorability of charts. In Elizabeth D. Mynatt, Don Schoner, Geraldine Fitzpatrick, Scott E. Hudson, W. Keith Edwards, and Tom Rodden, editors, *Proceedings of the 28th International Conference on Human Factors in Computing Systems, CHI*, pages 2573–2582. ACM, 2010.

[27] Patrick Baudisch and Ruth Rosenholtz. Halo: a technique for visualizing off-screen objects. In Gilbert Cockton and Panu Korhonen, editors, *Proceedings of the Conference on Human Factors in Computing Systems, CHI*, pages 481–488. ACM, 2003.

[28] Jeanette Bautista and Giuseppe Carenini. An empirical evaluation of interactive visualizations for preferential choice. In Stefano Levialdi, editor, *Proceedings of the working conference on Advanced Visual Interfaces, AVI*, pages 207–214. ACM Press, 2008.

[29] Fabian Beck, Michael Burch, and Stephan Diehl. Towards an aesthetic dimensions framework for dynamic graph visualisations. In *Proceedings of the 13th International Conference on Information Visualisation, IV*, pages 592–597. IEEE Computer Society, 2009.

[30] Fabian Beck, Michael Burch, Stephan Diehl, and Daniel Weiskopf. A taxonomy and survey of dynamic graph visualization. *Computer Graphics Forum*, 36(1):133–159, 2017.

[31] Fabian Beck, Michael Burch, Tanja Munz, Lorenzo Di Silvestro, and Daniel Weiskopf. Generalized pythagoras trees for visualizing hierarchies. In Robert S. Laramee, Andreas Kerren, and José Braz, editors, *Proceedings of the 5th International Conference on Information Visualization Theory and Applications, IVAPP*, pages 17–28. SciTePress, 2014.

[32] Fabian Beck, Michael Burch, Corinna Vehlow, Stephan Diehl, and Daniel Weiskopf. Rapid serial visual presentation in dynamic graph visualization. In Martin Erwig, Gem Stapleton, and Gennaro Costagliola, editors, *Proceedings of IEEE Symposium on Visual Languages and Human-Centric Computing, VL/HCC*, pages 185–192. IEEE, 2012.

[33] Roman Bednarik, Niko Myller, Erkki Sutinen, and Markku Tukiainen. Applying eye-movement tracking to program visualization. In

Proceedings of IEEE Symposium on Visual Languages and Human-Centric Computing, (VL/HCC, pages 302–304. IEEE Computer Society, 2005.

[34] Michael Behrisch, Benjamin Bach, Nathalie Henry Riche, Tobias Schreck, and Jean-Daniel Fekete. Matrix reordering methods for table and network visualization. *Computer Graphics Forum*, 35(3):693–716, 2016.

[35] David J. Berg, Susan E. Boehnke, Robert A. Marino, Douglas P. Munoz, and Laurent Itti. Free viewing of dynamic stimuli by humans and monkeys. *Journal of Vision*, 9(5):1–15, 2009.

[36] Regina Bernhaupt, Philippe A. Palanque, Marco Winckler, and David Navarre. Usability study of multi-modal interfaces using eye-tracking. In Maria Cecília Calani Baranauskas, Philippe A. Palanque, Julio Abascal, and Simone Diniz Junqueira Barbosa, editors, *Proceedings of the International Conference on Human-Computer Interaction - INTERACT*, volume 4663 of *Lecture Notes in Computer Science*, pages 412–424. Springer, 2007.

[37] Jacques Bertin. *Semiology of Graphics: Diagrams, Networks, Maps*. Wisconsin: University of Wisconsin Press, (first published in French in 1967 translated by William J. Berg in 1983), 1967.

[38] Jacques Bertin. *Graphics and Graphic Information Processing*. De Gruyter, Berlin. Translation:William J. Berg, Paul Scott, 1981.

[39] Jacques Bertin. *Semiology of Graphics - Diagrams, Networks, Maps*. ESRI, 2010.

[40] David Beymer and Daniel M. Russell. WebGazeAnalyzer: a system for capturing and analyzing web reading behavior using eye gaze. In Gerrit C. van der Veer and Carolyn Gale, editors, *Extended Abstracts Proceedings of the Conference on Human Factors in Computing Systems, CHI*, pages 1913–1916. ACM, 2005.

[41] David Beymer, Daniel M. Russell, and Peter Z. Orton. An eye tracking study of how font size and type influence online reading. In David England, editor, *Proceedings of the 22nd British HCI Group Annual Conference on HCI: People and Computers XXII: Culture, Creativity, Interaction - Volume 2, BCS HCI*, pages 15–18. BCS, 2008.

[42] Nikos Bikakis, George Papastefanatos, and Olga Papaemmanouil. Big data exploration, visualization and analytics. *Big Data Research*, 18, 2019.

[43] Pradipta Biswas and Jeevithashree D. V. Eye gaze controlled MFD for military aviation. In Shlomo Berkovsky, Yoshinori Hijikata,

Jun Rekimoto, Margaret M. Burnett, Mark Billinghurst, and Aaron Quigley, editors, *Proceedings of the 23rd International Conference on Intelligent User Interfaces, IUI*, pages 79–89. ACM, 2018.

[44] Tanja Blascheck, Michael Burch, Michael Raschke, and Daniel Weiskopf. Challenges and perspectives in big eye-movement data visual analytics. In *Big Data Visual Analytics, BDVA*, pages 17–24. IEEE, 2015.

[45] Tanja Blascheck and Thomas Ertl. Towards analyzing eye tracking data for evaluating interactive visualization systems. In Heidi Lam, Petra Isenberg, Tobias Isenberg, and Michael Sedlmair, editors, *Proceedings of the Fifth Workshop on Beyond Time and Errors: Novel Evaluation Methods for Visualization, BELIV*, pages 70–77. ACM, 2014.

[46] Tanja Blascheck, Markus John, Kuno Kurzhals, Steffen Koch, and Thomas Ertl. VA^2: A visual analytics approach for evaluating visual analytics applications. *IEEE Transactions on Visualization and Computer Graphics*, 22(1):61–70, 2016.

[47] Tanja Blascheck, Kuno Kurzhals, Michael Raschke, Michael Burch, Daniel Weiskopf, and Thomas Ertl. Visualization of eye tracking data: A taxonomy and survey. *Computer Graphics Forum*, 36(8):260–284, 2017.

[48] Tanja Blascheck, Michael Raschke, and Thomas Ertl. Circular heat map transition diagram. In *Proceedings of the Conference on Eye Tracking South Africa, ETSA*, pages 58–61, 2013.

[49] Pieter J. Blignaut. Visual span and other parameters for the generation of heatmaps. In Carlos Hitoshi Morimoto, Howell O. Istance, Aulikki Hyrskykari, and Qiang Ji, editors, *Proceedings of the Symposium on Eye-Tracking Research & Applications, ETRA*, pages 125–128. ACM, 2010.

[50] Agnieszka Bojko. Informative or misleading? heatmaps deconstructed. In Julie A. Jacko, editor, *Proceedings of the Conference on Human-Computer Interaction, HCI*, volume 5610 of *Lecture Notes in Computer Science*, pages 30–39. Springer, 2009.

[51] Alexandra Bonnici, Stefania Cristina, and Kenneth P. Camilleri. Preparation of music scores to enable hands-free page turning based on eye-gaze tracking. In Kenneth P. Camilleri and Alexandra Bonnici, editors, *Proceedings of the 2017 ACM Symposium on Document Engineering, DocEng*, pages 201–210. ACM, 2017.

[52] Rita Borgo, Luana Micallef, Benjamin Bach, Fintan McGee, and Bongshin Lee. Information visualization evaluation using crowdsourcing. *Computer Graphics Forum*, 37(3):573–595, 2018.

[53] Mina Shirvani Boroujeni, Sébastien Cuendet, Lorenzo Lucignano, Beat Adrian Schwendimann, and Pierre Dillenbourg. Screen or tabletop: An eye-tracking study of the effect of representation location in a tangible user interface system. In Gráinne Conole, Tomaz Klobucar, Christoph Rensing, Johannes Konert, and Élise Lavoué, editors, *Proceedings of 10th European Conference on Design for Teaching and Learning in a Networked World, EC-TEL*, volume 9307 of *Lecture Notes in Computer Science*, pages 473–478. Springer, 2015.

[54] Stephan A. Brandt and Lawrence W. Stark. Spontaneous eye movements during visual imagery reflect the content of the visual scene. *Journal of Cognitive Neuroscience*, 9(1):27 – 38, 1997.

[55] Adam Brasel and James Gips. Points of view: where do we look when we watch tv? *Perception*, 37(12):1890–1894, 2008.

[56] Matthew Brehmer, Bongshin Lee, Petra Isenberg, and Eun Kyoung Choe. A comparative evaluation of animation and small multiples for trend visualization on mobile phones. *IEEE Transactions on Visualization and Computer Graphics*, 26(1):364–374, 2020.

[57] Arthur Brisbane. Advice, Speakers Give Sound, Syracuse Post Standard, 1911.

[58] Daniel G. Brown and Alexander K. Hudek. New algorithms for multiple DNA sequence alignment. In Inge Jonassen and Junhyong Kim, editors, *Proceedings of 4th International Workshop on Algorithms in Bioinformatics, WABI*, volume 3240 of *Lecture Notes in Computer Science*, pages 314–325. Springer, 2004.

[59] Marc H. Brown. Exploring algorithms using balsa-ii. *Computer*, 21(5):14–36, 1988.

[60] Andreas Bulling. Human visual behaviour for collaborative human-machine interaction. In Kenji Mase, Marc Langheinrich, Daniel Gatica-Perez, Hans Gellersen, Tanzeem Choudhury, and Koji Yatani, editors, *Proceedings of the ACM International Joint Conference on Pervasive and Ubiquitous Computing and Proceedings of the ACM International Symposium on Wearable Computers, UbiComp/ISWC Adjunct*, pages 901–905. ACM, 2015.

[61] Andreas Bulling, Raimund Dachselt, Andrew T. Duchowski, Robert J. K. Jacob, Sophie Stellmach, and Veronica Sundstedt. Gaze interaction in the post-wimp world. In Wendy E. Mackay, Stephen A.

Brewster, and Susanne Bødker, editors, *Proceedings of ACM SIGCHI Conference on Human Factors in Computing Systems, CHI*, pages 3195–3198. ACM, 2013.

[62] Michael Burch. Visualizing software metrics in a software system hierarchy. In George Bebis, Richard Boyle, Bahram Parvin, Darko Koracin, Ioannis T. Pavlidis, Rogério Schmidt Feris, Tim McGraw, Mark Elendt, Regis Kopper, Eric D. Ragan, Zhao Ye, and Gunther H. Weber, editors, *Proceedings of 11th International Symposium on Advances in Visual Computing, ISVC*, volume 9475 of *Lecture Notes in Computer Science*, pages 733–744. Springer, 2015.

[63] Michael Burch. Mining and visualizing eye movement data. In Koji Koyamada and Puripant Ruchikachorn, editors, *Proceedings of SIGGRAPH ASIA - Symposium on Visualization*, pages 3:1–3:8. ACM, 2017.

[64] Michael Burch. Visual analytics of large dynamic digraphs. *Information Visualization*, 16(3):167–178, 2017.

[65] Michael Burch. Which symbols, features, and regions are visually attended in metro maps? In Ireneusz Czarnowski, Robert J. Howlett, and Lakhmi C. Jain, editors, *Proceedings of the 9th KES International Conference on Intelligent Decision Technologies (KES-IDT*, volume 73 of *Smart Innovation, Systems and Technologies*, pages 237–246. Springer, 2017.

[66] Michael Burch. Identifying similar eye movement patterns with t-sne. In Fabian Beck, Carsten Dachsbacher, and Filip Sadlo, editors, *Proceedings of the 23rd International Symposium on Vision, Modeling, and Visualization, VMV*, pages 111–118. Eurographics Association, 2018.

[67] Michael Burch. Property-driven dynamic call graph exploration. In Andreas Kerren, Karsten Klein, and Yi-Na Li, editors, *Proceedings of the 11th International Symposium on Visual Information Communication and Interaction, VINCI*, pages 72–79. ACM, 2018.

[68] Michael Burch. Interaction graphs: visual analysis of eye movement data from interactive stimuli. In Krzysztof Krejtz and Bonita Sharif, editors, *Proceedings of the 11th ACM Symposium on Eye Tracking Research & Applications, ETRA*, pages 89:1–89:5. ACM, 2019.

[69] Michael Burch. Graph-related properties for comparing dynamic call graphs. *Journal of Computer Languages*, 58, 2020.

[70] Michael Burch. The importance of requirements engineering for teaching large visualization courses. In *Proceedings of 4th*

International Workshop on Learning from Other Disciplines for Requirements Engineering, D4RE@RE, pages 6–10. IEEE, 2020.

[71] Michael Burch. Teaching eye tracking visual analytics in computer and data science bachelor courses. In Andreas Bulling, Anke Huckauf, Eakta Jain, Ralph Radach, and Daniel Weiskopf, editors, *Proceedings of the Symposium on Eye Tracking Research and Applications, ETRA*, pages 17:1–17:9. ACM, 2020.

[72] Michael Burch, Gennady L. Andrienko, Natalia V. Andrienko, Markus Höferlin, Michael Raschke, and Daniel Weiskopf. Visual task solution strategies in tree diagrams. In Sheelagh Carpendale, Wei Chen, and Seok-Hee Hong, editors, *Proceedings of IEEE Pacific Visualization Symposium, PacificVis*, pages 169–176. IEEE Computer Society, 2013.

[73] Michael Burch, Fabian Beck, Michael Raschke, Tanja Blascheck, and Daniel Weiskopf. A dynamic graph visualization perspective on eye movement data. In Pernilla Qvarfordt and Dan Witzner Hansen, editors, *Proceedings of the Symposium on Eye Tracking Research and Applications, ETRA*, pages 151–158. ACM, 2014.

[74] Michael Burch, Stephan Diehl, and Peter Weißgerber. Visual data mining in software archives. In Thomas L. Naps and Wim De Pauw, editors, *Proceedings of the ACM Symposium on Software Visualization*, pages 37–46. ACM, 2005.

[75] Michael Burch, Marcel Hlawatsch, and Daniel Weiskopf. Visualizing a sequence of a thousand graphs (or even more). *Computer Graphics Forum*, 36(3):261–271, 2017.

[76] Michael Burch, Weidong Huang, Mathew Wakefield, Helen C. Purchase, Daniel Weiskopf, and Jie Hua. The state of the art in empirical user evaluation of graph visualizations. *IEEE Access*, 9:4173–4198, 2021.

[77] Michael Burch, Andrei Jalba, and Carl van Dueren den Hollander. Convolutional neural networks for real-time eye tracking in interactive applications. In *Handbook of Research on Applied AI for International Business and Marketing Applications*, pages 455 – 473, 2021.

[78] Michael Burch, Natalia Konevtsova, Julian Heinrich, Markus Höferlin, and Daniel Weiskopf. Evaluation of traditional, orthogonal, and radial tree diagrams by an eye tracking study. *IEEE Transactions on Visualization and Computer Graphics*, 17(12):2440–2448, 2011.

[79] Michael Burch, Tos Kuipers, Chen Qian, and Fangqin Zhou. Comparing dimensionality reductions for eye movement data. In Michael Burch, Michel A. Westenberg, Quang Vinh Nguyen, and

Ying Zhao, editors, *Proceedings of the 13th International Symposium on Visual Information Communication and Interaction, VINCI*, pages 18:1–18:5. ACM, 2020.

[80] Michael Burch, Andreas Kull, and Daniel Weiskopf. AOI rivers for visualizing dynamic eye gaze frequencies. *Computer Graphics Forum*, 32(3):281–290, 2013.

[81] Michael Burch, Ayush Kumar, and Klaus Mueller. The hierarchical flow of eye movements. In Lewis L. Chuang, Michael Burch, and Kuno Kurzhals, editors, *Proceedings of the 3rd Workshop on Eye Tracking and Visualization, ETVIS@ETRA*, pages 3:1–3:5. ACM, 2018.

[82] Michael Burch, Ayush Kumar, and Neil Timmermans. An interactive web-based visual analytics tool for detecting strategic eye movement patterns. In Krzysztof Krejtz and Bonita Sharif, editors, *Proceedings of the 11th ACM Symposium on Eye Tracking Research & Applications, ETRA*, pages 93:1–93:5. ACM, 2019.

[83] Michael Burch and Kuno Kurzhals. Visual analysis of eye movements during game play. In Andreas Bulling, Anke Huckauf, Eakta Jain, Ralph Radach, and Daniel Weiskopf, editors, *Proceedings of the Symposium on Eye Tracking Research and Applications, ETRA, Short Papers*, pages 59:1–59:5. ACM, 2020.

[84] Michael Burch, Kuno Kurzhals, Niklas Kleinhans, and Daniel Weiskopf. EyeMSA: exploring eye movement data with pairwise and multiple sequence alignment. In Bonita Sharif and Krzysztof Krejtz, editors, *Proceedings of the ACM Symposium on Eye Tracking Research & Applications, ETRA*, pages 52:1–52:5. ACM, 2018.

[85] Michael Burch, Kuno Kurzhals, and Daniel Weiskopf. Visual task solution strategies in public transport maps. In Peter Kiefer, Ioannis Giannopoulos, Martin Raubal, and Antonio Krüger, editors, *Proceedings of the 2nd International Workshop on Eye Tracking for Spatial Research co-located with the 8th International Conference on Geographic Information Science, ET4S@GIScience*, volume 1241 of *CEUR Workshop Proceedings*, pages 32–36. CEUR-WS.org, 2014.

[86] Michael Burch, Steffen Lohmann, Daniel Pompe, and Daniel Weiskopf. Prefix tag clouds. In *Proceedings of 17th International Conference on Information Visualisation, IV*, pages 45–50. IEEE Computer Society, 2013.

[87] Michael Burch and Elisabeth Melby. What more than a hundred project groups reveal about teaching visualization. *Journal of Visualization*, 23(5):895–911, 2020.

[88] Michael Burch, Christoph Müller, Guido Reina, Hansjörg Schmauder, Miriam Greis, and Daniel Weiskopf. Visualizing dynamic call graphs. In Michael Goesele, Thorsten Grosch, Holger Theisel, Klaus D. Tönnies, and Bernhard Preim, editors, *Proceedings of 17th International Workshop on Vision, Modeling, and Visualization, VMV*, pages 207–214. Eurographics Association, 2012.

[89] Michael Burch, Tanja Munz, Fabian Beck, and Daniel Weiskopf. Visualizing work processes in software engineering with developer rivers. In *Proceedings of 3rd IEEE Working Conference on Software Visualization, VISSOFT*, pages 116–124. IEEE Computer Society, 2015.

[90] Michael Burch, Daniel Pompe, and Daniel Weiskopf. An analysis and visualization tool for DBLP data. In *Proceedings of 19th International Conference on Information Visualisation, IV*, pages 163–170. IEEE Computer Society, 2015.

[91] Michael Burch, Michael Raschke, and Daniel Weiskopf. Indented pixel tree plots. In *Proceedings of 6th International Symposium on Advances in Visual Computing, ISVC*, volume 6453 of *Lecture Notes in Computer Science*, pages 338–349. Springer, 2010.

[92] Michael Burch, Abdullah Saeed, Alina Vorobiova, Armin Memar Zahedani, Linus Hafkemeyer, and Marco Palazzo. eDBLP: Visualizing scientific publications. In *Proceedings of the 13th International Symposium on Visual Information Communication and Interaction, VINCI*, 2020.

[93] Michael Burch and Hansjörg Schmauder. Challenges and perspectives of interacting with hierarchy visualizations on large-scale displays. In Andreas Kerren, Karsten Klein, and Yi-Na Li, editors, *Proceedings of the 11th International Symposium on Visual Information Communication and Interaction, VINCI*, pages 33–40. ACM, 2018.

[94] Michael Burch, Hansjörg Schmauder, Michael Raschke, and Daniel Weiskopf. Saccade plots. In Pernilla Qvarfordt and Dan Witzner Hansen, editors, *Proceedings of the Symposium on Eye Tracking Research and Applications, ETRA*, pages 307–310. ACM, 2014.

[95] Michael Burch, Hansjörg Schmauder, and Daniel Weiskopf. Indented pixel tree browser for exploring huge hierarchies. In *Proceedings of the 7th International Symposium on Advances in Visual Computing, ISVC*, volume 6938 of *Lecture Notes in Computer Science*, pages 301–312. Springer, 2011.

[96] Michael Burch, Yves Staudt, Sina Frommer, Janis Uttenweiler, Peter Grupp, Steffen Hähnle, Josia Scheytt, and Uwe Kloos. Pasvis: enhancing public transport maps with interactive passenger data visualizations. In Michael Burch, Michel A. Westenberg, Quang Vinh Nguyen, and Ying Zhao, editors, *Proceedings of the 13th International Symposium on Visual Information Communication and Interaction, VINCI*, pages 13:1–13:8. ACM, 2020.

[97] Michael Burch, Corinna Vehlow, Natalia Konevtsova, and Daniel Weiskopf. Evaluating partially drawn links for directed graph edges. In Marc J. van Kreveld and Bettina Speckmann, editors, *Proceedings of the 19th International Symposium on Graph Drawing, GD*, volume 7034 of *Lecture Notes in Computer Science*, pages 226–237. Springer, 2011.

[98] Michael Burch, Alberto Veneri, and Bangjie Sun. Exploring eye movement data with image-based clustering. *Journal of Visualization*, 23(4):677–694, 2020.

[99] Michael Burch, Adrian Vramulet, Alex Thieme, Alina Vorobiova, Denis Shehu, Mara Miulescu, Mehrdad Farsadyar, and Tar van Krieken. Vizwick: a multiperspective view of hierarchical data. In Michael Burch, Michel A. Westenberg, Quang Vinh Nguyen, and Ying Zhao, editors, *Proceedings of the 13th International Symposium on Visual Information Communication and Interaction, VINCI*, pages 23:1–23:5. ACM, 2020.

[100] Michael Burch, Günter Wallner, Nick Broeks, Lulof Piree, Nynke Boonstra, Paul Vlaswinkel, Silke Franken, and Vince van Wijk. The power of linked eye movement data visualization. In *Proceedings of the Symposium on Eye Tracking Research and Applications, ETRA*. ACM, 2021.

[101] Michael Burch and Daniel Weiskopf. Visualizing dynamic quantitative data in hierarchies - timeedgetrees: Attaching dynamic weights to tree edges. In Gabriela Csurka, Martin Kraus, and José Braz, editors, *Proceedings of the International Conference on Imaging Theory and Applications and International Conference on Information Visualization Theory and Applications*, pages 177–186. SciTePress, 2011.

[102] Michael Burch and Daniel Weiskopf. *On the Benefits and Drawbacks of Radial Diagrams*, pages 429–451. Springer New York, New York, NY, 2014.

[103] Michael Burch, Robin Woods, Rudolf Netzel, and Daniel Weiskopf. The challenges of designing metro maps. In Nadia Magnenat-Thalmann, Paul Richard, Lars Linsen, Alexandru C. Telea, Sebastiano Battiato, Francisco H. Imai, and José Braz, editors, *Proceedings of the 11th Joint Conference on Computer Vision, Imaging and Computer Graphics Theory and Applications, (VISIGRAPP*, pages 197–204. SciTePress, 2016.

[104] Wolfgang Büschel, Patrick Reipschläger, Ricardo Langner, and Raimund Dachselt. Investigating the use of spatial interaction for 3d data visualization on mobile devices. In Sriram Subramanian, Jürgen Steimle, Raimund Dachselt, Diego Martínez Plasencia, and Tovi Grossman, editors, *Proceedings of the Interactive Surfaces and Spaces, ISS*, pages 62–71. ACM, 2017.

[105] Michael D. Byrne, John R. Anderson, Scott Douglass, and Michael Matessa. Eye tracking the visual search of click-down menus. In Marian G. Williams and Mark W. Altom, editors, *Proceedings of the CHI Conference on Human Factors in Computing Systems*, pages 402–409. ACM, 1999.

[106] Bram C. M. Cappers, Paulus N. Meessen, Sandro Etalle, and Jarke J. van Wijk. Eventpad: Rapid malware analysis and reverse engineering using visual analytics. In Diane Staheli, Celeste Lyn Paul, Jörn Kohlhammer, Daniel M. Best, Stoney Trent, Nicolas Prigent, Robert Gove, and Graig Sauer, editors, *Proceedings of IEEE Symposium on Visualization for Cyber Security, VizSec*, pages 1–8. IEEE, 2018.

[107] Stuart Card. *Information visualization, in A. Sears and J.A. Jacko (eds.), The Human-Computer Interaction Handbook: Fundamentals, Evolving Technologies, and Emerging Applications.* Lawrence Erlbaum Assoc Inc, 2007.

[108] Stuart K. Card, Jock D. Mackinlay, and Ben Shneiderman. *Readings in information visualization - using vision to think.* Academic Press, 1999.

[109] Stuart K. Card, Thomas P. Moran, and Allen Newell. *The psychology of human-computer interaction.* Erlbaum, 1983.

[110] Jorge C. S. Cardoso. Comparison of gesture, gamepad, and gaze-based locomotion for VR worlds. In Dieter Kranzlmüller and Gudrun Klinker, editors, *Proceedings of the 22nd ACM Conference on Virtual Reality Software and Technology, VRST*, pages 319–320. ACM, 2016.

[111] Sheelagh Carpendale. Evaluating information visualizations. In Andreas Kerren, John T. Stasko, Jean-Daniel Fekete, and Chris

North, editors, *Information Visualization - Human-Centered Issues and Perspectives*, volume 4950 of *Lecture Notes in Computer Science*, pages 19–45. Springer, 2008.

[112] Thomas Carroll, Aaron Rogers, Dimitrios Charalampidis, and Huimin Chen. Eye tracking and its application in human computer interfaces. In Zia-ur Rahman, Stephen E. Reichenbach, and Mark A. Neifeld, editors, *Proceedings of Visual Information Processing*, volume 8056 of *SPIE Proceedings*, page 805607. SPIE, 2011.

[113] Monica S. Castelhano and Paul Muter. Optimizing the reading of electronic text using rapid serial visual presentation. *Behaviour & Information Technology*, 20(4):237–247, 2001.

[114] Hubert Cecotti, Yogesh Kumar Meena, and Girijesh Prasad. A multimodal virtual keyboard using eye-tracking and hand gesture detection. In *Proceedings of the 40th Annual International Conference of the IEEE Engineering in Medicine and Biology Society, EMBC*, pages 3330–3333. IEEE, 2018.

[115] Xiaolei Cha, Xiaohui Yang, Yingji Zhang, Zhiquan Feng, Tao Xu, and Xue Fan. Eye tracking in driving environment based on multichannel convolutional neural network. In *Proceedings of the 4th International Conference on Digital Signal Processing, ICDSP*, pages 141–144. ACM, 2020.

[116] Chin-Sheng Chang, Chih-Ming Chen, and Yu-Chieh Lin. A visual interactive reading system based on eye tracking technology to improve digital reading performance. In *Proceedings of 7th International Congress on Advanced Applied Informatics, IIAI-AAI*, pages 182–187. IEEE, 2018.

[117] Remco Chang, Mohammad Ghoniem, Robert Kosara, William Ribarsky, Jing Yang, Evan A. Suma, Caroline Ziemkiewicz, Daniel A. Kern, and Agus Sudjianto. Wirevis: Visualization of categorical, time-varying data from financial transactions. In *Proceedings of 2nd IEEE Symposium on Visual Analytics Science and Technology, VAST*, pages 155–162. IEEE Computer Society, 2007.

[118] Monchu Chen, Nelson Alves, and Ricardo Sol. Combining spatial and temporal information of eye movements in goal-oriented tasks. In Andreas Holzinger, Martina Ziefle, Martin Hitz, and Matjaz Debevc, editors, *Proceedings of the First International Conference on Human Factors in Computing and Informatics, SouthCHI*, volume 7946 of *Lecture Notes in Computer Science*, pages 827–830. Springer, 2013.

[119] Shiwei Cheng and Ying Liu. Eye-tracking based adaptive user interface: implicit human-computer interaction for preference indication. *Journal on Multimodal User Interfaces*, 5(1-2):77–84, 2012.

[120] Dimitri A. Chernyak and Lawrence W. Stark. Top-down guided eye movements: peripheral model. In Bernice E. Rogowitz and Thrasyvoulos N. Pappas, editors, *Proceedings of Human Vision and Electronic Imaging VI*, volume 4299 of *SPIE Proceedings*, pages 349–360. SPIE, 2001.

[121] Ed Huai-hsin Chi, Michelle Gumbrecht, and Lichan Hong. Visual foraging of highlighted text: An eye-tracking study. In Julie A. Jacko, editor, *Proceedings of 12th International Conference on Intelligent Multimodal Interaction Environments*, volume 4552 of *Lecture Notes in Computer Science*, pages 589–598. Springer, 2007.

[122] Kang-A. Choi, Chunfei Ma, and Sung-Jea Ko. Improving the usability of remote eye gaze tracking for human-device interaction. *IEEE Transactions on Consumer Electronics*, 60(3):493–498, 2014.

[123] Jaegul Choo and Shixia Liu. Visual analytics for explainable deep learning. *IEEE Computer Graphics and Applications*, 38(4):84–92, 2018.

[124] William S. Cleveland and Robert McGill. An experiment in graphical perception. *International Journal of Man-Machine Studies*, 25(5):491–501, 1986.

[125] Jean Clottes. *Chauvet Cave: The Art of Earliest Times*. University of Utah Press, 2003.

[126] Moreno I. Coco. The statistical challenge of scan-path analysis. In *Proceedings of the 2nd conference on Human System Interactions, HSI*, pages 369–372, 2009.

[127] Christopher Collins, Gerald Penn, and Sheelagh Carpendale. Bubble sets: Revealing set relations with isocontours over existing visualizations. *IEEE Transactions on Visualization and Computer Graphics*, 15(6):1009–1016, 2009.

[128] Cristina Conati, Christina Merten, Saleema Amershi, and Kasia Muldner. Using eye-tracking data for high-level user modeling in adaptive interfaces. In *Proceedings of the Twenty-Second AAAI Conference on Artificial Intelligence*, pages 1614–1617. AAAI Press, 2007.

[129] Gregorio Convertino, Jian Chen, Beth Yost, Young-Sam Ryu, and Chris North. Exploring context switching and cognition in dual-view

coordinated visualizations. *Proceedings International Conference on Coordinated and Multiple Views in Exploratory Visualization - CMV 2003 -*, pages 55–62, 2003.

[130] Edward Tyas Cook. *The Life of Florence Nightingale: 1820-1861*. Macmillan, University of Wisconsin - Madison, 1 edition, 1914.

[131] Maia B. Cook and Harvey S. Smallman. Human factors of the confirmation bias in intelligence analysis: Decision support from graphical evidence landscapes. *Human Factors*, 50(5):745–754, 2008.

[132] Thomas H. Cormen, Charles E. Leiserson, and Ronald L. Rivest. *Introduction to Algorithms*. The MIT Press and McGraw-Hill Book Company, 1989.

[133] Marcella Cornia, Lorenzo Baraldi, Giuseppe Serra, and Rita Cucchiara. Predicting human eye fixations via an lstm-based saliency attentive model. *IEEE Transactions on Image Processing*, 27(10):5142–5154, 2018.

[134] Alberto Corvò, Marc A. van Driel, and Michel A. Westenberg. Pathova: A visual analytics tool for pathology diagnosis and reporting. In *Proceedings of IEEE Workshop on Visual Analytics in Healthcare, VAHC*, pages 77–83. IEEE, 2017.

[135] Alberto Corvò, Michel A. Westenberg, Reinhold Wimberger-Friedl, Stephan Fromme, Michel M. R. Peeters, Marc A. van Driel, and Jarke J. van Wijk. Visual analytics in digital pathology: Challenges and opportunities. In Barbora Kozlíková and Renata Georgia Raidou, editors, *Proceedings of the Eurographics Workshop on Visual Computing for Biology and Medicine, VCBM*, pages 129–143. Eurographics Association, 2019.

[136] Marshall Couch. Effects of curved lines on force-directed graphs. Undergraduate Honors Theses. Paper 64, 2013.

[137] Chris Creed, Maite Frutos Pascual, and Ian Williams. Multimodal gaze interaction for creative design. In Regina Bernhaupt, Florian 'Floyd' Mueller, David Verweij, Josh Andres, Joanna McGrenere, Andy Cockburn, Ignacio Avellino, Alix Goguey, Pernille Bjøn, Shengdong Zhao, Briane Paul Samson, and Rafal Kocielnik, editors, *Proceedings of the Conference on Human Factors in Computing Systems, CHI*, pages 1–13. ACM, 2020.

[138] Eric C. Crowe and N. Hari Narayanan. Comparing interfaces based on what users watch and do. In Andrew T. Duchowski, editor, *Proceedings of the Eye Tracking Research & Application Symposium, ETRA*, pages 29–36. ACM, 2000.

[139] Aaron Crug and Rochel Gelman. Counting is a piece of cake: Yet another counting task. In Laura A. Carlson, Christoph Hölscher, and Thomas F. Shipley, editors, *Proceedings of the 33th Annual Meeting of the Cognitive Science Society, CogSci*. cognitivesciencesociety.org, 2011.

[140] James E. Cutting. How the eye measures reality and virtual reality. *Behavior Research Methods, Instruments, and Computers*, 29:27–36, 1997.

[141] Filip Dabek and Jesus J. Caban. A grammar-based approach for modeling user interactions and generating suggestions during the data exploration process. *IEEE Transactions on Visualization and Computer Graphics*, 23(1):41–50, 2017.

[142] Robertas Damasevicius, Jevgenijus Toldinas, Algimantas Venckauskas, Sarunas Grigaliunas, Nerijus Morkevicius, and Vaidas Jukavicius. Visual analytics for cyber security domain: State-of-the-art and challenges. In Robertas Damasevicius and Giedre Vasiljeviene, editors, *Proceedings of 25th International Conference on Information and Software Technologies, ICIST*, volume 1078 of *Communications in Computer and Information Science*, pages 256–270. Springer, 2019.

[143] Subhrajit Das, Tom McEwan, and Donna Douglas. Using eye-tracking to evaluate label alignment in online forms. In Agneta Gulz, Charlotte Magnusson, Lone Malmborg, Håkan Eftring, Bodil Jönsson, and Konrad Tollmar, editors, *Proceedings of the 5th Nordic Conference on Human-Computer Interaction*, volume 358 of *ACM International Conference Proceeding Series*, pages 451–454. ACM, 2008.

[144] Alan Davies, Markel Vigo, Simon Harper, and Caroline Jay. The visualisation of eye-tracking scanpaths: what can they tell us about how clinicians view electrocardiograms? In Michael Burch, Lewis L. Chuang, and Andrew T. Duchowski, editors, *Proceedings of IEEE Workshop on Eye Tracking and Visualization, ETVIS*, pages 79–83. IEEE Computer Society, 2016.

[145] Marcelo de Paiva Guimarães, Diego Roberto Colombo Dias, José Hamilton Mota, Bruno Barberi Gnecco, Vinicius Humberto Serapilha Durelli, and Luis Carlos Trevelin. Immersive and interactive virtual reality applications based on 3d web browsers. *Multimedia Tools and Applications*, 77(1):347–361, 2018.

[146] Thomas A. DeFanti, Maxine D. Brown, and Bruce H. McCormick. Visualization: Expanding scientific and engineering research opportunities. *Computer*, 22(8):12–25, 1989.

[147] Philippe Dessus, Olivier Cosnefroy, and Vanda Luengo. "keep your eyes on 'em all!": A mobile eye-tracking analysis of teachers' sensitivity to students. In Katrien Verbert, Mike Sharples, and Tomaz Klobucar, editors, *Proceedings of the 11th European Conference on Technology Enhanced Learning, Adaptive and Adaptable Learning , EC-TEL*, volume 9891 of *Lecture Notes in Computer Science*, pages 72–84. Springer, 2016.

[148] Richard Dewhurst, Marcus Nyström, Halszka Jarodzka, Tom Foulsham, Roger Johansson, and Kenneth Holmqvist. It depends on how you look at it: Scanpath comparison in multiple dimensions with MultiMatch, a vector-based approach. *Behavior Research Methods*, 44:1079 – 1100, 2012.

[149] Stephan Diehl. *Software Visualization - Visualizing the Structure, Behaviour, and Evolution of Software*. Springer, 2007.

[150] Stephan Diehl, Fabian Beck, and Michael Burch. Uncovering strengths and weaknesses of radial visualizations—an empirical approach. *IEEE Transactions on Visualization and Computer Graphics*, 16(6):935–942, 2010.

[151] Stephan Diehl and Carsten Görg. Graphs, they are changing. In Stephen G. Kobourov and Michael T. Goodrich, editors, *Proceedings of the 10th International Symposium on Graph Drawing, GD*, volume 2528 of *Lecture Notes in Computer Science*, pages 23–30. Springer, 2002.

[152] Evanthia Dimara and Charles Perin. What is interaction for data visualization? *IEEE Transactions on Visualization and Computer Graphics*, 26(1):119–129, 2020.

[153] Dimitra Dimitrakopoulou, Evanthia Faliagka, and Maria Rigou. An eye-tracking study of web search interaction design patterns. In *Proceedings of the 20th Pan-Hellenic Conference on Informatics*, page 22. ACM, 2016.

[154] Alan J. Dix. Designing for appropriation. In Thomas C. Ormerod and Corina Sas, editors, *Proceedings of the 21st British HCI Group Annual Conference on HCI 2007: HCI...but not as we know it - Volume 2, BCS HCI*, pages 27–30. BCS, 2007.

[155] Timothy D. Dixon, Jian Li, Jan M. Noyes, Tom Troscianko, Stavri G. Nikolov, John J. Lewis, Eduardo Fernández Canga, David R. Bull, and Cedric Nishan Canagarajah. Scanpath analysis of fused multi-sensor images with luminance change: A pilot study. In *Proceedings of 9th*

International Conference on Information Fusion, FUSION, pages 1–8. IEEE, 2006.

[156] Soussan Djamasbi, Marisa Siegel, and Tom Tullis. Visual hierarchy and viewing behavior: An eye tracking study. In Julie A. Jacko, editor, *Proceedings of International Conference on Human-Computer Interaction, HCI*, volume 6761 of *Lecture Notes in Computer Science*, pages 331–340. Springer, 2011.

[157] Weihua Dong, Shengkai Wang, Yizhou Chen, and Liqiu Meng. Using eye tracking to evaluate the usability of flow maps. *ISPRS International Journal of Geo Information*, 7(7):281, 2018.

[158] Michael Dorr, Thomas Martinetz, Karl R. Gegenfurtner, and Erhardt Barth. Variability of eye movements when viewing dynamic natural scenes. *Journal of Vision*, 10(10):1–17, 2010.

[159] Wenwen Dou, Dong Hyun Jeong, Felesia Stukes, William Ribarsky, Heather Richter Lipford, and Remco Chang. Recovering reasoning processes from user interactions. *IEEE Computer Graphics and Applications*, 29(3):52–61, 2009.

[160] Aaron Duane and Cathal Gurrin. Pilot study to investigate feasibility of visual lifelog exploration in virtual reality. In Cathal Gurrin, Xavier Giró-i-Nieto, Petia Radeva, Duc-Tien Dang-Nguyen, Mariella Dimiccoli, and Hideo Joho, editors, *Proceedings of the 2nd Workshop on Lifelogging Tools and Applications, LTA@MM*, pages 29–32. ACM, 2017.

[161] Andrew T. Duchowski. *Eye Tracking Methodology - Theory and Practice, Third Edition*. Springer, 2017.

[162] Andrew T. Duchowski, Jason Driver, Sheriff Jolaoso, William Tan, Beverly N. Ramey, and Ami Robbins. Scanpath comparison revisited. In Carlos Hitoshi Morimoto, Howell O. Istance, Aulikki Hyrskykari, and Qiang Ji, editors, *Proceedings of the 2010 Symposium on Eye-Tracking Research & Applications, ETRA*, pages 219–226. ACM, 2010.

[163] Andrew T. Duchowski, Donald H. House, Jordan Gestring, Robert Congdon, Lech Swirski, Neil A. Dodgson, Krzysztof Krejtz, and Izabela Krejtz. Comparing estimated gaze depth in virtual and physical environments. In Pernilla Qvarfordt and Dan Witzner Hansen, editors, *Proceedings of the Symposium on Eye Tracking Research and Applications, ETRA*, pages 103–110. ACM, 2014.

[164] Andrew T. Duchowski and Bruce H. McCormick. Gaze-contingent video resolution degradation. In Bernice E. Rogowitz and

Thrasyvoulos N. Pappas, editors, *Proceedings of Human Vision and Electronic Imaging III*, volume 3299 of *SPIE Proceedings*, pages 318–329. SPIE, 1998.

[165] Andrew T. Duchowski, Eric Medlin, Nathan Cournia, Anand K. Gramopadhye, Brian J. Melloy, and Santosh Nair. 3d eye movement analysis for VR visual inspection training. In Andrew T. Duchowski, Roel Vertegaal, and John W. Senders, editors, *Proceedings of the Eye Tracking Research & Application Symposium, ETRA*, pages 103–110. ACM, 2002.

[166] Andrew T. Duchowski, Brandon Pelfrey, Donald H. House, and Rui I. Wang. Measuring gaze depth with an eye tracker during stereoscopic display. In Rachel McDonnell, Simon J. Thorpe, Stephen N. Spencer, Diego Gutierrez, and Martin A. Giese, editors, *Proceedings of the 8th Symposium on Applied Perception in Graphics and Visualization, APGV*, pages 15–22. ACM, 2011.

[167] Andrew T. Duchowski, Margaux M. Price, Miriah D. Meyer, and Pilar Orero. Aggregate gaze visualization with real-time heatmaps. In Carlos Hitoshi Morimoto, Howell O. Istance, Stephen N. Spencer, Jeffrey B. Mulligan, and Pernilla Qvarfordt, editors, *Proceedings of the Symposium on Eye-Tracking Research and Applications, ETRA*, pages 13–20. ACM, 2012.

[168] Tim Dwyer, Kim Marriott, Tobias Isenberg, Karsten Klein, Nathalie Henry Riche, Falk Schreiber, Wolfgang Stuerzlinger, and Bruce H. Thomas. Immersive analytics: An introduction. In Kim Marriott, Falk Schreiber, Tim Dwyer, Karsten Klein, Nathalie Henry Riche, Takayuki Itoh, Wolfgang Stuerzlinger, and Bruce H. Thomas, editors, *Immersive Analytics*, volume 11190 of *Lecture Notes in Computer Science*, pages 1–23. Springer, 2018.

[169] Yuka Egusa, Masao Takaku, Hitoshi Terai, Hitomi Saito, Noriko Kando, and Makiko Miwa. Visualization of user eye movements for search result pages. In Tetsuya Sakai, Mark Sanderson, and Noriko Kando, editors, *Proceedings of the 2nd International Workshop on Evaluating Information Access, EVIA*. National Institute of Informatics (NII), 2008.

[170] Abdelaziz Elfadaly, Mohamed A. R. Abouarab, Radwa R. M. El Shabrawy, Wael Mostafa, Penelope Wilson, Christophe Morhange, Jay Silverstein, and Rosa Lasaponara. Discovering potential settlement areas around archaeological tells using the integration between historic

topographic maps, optical, and radar data in the northern nile delta, egypt. *Remote Sensing*, 11(24):3039, 2019.

[171] Geoffrey P. Ellis and Alan J. Dix. An explorative analysis of user evaluation studies in information visualisation. In Enrico Bertini, Catherine Plaisant, and Giuseppe Santucci, editors, *Proceedings of the AVI Workshop on BEyond time and errors: novel evaluation methods for information visualization, BELIV*, pages 1–7. ACM Press, 2006.

[172] Niklas Elmqvist, Pierre Dragicevic, and Jean-Daniel Fekete. Rolling the dice: Multidimensional visual exploration using scatterplot matrix navigation. *IEEE Transactions on Visualization and Computer Graphics*, 14(6):1539–1148, 2008.

[173] Alex Endert, Patrick Fiaux, and Chris North. Semantic interaction for sensemaking: Inferring analytical reasoning for model steering. *IEEE Transactions on Visualization and Computer Graphics*, 18(12):2879–2888, 2012.

[174] Alex Endert, Mahmud Shahriar Hossain, Naren Ramakrishnan, Chris North, Patrick Fiaux, and Christopher Andrews. The human is the loop: new directions for visual analytics. *Journal of Intelligent Information Systems*, 43(3):411–435, 2014.

[175] Sukru Eraslan, Yeliz Yesilada, and Simon Harper. Scanpath trend analysis on web pages: Clustering eye tracking scanpaths. *ACM Transactions on the Web*, 10(4):20:1–20:35, 2016.

[176] Mateus Espadoto, Rafael Messias Martins, Andreas Kerren, Nina S. T. Hirata, and Alexandru C. Telea. Toward a quantitative survey of dimension reduction techniques. *IEEE Transactions on Visualization and Computer Graphics*, 27(3):2153–2173, 2021.

[177] Augusto Esteves, Eduardo Velloso, Andreas Bulling, and Hans Gellersen. Orbits: Gaze interaction for smart watches using smooth pursuit eye movements. In Celine Latulipe, Bjoern Hartmann, and Tovi Grossman, editors, *Proceedings of the 28th Annual ACM Symposium on User Interface Software & Technology, UIST*, pages 457–466. ACM, 2015.

[178] Ronak Etemadpour, Bettina Olk, and Lars Linsen. Eye-tracking investigation during visual analysis of projected multidimensional data with 2d scatterplots. In Robert S. Laramee, Andreas Kerren, and José Braz, editors, *Proceedings of the 5th International Conference on Information Visualization Theory and Applications, IVAPP*, pages 233–246. SciTePress, 2014.

[179] Leonard Euler. Solutio problematis ad geometriam situs pertinentis. *Commentarii Academiae Scientiarum Petropolitanae*, 8:128–140, 1741.

[180] Adil Fahad, Najlaa Alshatri, Zahir Tari, Abdullah Alamri, Ibrahim Khalil, Albert Y. Zomaya, Sebti Foufou, and Abdelaziz Bouras. A survey of clustering algorithms for big data: Taxonomy and empirical analysis. *IEEE Transactions on Emerging Topics in Computing*, 2(3):267–279, 2014.

[181] Paolo Federico and Silvia Miksch. Evaluation of two interaction techniques for visualization of dynamic graphs. In Yifan Hu and Martin Nöllenburg, editors, *Proceedings of 24th International Symposium on Graph Drawing and Network Visualization, GD*, volume 9801 of *Lecture Notes in Computer Science*, pages 557–571. Springer, 2016.

[182] Jean-Daniel Fekete, Jarke J. van Wijk, John T. Stasko, and Chris North. The value of information visualization. In Andreas Kerren, John T. Stasko, Jean-Daniel Fekete, and Chris North, editors, *Information Visualization - Human-Centered Issues and Perspectives*, volume 4950 of *Lecture Notes in Computer Science*, pages 1–18. Springer, 2008.

[183] Ronald A. Fisher. On the "probable error" of a coefficient of correlation deduced from a small sample. *Metron*, 1:3–32, 1921.

[184] Paul M. Fitts, Richard E. Jones, and John L. Milton. Eye movement of aircraft pilots during instrument-landing approaches. *Ergonomics: Psychological Mechanisms and Models in Ergonomics*, 3:56–66, 2005.

[185] Carla M. D. S. Freitas, Paulo Roberto Gomes Luzzardi, Ricardo Andrade Cava, Marco Winckler, Marcelo Soares Pimenta, and Luciana P. Nedel. On evaluating information visualization techniques. In Maria De Marsico, Stefano Levialdi, and Emanuele Panizzi, editors, *Proceedings of the Working Conference on Advanced Visual Interfaces, AVI*, pages 373–374. ACM, 2002.

[186] Johann Christoph Freytag, Raghu Ramakrishnan, and Rakesh Agrawal. Data mining: The next generation. *it - Information Technology*, 47(5):308–312, 2005.

[187] Milton Friedman. The use of ranks to avoid the assumption of normality implicit in the analysis of variance. *Journal of the American Statistical Association*, 200(32):675–701, 1937.

[188] Michael Friendly. Milestones in the history of data visualization: A case study in statistical historiography. In Claus Weihs and Wolfgang Gaul, editors, *Proceedings of the Annual Conference on Classification*

- *the Ubiquitous Challenge*, Studies in Classification, Data Analysis, and Knowledge Organization, pages 34–52. Springer, 2004.

[189] Michael Friendly and Dan Michael. The early origins and development of the scatterplot. *Journal of the History of the Behavioral Sciences*, 2(41):103–130, 2005.

[190] Yaniv Frishman and Ayellet Tal. Dynamic drawing of clustered graphs. In Matthew O. Ward and Tamara Munzner, editors, *Proceedings of 10th IEEE Symposium on Information Visualization (InfoVis)*, pages 191–198. IEEE Computer Society, 2004.

[191] Bernd Fröhlich, Jan Hochstrate, Alexander Kulik, and Anke Huckauf. On 3d input devices. *IEEE Computer Graphics and Applications*, 26(2):15–19, 2006.

[192] Bo Fu, Natalya Fridman Noy, and Margaret-Anne D. Storey. Indented tree or graph? A usability study of ontology visualization techniques in the context of class mapping evaluation. In Harith Alani, Lalana Kagal, Achille Fokoue, Paul Groth, Chris Biemann, Josiane Xavier Parreira, Lora Aroyo, Natasha F. Noy, Chris Welty, and Krzysztof Janowicz, editors, *Proceedings of the 12th International Semantic Web Conference, ISWC*, volume 8218 of *Lecture Notes in Computer Science*, pages 117–134. Springer, 2013.

[193] Bo Fu, Natalya Fridman Noy, and Margaret-Anne D. Storey. Eye tracking the user experience - an evaluation of ontology visualization techniques. *Semantic Web*, 8(1):23–41, 2017.

[194] Ujwal Gadiraju, Sebastian Möller, Martin Nöllenburg, Dietmar Saupe, Sebastian Egger-Lampl, Daniel W. Archambault, and Brian D. Fisher. Crowdsourcing versus the laboratory: Towards human-centered experiments using the crowd. In Daniel W. Archambault, Helen C. Purchase, and Tobias Hoßfeld, editors, *Evaluation in the Crowd. Crowdsourcing and Human-Centered Experiments - Dagstuhl Seminar 15481, Dagstuhl Castle, Germany*, volume 10264 of *Lecture Notes in Computer Science*, pages 6–26. Springer, 2015.

[195] Michael R. Garey and David S. Johnson. *Computers and Intractability: A Guide to the Theory of NP-Completeness*. W. H. Freeman, 1979.

[196] Christine Garhart and Vasudevan Lakshminarayanan. Anatomy of the eye. In Janglin Chen, Wayne Cranton, and Mark Fihn, editors, *Handbook of Visual Display Technology*, pages 73–83. Springer, 2012.

[197] Péter Gáspár, Michal Kompan, Jakub Simko, and Mária Bieliková. Analysis of user behavior in interfaces with recommended items: An eye-tracking study. In Peter Brusilovsky, Marco de Gemmis, Alexander

Felfernig, Pasquale Lops, John O'Donovan, Giovanni Semeraro, and Martijn C. Willemsen, editors, *Proceedings of the 5th Joint Workshop on Interfaces and Human Decision Making for Recommender Systems, IntRS*, volume 2225 of *CEUR Workshop Proceedings*, pages 32–36. CEUR-WS.org, 2018.

[198] Alexander Gee, Min Yu, and Georges Grinstein. Dynamic and interactive dimensional anchors for spring-based visualizations. Technical report, University of Massachussetts Lowell, 2005.

[199] Andreas Gegenfurtner and Marko Seppänen. Transfer of expertise: An eye tracking and think aloud study using dynamic medical visualizations. *Computers & Education*, 63:393–403, 2013.

[200] Mohammad Ghoniem, Jean-Daniel Fekete, and Philippe Castagliola. On the readability of graphs using node-link and matrix-based representations: a controlled experiment and statistical analysis. *Information Visualization*, 4(2):114–135, 2005.

[201] Sophie Giesa, Manuel Heinzig, Robert Manthey, Christian Roschke, Rico Thomanek, and Marc Ritter. An exploratory study on the perception of optical illusions in real world and virtual environments. In Don Harris and Wen-Chin Li, editors, *Proceedings of Engineering Psychology and Cognitive Ergonomics. Mental Workload, Human Physiology, and Human Energy - 17th International Conference, EPCE*, volume 12186 of *Lecture Notes in Computer Science*, pages 161–170. Springer, 2020.

[202] Joseph H. Goldberg and Jonathan I. Helfman. Scanpath clustering and aggregation. In Carlos Hitoshi Morimoto, Howell O. Istance, Aulikki Hyrskykari, and Qiang Ji, editors, *Proceedings of the Symposium on Eye-Tracking Research & Applications, ETRA*, pages 227–234. ACM, 2010.

[203] Joseph H. Goldberg and Jonathan I. Helfman. Visual scanpath representation. In *Proceedings of the Symposium on Eye-Tracking Research & Applications, ETRA*, pages 203–210, 2010.

[204] Joseph H. Goldberg and Jonathan I. Helfman. Eye tracking for visualization evaluation: Reading values on linear versus radial graphs. *Information Visualization*, 10(3):182–195, 2011.

[205] Joseph H. Goldberg and Xerxes P. Kotval. Computer interface evaluation using eye movements: methods and constructs. *International Journal of Industrial Ergonomics*, 24(6):631–645, 1999.

[206] Robert B. Goldstein, Russell L. Woods, and Eli Peli. Where people look when watching movies: Do all viewers look at the same place? *Computers in Biology and Medicine*, 37(7):957–964, 2007.

[207] David Gotz, Shun Sun, and Nan Cao. Adaptive contextualization: Combating bias during high-dimensional visualization and data selection. In Jeffrey Nichols, Jalal Mahmud, John O'Donovan, Cristina Conati, and Massimo Zancanaro, editors, *Proceedings of the 21st International Conference on Intelligent User Interfaces, IUI*, pages 85–95. ACM, 2016.

[208] Natalia Grabar, Emmanuel Farce, and Laurent Sparrow. Study of readability of health documents with eye-tracking methods. In Pascale Sébillot and Vincent Claveau, editors, *Actes de la Conférence, CORIA-TALN-RJC*, pages 3–18. ATALA, 2018.

[209] Thomas Grindinger, Andrew T. Duchowski, and Michael W. Sawyer. Group-wise similarity and classification of aggregate scanpaths. In Carlos Hitoshi Morimoto, Howell O. Istance, Aulikki Hyrskykari, and Qiang Ji, editors, *Proceedings of the Symposium on Eye-Tracking Research & Applications, ETRA*, pages 101–104. ACM, 2010.

[210] Martin Groen and Jan Noyes. Using eye tracking to evaluate usability of user interfaces: Is it warranted? In Frédéric Vanderhaegen, editor, *Proceedings of 11th IFAC/IFIP/IFORS/IEA Symposium on Analysis, Design, and Evaluation of Human-Machine Systems*, pages 489–493. International Federation of Automatic Control, 2010.

[211] Yi Gu, Chaoli Wang, Robert Bixler, and Sidney K. D'Mello. Etgraph: A graph-based approach for visual analytics of eye-tracking data. *Computers & Graphics*, 62:1–14, 2017.

[212] Drew T. Guarnera. Enhancing eye tracking of source code: A specialized fixation filter for source code. In *Proceedings of the IEEE International Conference on Software Maintenance and Evolution, ICSME*, pages 615–618. IEEE, 2019.

[213] Egon Guba, Willavene Wolf, Sybil de Groot, Manfred Knemeyer, Ralph van Atta, and Larry Light. Eye movements and tv viewing in children. *AV Communication Review*, 12(4):386 – 401, 1964.

[214] Wei Guo and Shiwei Cheng. An approach to reading assistance with eye tracking data and text features. In *Proceedings of Adjunct of the 2019 International Conference on Multimodal Interaction, ICMI*, pages 7:1–7:7. ACM, 2019.

[215] Jacek Gwizdka and Michael J. Cole. Does interactive search results overview help?: an eye tracking study. In Wendy E. Mackay,

Stephen A. Brewster, and Susanne Bødker, editors, *Proceedings of ACM SIGCHI Conference on Human Factors in Computing Systems, CHI*, pages 1869–1874. ACM, 2013.

[216] Kunhee Ha, Il-Hyun Jo, Sohye Lim, and Yeonjeong Park. Tracking students' eye-movements on visual dashboard presenting their online learning behavior patterns. In Guang Chen, Vive Kumar, Kinshuk, Ronghuai Huang, and Siu Cheung Kong, editors, *Proceedings of International Conference on Smart Learning Environments, ICSLE*, Lecture Notes in Educational Technology, pages 371–376. Springer, 2014.

[217] Katarzyna Harezlak and Pawel Kasprowski. Application of eye tracking in medicine: A survey, research issues and challenges. *Computerized Medical Imaging and Graphics*, 65:176–190, 2018.

[218] Susan Havre, Elizabeth G. Hetzler, Paul Whitney, and Lucy T. Nowell. Themeriver: Visualizing thematic changes in large document collections. *IEEE Transactions on Visualization and Computer Graphics*, 8(1):9–20, 2002.

[219] Christopher G. Healey and James T. Enns. Attention and visual memory in visualization and computer graphics. *IEEE Transactions on Visualization and Computer Graphics*, 18(7):1170–1188, 2012.

[220] Marti A. Hearst and Daniela Karin Rosner. Tag clouds: Data analysis tool or social signaller? In *Proceedings of 41st Hawaii International Conference on Systems Science HICSS*, page 160. IEEE Computer Society, 2008.

[221] Jeffrey Heer and Michael Bostock. Crowdsourcing graphical perception: using mechanical turk to assess visualization design. In Elizabeth D. Mynatt, Don Schoner, Geraldine Fitzpatrick, Scott E. Hudson, W. Keith Edwards, and Tom Rodden, editors, *Proceedings of the 28th International Conference on Human Factors in Computing Systems, CHI*, pages 203–212. ACM, 2010.

[222] Birte Heinemann, Matthias Ehlenz, and Ulrik Schroeder. Eye-tracking in educational multi-touch games: Design-based (interaction) research and great visions. In Andreas Bulling, Anke Huckauf, Eakta Jain, Ralph Radach, and Daniel Weiskopf, editors, *Proceedings of the Symposium on Eye Tracking Research and Applications, Short Papers*, pages 58:1–58:5. ACM, 2020.

[223] Julian Heinrich. *Visualization techniques for parallel coordinates*. PhD thesis, University of Stuttgart, 2013.

[224] Julian Heinrich and Daniel Weiskopf. State of the art of parallel coordinates. In Mateu Sbert and László Szirmay-Kalos, editors, *Proceedings of 34th Annual Conference of the European Association for Computer Graphics, Eurographics - State of the Art Reports*, pages 95–116. Eurographics Association, 2013.

[225] Helene Hembrooke, Matthew K. Feusner, and Geri Gay. Averaging scan patterns and what they can tell us. In Kari-Jouko Räihä and Andrew T. Duchowski, editors, *Proceedings of the Eye Tracking Research & Application Symposium, ETRA*, page 41. ACM, 2006.

[226] Bradley M. Hemminger, Trish Losi, and Anne Bauers. Survey of bioinformatics programs in the united states. *Journal of the American Society for Information Science and Technology (JASIST)*, 56(5):529–537, 2005.

[227] Nathalie Henry, Jean-Daniel Fekete, and Michael J. McGuffin. Nodetrix: a hybrid visualization of social networks. *IEEE Transactions on Visualization and Computer Graphics*, 13(6):1302–1309, 2007.

[228] Elad Hirsch and Ayellet Tal. Color visual illusions: A statistics-based computational model. In Hugo Larochelle, Marc'Aurelio Ranzato, Raia Hadsell, Maria-Florina Balcan, and Hsuan-Tien Lin, editors, *Proceedings of Advances in Neural Information Processing Systems 33: Annual Conference on Neural Information Processing Systems, NeurIPS*, 2020.

[229] Marcel Hlawatsch, Michael Burch, and Daniel Weiskopf. Visual adjacency lists for dynamic graphs. *IEEE Transactions on Visualization and Computer Graphics*, 20(11):1590–1603, 2014.

[230] Marcel Hlawatsch, Filip Sadlo, Michael Burch, and Daniel Weiskopf. Scale-stack bar charts. *Computer Graphics Forum*, 32(3):181–190, 2013.

[231] Hsin Yang Ho, I-Cheng Yeh, Yu-Chi Lai, Wen-Chieh Lin, and Fu-Yin Cherng. Evaluating 2d flow visualization using eye tracking. *Computer Graphics Forum*, 34(3):501–510, 2015.

[232] Benjamin Höferlin. *Scalable Visual Analytics for Video Surveillance*. PhD thesis, University of Stuttgart, 2014.

[233] Benjamin Höferlin, Markus Höferlin, Gunther Heidemann, and Daniel Weiskopf. Scalable video visual analytics. *Information Visualization*, 14(1):10–26, 2015.

[234] Benjamin Höferlin, Rudolf Netzel, Markus Höferlin, Daniel Weiskopf, and Gunther Heidemann. Inter-active learning of ad-hoc classifiers for video visual analytics. In *Proceedings of 7th IEEE Conference on*

Visual Analytics Science and Technology, IEEE VAST, pages 23–32. IEEE Computer Society, 2012.

[235] Kenneth Holmqvist. *Eye tracking: a comprehensive guide to methods and measures*. Oxford University Press, 2011.

[236] Danny Holten. Hierarchical edge bundles: Visualization of adjacency relations in hierarchical data. *IEEE Transactions on Visualization and Computer Graphics*, 12(5):741–748, 2006.

[237] Danny Holten, Petra Isenberg, Jarke J. van Wijk, and Jean-Daniel Fekete. An extended evaluation of the readability of tapered, animated, and textured directed-edge representations in node-link graphs. In Giuseppe Di Battista, Jean-Daniel Fekete, and Huamin Qu, editors, *Proceedings of the IEEE Pacific Visualization Symposium, PacificVis*, pages 195–202. IEEE Computer Society, 2011.

[238] Danny Holten, Petra Isenberg, Jarke J. van Wijk, and Jean-Daniel Fekete. An extended evaluation of the readability of tapered, animated, and textured directed-edge representations in node-link graphs. In Giuseppe Di Battista, Jean-Daniel Fekete, and Huamin Qu, editors, *Proceedings of the IEEE Pacific Visualization Symposium, PacificVis*, pages 195–202. IEEE Computer Society, 2011.

[239] Danny Holten and Jarke J. van Wijk. Force-directed edge bundling for graph visualization. *Computer Graphics Forum*, 28(3):983–990, 2009.

[240] Anthony Hornof and Tim Halverson. Cleaning up systematic error in eye-tracking data by using required fixation locations. *Behavior Research Methods, Instruments, and Computers*, 34:592–604, 2002.

[241] Harold Hotelling. Analysis of a complex of statistical variables into principal components. *Journal of Educational Psychology*, 24:417–441, 1933.

[242] Ya Ting Hu, Michael Burch, and Huub van de Wetering. Visualizing dynamic graphs with heat triangles. In Michael Burch, Michel A. Westenberg, Quang Vinh Nguyen, and Ying Zhao, editors, *Proceedings of the 13th International Symposium on Visual Information Communication and Interaction, VINCI*, pages 7:1–7:8. ACM, 2020.

[243] Yifan Hu, Emden R. Gansner, and Stephen G. Kobourov. Visualizing graphs and clusters as maps. *IEEE Computer Graphics and Applications*, 30(6):54–66, 2010.

[244] Yiling Hu, Bian Wu, and Xiaoqing Gu. An eye tracking study of high- and low-performing students in solving interactive and analytical

problems. *Journal on Educational Technology and Society*, 20(4):300–311, 2017.

[245] Weidong Huang. An eye tracking study into the effects of graph layout. *CoRR*, abs/0810.4431, 2008.

[246] Weidong Huang. Window to the soul: Tracking eyes to inform the design of visualizations. In *Proceedings of Ninth International Conference on Computer Graphics, Imaging and Visualization, CGIV*, page 61. IEEE Computer Society, 2012.

[247] Weidong Huang, Seok-Hee Hong, and Peter Eades. Effects of sociogram drawing conventions and edge crossings in social network visualization. *Journal on Graph Algorithms and Applications*, 11(2):397–429, 2007.

[248] Weidong Huang, Seok-Hee Hong, and Peter Eades. Effects of crossing angles. In *Proceedings of IEEE VGTC Pacific Visualization Symposium, PacificVis*, pages 41–46. IEEE Computer Society, 2008.

[249] Xun Huang, Chengyao Shen, Xavier Boix, and Qi Zhao. SALICON: reducing the semantic gap in saliency prediction by adapting deep neural networks. In *Proceedings of the IEEE International Conference on Computer Vision, ICCV*, pages 262–270. IEEE Computer Society, 2015.

[250] Edmund Burke Huey. *The psychology and pedagogy of reading. With a review of the history of reading and writing and of methods, texts, and hygiene in reading*. MIT Psychology, Band 86, MIT Press, Cambridge MA, 1968.

[251] Christophe Hurter, Ozan Ersoy, Sara Irina Fabrikant, Tijmen R. Klein, and Alexandru C. Telea. Bundled visualization of dynamicgraph and trail data. *IEEE Transactions on Visualization and Computer Graphics*, 20(8):1141–1157, 2014.

[252] Stephanie Hüttermann, Benjamin Noël, and Daniel Memmert. Eye tracking in high-performance sports: Evaluation of its application in expert athletes. *International Journal of Computer Science in Sport*, 17(2):182–203, 2018.

[253] Keiichiro Inagaki, Takayuki Kannon, Yoshimi Kamiyama, and Shiro Usui. Effect of fixational eye movement on signal processing of retinal photoreceptor: A computational study. *IEICE Transactions on Information and Systems*, 103-D(7):1753–1759, 2020.

[254] Ohad Inbar, Noam Tractinsky, and Joachim Meyer. Minimalism in information visualization: attitudes towards maximizing the data-ink ratio. In Willem-Paul Brinkman, Dong-Han Ham, and B. L. William

Wong, editors, *Proceedings of the 14th European Conference on Cognitive Ergonomics: invent! explore!, ECCE*, volume 250 of *ACM International Conference Proceeding Series*, pages 185–188. ACM, 2007.

[255] Alfred Inselberg and Bernard Dimsdale. Parallel coordinates: A tool for visualizing multi-dimensional geometry. In Arie E. Kaufman, editor, *Proceedings of 1st IEEE Visualization Conference, IEEE Vis*, pages 361–378. IEEE Computer Society Press, 1990.

[256] Petra Isenberg and Danyel Fisher. Collaborative brushing and linking for co-located visual analytics of document collections. *Computer Graphics Forum*, 28(3):1031–1038, 2009.

[257] Tobias Isenberg, Petra Isenberg, Jian Chen, Michael Sedlmair, and Torsten Möller. A systematic review on the practice of evaluating visualization. *IEEE Transactions on Visualization and Computer Graphics*, 19(12):2818–2827, 2013.

[258] Shinobu Ishihara. *Tests for color-blindness*. Handaya, Tokyo, Hongo Harukicho, 1917.

[259] Kirill V. Istomin, Jaroslava Panáková, and Patrick Heady. Culture, perception, and artistic visualization: A comparative study of children's drawings in three siberian cultural groups. *Cognitive Science*, 38(1):76–100, 2014.

[260] Robert J. K. Jacob. What you look at is what you get: eye movement-based interaction techniques. In *Proceedings of the SIGCHI conference on Human factors in computing systems*, pages 11–18, 1990.

[261] Robert J.K. Jacob and Keith S. Karn. Commentary on section 4 - eye tracking in human-computer interaction and usability research: Ready to deliver the promises. In J. Hyönä, R. Radach, and H. Deubel, editors, *The Mind's Eye*, pages 573–605. North-Holland, 2003.

[262] Monique W. M. Jaspers, Thiemo Steen, Cor van den Bos, and Maud M. Geenen. The think aloud method: a guide to user interface design. *International Journal of Medical Informatics*, 73(11-12):781–795, 2004.

[263] Louis Emile Javal. Essai sur la physiologie de la lecture. In *in Annales d'ocullistique 80*, pages 61–73, 1878.

[264] Gavindya Jayawardena and Sampath Jayarathna. Automated filtering of eye gaze metrics from dynamic areas of interest. In *Proceedings of 21st International Conference on Information Reuse and Integration for Data Science, IRI*, pages 67–74. IEEE, 2020.

[265] Jimmy Johansson and Camilla Forsell. Evaluation of parallel coordinates: Overview, categorization and guidelines for future research. *IEEE Transactions on Visualization and Computer Graphics*, 22(1):579–588, 2016.

[266] Brian Johnson and Ben Shneiderman. Tree maps: A space-filling approach to the visualization of hierarchical information structures. In Gregory M. Nielson and Lawrence J. Rosenblum, editors, *Proceedings of 2nd IEEE Visualization Conference, IEEE*, pages 284–291. IEEE Computer Society Press, 1991.

[267] Patrick Jungkunz and Christian J. Darken. A computational model for human eye-movements in military simulations. *Computational & Mathematical Organization Theory*, 17(3):229–250, 2011.

[268] Marcel Adam Just and Patricia A. Carpenter. Eye fixations and cognitive processes. *Cognitive Psychology*, 8(4):441–480, 1976.

[269] Marcel Adam Just and Patricia A. Carpenter. A theory of reading: From eye fixations to comprehension. *Psychological Review*, 87(4):329–354, 1980.

[270] Rudolf Kajan, Adam Herout, Roman Bednarik, and Filip Povolný. Peeplist: Adapting ex-post interaction with pervasive display content using eye tracking. *Pervasive and Mobile Computing*, 30:71–83, 2016.

[271] Jiannan Kang, Xiaoya Han, Jiajia Song, Zikang Niu, and Xiaoli Li. The identification of children with autism spectrum disorder by SVM approach on EEG and eye-tracking data. *Computers in Biology and Medicine*, 120:103722, 2020.

[272] Xin Kang. The effect of color on short-term memory in information visualization. In Andreas Kerren and Kang Zhang, editors, *Proceedings of the 9th International Symposium on Visual Information Communication and Interaction, VINCI*, pages 144–145. ACM, 2016.

[273] Ziho Kang and Steven J. Landry. An eye movement analysis algorithm for a multielement target tracking task: Maximum transition-based agglomerative hierarchical clustering. *IEEE Transactions on Human-Machine Systems*, 45(1):13–24, 2015.

[274] Valeria Karpinskaia, Vsevolod Lyakhovetskii, Alla Cherniavskaia, and Yuri Shilov. The aftereffects of visual illusions (ponzo and müller-lyer): Hand-dependent effects in sensorimotor domain. In Tingwen Huang, Jiancheng Lv, Changyin Sun, and Alexander V. Tuzikov, editors, *Proceedings of Advances in Neural Networks - ISNN - 15th International Symposium on Neural Networks, ISNN*, volume 10878 of *Lecture Notes in Computer Science*, pages 800–806. Springer, 2018.

[275] Pawel Kasprowski, Katarzyna Harezlak, Pawel Fudalej, and Piotr Fudalej. Examining the impact of dental imperfections on scan-path patterns. In Ireneusz Czarnowski, Robert J. Howlett, and Lakhmi C. Jain, editors, *Intelligent Decision Technologies 2017*, pages 278–286. Springer International Publishing, 2018.

[276] Daniel F. Keefe and David H. Laidlaw. Virtual reality data visualization for team-based STEAM education: Tools, methods, and lessons learned. In Randall Shumaker, editor, *Proceedings of 5th International Conference on Virtual, Augmented and Mixed Reality. Systems and Applications, VAMR*, volume 8022 of *Lecture Notes in Computer Science*, pages 179–187. Springer, 2013.

[277] Daniel A. Keim. Solving problems with visual analytics: Challenges and applications. In *Proceedings of Machine Learning and Knowledge Discovery in Databases - European Conference*, pages 5–6, 2012.

[278] Daniel A. Keim, Florian Mansmann, Jörn Schneidewind, Jim Thomas, and Hartmut Ziegler. Visual analytics: Scope and challenges. In Simeon J. Simoff, Michael H. Böhlen, and Arturas Mazeika, editors, *Visual Data Mining - Theory, Techniques and Tools for Visual Analytics*, volume 4404 of *Lecture Notes in Computer Science*, pages 76–90. Springer, 2008.

[279] Daniel A. Keim, Florian Mansmann, Jörn Schneidewind, and Hartmut Ziegler. Challenges in visual data analysis. In *Proceedings of 10th International Conference on Information Visualisation, IV*, pages 9–16. IEEE Computer Society, 2006.

[280] René Keller, Claudia M. Eckert, and P. John Clarkson. Matrices or node-link diagrams: which visual representation is better for visualising connectivity models? *Information Visualization*, 5(1):62–76, 2006.

[281] Andreas Kerren and Falk Schreiber. Why integrate infovis and scivis?: An example from systems biology. *IEEE Computer Graphics and Applications*, 34(6):69–73, 2014.

[282] Andreas Kerren and John T. Stasko. Chapter 1 algorithm animation. In Stephan Diehl, editor, *Software Visualization*, pages 1–15, Berlin, Heidelberg, 2002. Springer Berlin Heidelberg.

[283] Mohamed Khamis, Axel Hoesl, Alexander Klimczak, Martin Reiss, Florian Alt, and Andreas Bulling. Eyescout: Active eye tracking for position and movement independent gaze interaction with large public displays. In Krzysztof Gajos, Jennifer Mankoff, and Chris Harrison, editors, *Proceedings of the 30th Annual ACM Symposium on User*

Interface Software and Technology, UIST, pages 155–166. ACM, 2017.

[284] Mohamed Khamis, Ludwig Trotter, Ville Mäkelä, Emanuel von Zezschwitz, Jens Le, Andreas Bulling, and Florian Alt. Cueauth: Comparing touch, mid-air gestures, and gaze for cue-based authentication on situated displays. *Proceedings of the ACM on Interactive, Mobile, Wearable and Ubiquitous Technoligies*, 2(4):174:1–174:22, 2018.

[285] M. Imran Khan and Atif Bin Mansoor. Real time eyes tracking and classification for driver fatigue detection. In Aurélio C. Campilho and Mohamed S. Kamel, editors, *Proceedings of 5th International Conference on Image Analysis and Recognition, ICIAR*, volume 5112 of *Lecture Notes in Computer Science*, pages 729–738. Springer, 2008.

[286] Hansol Kim, Kun Ha Suh, and Eui Chul Lee. Multi-modal user interface combining eye tracking and hand gesture recognition. *Journal on Multimodal User Interfaces*, 11(3):241–250, 2017.

[287] Minseok Kim, Kyeong-Beom Park, Sung Ho Choi, and Jae Yeol Lee. Inside-reachable and see-through augmented reality shell for 3d visualization and tangible interaction. *Multimedia Tools and Applications*, 79(9-10):5941–5963, 2020.

[288] Seondae Kim, Yeongil Ryu, Jinsoo Cho, and Eun-Seok Ryu. Towards tangible vision for the visually impaired through 2d multiarray braille display. *Sensors*, 19(23):5319, 2019.

[289] Jeff Klingner, Rakshit Kumar, and Pat Hanrahan. Measuring the task-evoked pupillary response with a remote eye tracker. In Kari-Jouko Räihä and Andrew T. Duchowski, editors, *Proceedings of the Eye Tracking Research & Application Symposium, ETRA*, pages 69–72. ACM, 2008.

[290] Hiromi Kobayashi, , and Shiro Kohshima. Unique morphology of the human eye and its adaptive meaning: comparative studies on external morphology of the primate eye. *Journal of Human Evolution*, 40(5):419–435, 2001.

[291] Alfred Kobsa. User experiments with tree visualization systems. In Matthew O. Ward and Tamara Munzner, editors, *Proceedings of 10th IEEE Symposium on Information Visualization (InfoVis)*, pages 9–16. IEEE Computer Society, 2004.

[292] Kurt Koffka. *Principles of Gestalt Psychology*. New York: Harcourt, Brace, 1935.

[293] Menno-Jan Kraak. The space-time cube revisited from a geovisualization perspective. In *Proceedings of the 21st International Cartographic Conference*, pages 1988 – 1995, 2003.

[294] Michael Krone, Sebastian Grottel, Guido Reina, Christoph Müller, and Thomas Ertl. 10 years of megamol: The pain and gain of creating your own visualization framework. *IEEE Computer Graphics and Applications*, 38(1):109–114, 2018.

[295] Joseph Kruskal and James Landwehr. Icicle plots: Better displays for hierarchical clustering. *The American Statistician*, 37(2):162–168, 1983.

[296] William H. Kruskal and Wilson A. Wallis. Use of ranks in one-criterion variance analysis. *Journal of the American Statistical Association*, 160(47):583–621, 1952.

[297] Thomas C. Kübler, Enkelejda Kasneci, and Wolfgang Rosenstiel. Subsmatch: scanpath similarity in dynamic scenes based on subsequence frequencies. In Pernilla Qvarfordt and Dan Witzner Hansen, editors, *Proceedings of the Symposium on Eye Tracking Research and Applications, ETRA*, pages 319–322. ACM, 2014.

[298] Ayush Kumar, Michael Burch, and Klaus Mueller. Visually comparing eye movements over space and time. In Krzysztof Krejtz and Bonita Sharif, editors, *Proceedings of the 11th ACM Symposium on Eye Tracking Research & Applications, ETRA*, pages 81:1–81:9. ACM, 2019.

[299] Ayush Kumar, Rudolf Netzel, Michael Burch, Daniel Weiskopf, and Klaus Mueller. Visual multi-metric grouping of eye-tracking data. *Journal of Eye Movement Research*, 10(5), 2018.

[300] Ayush Kumar, Neil Timmermans, Michael Burch, and Klaus Mueller. Clustered eye movement similarity matrices. In Krzysztof Krejtz and Bonita Sharif, editors, *Proceedings of the 11th ACM Symposium on Eye Tracking Research & Applications, ETRA*, pages 82:1–82:9. ACM, 2019.

[301] Shailesh Kumar, Shashwat Pathak, and Basant Kumar. Automated detection of eye related diseases using digital image processing. In Amit Kumar Singh and Anand Mohan, editors, *Handbook of Multimedia Information Security: Techniques and Applications*, pages 513–544. Springer, 2019.

[302] Matthias Kümmerer, Thomas S. A. Wallis, and Matthias Bethge. Deepgaze II: reading fixations from deep features trained on object recognition. *CoRR*, abs/1610.01563, 2016.

[303] Eugenijus Kurilovas. Evaluation of quality and personalisation of VR/AR/MR learning systems. *Behaviour & Information Technology*, 35(11):998–1007, 2016.

[304] Kuno Kurzhals, Michael Burch, Tanja Blascheck, Gennady Andrienko, Natalia Andrienko, and Daniel Weiskopf. A task-based view on the visual analysis of eye-tracking data. In Michael Burch, Lewis Chuang, Brian Fisher, Albrecht Schmidt, and Daniel Weiskopf, editors, *Eye Tracking and Visualization*, pages 3–22. Springer International Publishing, 2017.

[305] Kuno Kurzhals, Michael Burch, and Daniel Weiskopf. What we see and what we get from visualization: Eye tracking beyond gaze distributions and scanpaths. *CoRR*, abs/2009.14515, 2020.

[306] Kuno Kurzhals, Brian D. Fisher, Michael Burch, and Daniel Weiskopf. Evaluating visual analytics with eye tracking. In Heidi Lam, Petra Isenberg, Tobias Isenberg, and Michael Sedlmair, editors, *Proceedings of the Fifth Workshop on Beyond Time and Errors: Novel Evaluation Methods for Visualization, BELIV*, pages 61–69. ACM, 2014.

[307] Kuno Kurzhals, Brian D. Fisher, Michael Burch, and Daniel Weiskopf. Eye tracking evaluation of visual analytics. *Information Visualization*, 15(4):340–358, 2016.

[308] Kuno Kurzhals, Florian Heimerl, and Daniel Weiskopf. ISeeCube: visual analysis of gaze data for video. In Pernilla Qvarfordt and Dan Witzner Hansen, editors, *Proceedings of Symposium on Eye Tracking Research and Applications, ETRA*, pages 43–50. ACM, 2014.

[309] Kuno Kurzhals, Marcel Hlawatsch, Florian Heimerl, Michael Burch, Thomas Ertl, and Daniel Weiskopf. Gaze stripes: Image-based visualization of eye tracking data. *IEEE Transactions on Visualization and Computer Graphics*, 22(1):1005–1014, 2016.

[310] Kuno Kurzhals, Markus Höferlin, and Daniel Weiskopf. Evaluation of attention-guiding video visualization. *Computer Graphics Forum*, 32(3):51–60, 2013.

[311] Kuno Kurzhals and Daniel Weiskopf. Space-time visual analytics of eye-tracking data for dynamic stimuli. *IEEE Transactions on Visualization and Computer Graphics*, 19(12):2129–2138, 2013.

[312] Kuno Kurzhals and Daniel Weiskopf. Eye tracking for personal visual analytics. *IEEE Computer Graphics and Applications*, 35(4):64–72, 2015.

[313] Bum Chul Kwon, Brian D. Fisher, and Ji Soo Yi. Visual analytic roadblocks for novice investigators. In *Proceedings of 6th IEEE*

Conference on Visual Analytics Science and Technology, VAST, pages 3–11. IEEE Computer Society, 2011.

[314] Vasudevan Lakshminarayanan. Visual acuity. In Janglin Chen, Wayne Cranton, and Mark Fihn, editors, *Handbook of Visual Display Technology*, pages 93–99. Springer, 2012.

[315] Heidi Lam, Enrico Bertini, Petra Isenberg, Catherine Plaisant, and Sheelagh Carpendale. Empirical studies in information visualization: Seven scenarios. *IEEE Transactions on Visualization and Computer Graphics*, 18(9):1520–1536, 2012.

[316] Edmund Landau. *Handbuch der Lehre von der Verteilung der Primzahlen*. Leipzig B.G. Teubner, 1909.

[317] Ulrich Lang, Georges G. Grinstein, and R. Daniel Bergeron. Visualization related metadata. In Andreas Wierse, Georges G. Grinstein, and Ulrich Lang, editors, *Proceedings of Database Issues for Data Visualization, IEEE Visualization Workshop*, volume 1183 of *Lecture Notes in Computer Science*, pages 26–34. Springer, 1995.

[318] Chris Lankford. Gazetracker: software designed to facilitate eye movement analysis. In Andrew T. Duchowski, editor, *Proceedings of the Eye Tracking Research & Application Symposium, ETRA*, pages 51–55. ACM, 2000.

[319] Sven Laqua, Gemini Pate, and Martina Angela Sasse. Personalised focus-metaphor interfaces: An eye tracking study on user confusion. In Andreas M. Heinecke and Hansjürgen Paul, editors, *Mensch und Computer 2006: Mensch und Computer im Strukturwandel*, pages 175–184. Oldenbourg Verlag, 2006.

[320] Ethan P. Larsen, Jacob M. Kolman, Faisal N. Masud, and Farzan Sasangohar. Ethical considerations when using a mobile eye tracker in a patient-facing area: Lessons from an intensive care unit observational protocol. *Ethics & Human Research*, 2020.

[321] Cyril Robert Latimer. Eye-movement data: Cumulative fixation time and cluster analysis. *Behavior Research Methods*, 20(5):437–470, 1988.

[322] Alexandra Lee, Daniel Archambault, and Miguel A. Nacenta. The effectiveness of interactive visualization techniques for time navigation of dynamic graphs on large displays. *IEEE Transactions on Visualization and Computer Graphics*, 27(2):528–538, 2021.

[323] Bongshin Lee, Catherine Plaisant, Cynthia Sims Parr, Jean-Daniel Fekete, and Nathalie Henry. Task taxonomy for graph visualization. In Enrico Bertini, Catherine Plaisant, and Giuseppe Santucci, editors,

Proceedings of the 2006 AVI Workshop on BEyond time and errors: novel evaluation methods for information visualization, BELIV, pages 1–5. ACM Press, 2006.

[324] Bongshin Lee, Nathalie Henry Riche, Petra Isenberg, and Sheelagh Carpendale. More than telling a story: Transforming data into visually shared stories. *IEEE Computer Graphics and Applications*, 35(5):84–90, 2015.

[325] Michael D. Lee and Rachel E. Reilly. An empirical evaluation of chernoff faces, star glyphs, and spatial visualisations for binary data. In Tim Pattison and Bruce H. Thomas, editors, *Proceedings of Australasian Symposium on Information Visualisation, InVis.au*, volume 24 of *CRPIT*, pages 1–10. Australian Computer Society, 2003.

[326] Raymond S. T. Lee. *Artificial Intelligence in Daily Life*. Springer, 2020.

[327] Michael Ley. The DBLP computer science bibliography: Evolution, research issues, perspectives. In Alberto H. F. Laender and Arlindo L. Oliveira, editors, *String Processing and Information Retrieval, 9th International Symposium, SPIRE*, volume 2476 of *Lecture Notes in Computer Science*, pages 1–10. Springer, 2002.

[328] Michael Ley. DBLP - some lessons learned. *Proceddings of the VLDB Endowment*, 2(2):1493–1500, 2009.

[329] Rui Li, Jeff B. Pelz, Pengcheng Shi, Cecilia Ovesdotter Alm, and Anne R. Haake. Learning eye movement patterns for characterization of perceptual expertise. In Carlos Hitoshi Morimoto, Howell O. Istance, Stephen N. Spencer, Jeffrey B. Mulligan, and Pernilla Qvarfordt, editors, *Proceedings of the Symposium on Eye-Tracking Research and Applications, ETRA*, pages 393–396. ACM, 2012.

[330] Xia Li, Arzu Çöltekin, and Menno-Jan Kraak. Visual exploration of eye movement data using the space-time-cube. In Sara Irina Fabrikant, Tumasch Reichenbacher, Marc J. van Kreveld, and Christoph Schlieder, editors, *Proceedings of 6th International Conference on Geographic Information Science, GIScience*, volume 6292 of *Lecture Notes in Computer Science*, pages 295–309. Springer, 2010.

[331] Jie Liang, Mao Lin Huang, and Quang Vinh Nguyen. Perceptual user study for combined treemap. In *Proceedings of 11th International Conference on Machine Learning and Applications, ICMLA*, pages 300–305. IEEE, 2012.

[332] Jia Zheng Lim, James Mountstephens, and Jason Teo. Emotion recognition using eye-tracking: Taxonomy, review and current challenges. *Sensors*, 20(8):2384, 2020.

[333] Daniel Limberger, Carolin Fiedler, Sebastian Hahn, Matthias Trapp, and Jürgen Döllner. Evaluation of sketchiness as a visual variable for 2.5d treemaps. In Ebad Banissi et al., editor, *Proceedings of 20th International Conference Information Visualisation, IV*, pages 183–189. IEEE Computer Society, 2016.

[334] Heather Richter Lipford, Felesia Stukes, Wenwen Dou, Matthew E. Hawkins, and Remco Chang. Helping users recall their reasoning process. In *Proceedings of 5th IEEE Conference on Visual Analytics Science and Technology, VAST*, pages 187–194. IEEE Computer Society, 2010.

[335] Bing Liu, Weihua Dong, and Liqiu Meng. Using eye tracking to explore the guidance and constancy of visual variables in 3d visualization. *ISPRS International Journal of Geo Information*, 6(9):274, 2017.

[336] Xiaohui Liu, Gongxian Cheng, and John Xingwang Wu. AI for public health: Self-screening for eye diseases. *IEEE Intelligent Systems and Their Applications*, 13(5):28–35, 1998.

[337] Zhicheng Liu and Jeffrey Heer. The effects of interactive latency on exploratory visual analysis. *IEEE Transactions on Visualization and Computer Graphics*, 20(12):2122–2131, 2014.

[338] Angela Locoro, Federico Cabitza, Rossana Actis-Grosso, and Carlo Batini. Static and interactive infographics in daily tasks: A value-in-use and quality of interaction user study. *Computers in Human Behavior*, 71:240–257, 2017.

[339] Christophe Antony Lounis, Vsevolod Peysakhovich, and Mickaël Causse. Intelligent cockpit: eye tracking integration to enhance the pilot-aircraft interaction. In Bonita Sharif and Krzysztof Krejtz, editors, *Proceedings of the 2018 ACM Symposium on Eye Tracking Research & Applications, ETRA*, pages 74:1–74:3. ACM, 2018.

[340] Shang Lu, Yerly Paola Sanchez Perdomo, Xianta Jiang, and Bin Zheng. Integrating eye-tracking to augmented reality system for surgical training. *Journal of Medical Systems*, 44(11):192, 2020.

[341] Gerard A. Lutty. Effects of diabetes on the eye. *Investigative ophthalmology & visual science*, 54(14), 2013.

[342] Alan M. MacEachren. *How Maps Work: Representation, Visualization and Design*. Guilford Press, 1995.

[343] Alan M. MacEachren, Francis P. Boscoe, Daniel Haug, and Linda Pickle. Geographic visualization: Designing manipulable maps for exploring temporally varying georeferenced statistics. In *Proceedings*

of IEEE Symposium on Information Visualization (InfoVis '98), pages 87–94. IEEE Computer Society, 1998.

[344] Jock D. Mackinlay. Automating the design of graphical presentations of relational information. *ACM Transactions on Graphics, TOG*, 5(2):110–141, 1986.

[345] Jane F. Mackworth and Norman H. Mackworth. Eye fixations recorded on changing visual scenes by the television eye-marker. *Journal of the Optical Society of America*, 48(7):439–444, 1958.

[346] Azam Majooni, Mona Masood, and Amir Akhavan. An eye-tracking study on the effect of infographic structures on viewer's comprehension and cognitive load. *Information Visualization*, 17(3):257–266, 2018.

[347] Kim Marriott, Falk Schreiber, Tim Dwyer, Karsten Klein, Nathalie Henry Riche, Takayuki Itoh, Wolfgang Stuerzlinger, and Bruce H. Thomas, editors. *Immersive Analytics*, volume 11190 of *Lecture Notes in Computer Science*. Springer, 2018.

[348] Curtis E. Martin, J. O. Keller, Steven K. Rogers, and Matthew Kabrisky. Color blindness and a color human visual system model. *IEEE Transactions on Systems, Man, and Cybernetics Part A*, 30(4):494–500, 2000.

[349] Bruce Howard McCormick, Thomas A. DeFanti, and Maxine D. Brown. Visualization in scientific computing. *Computer Graphics*, 21(6), 1987.

[350] Michael J. McGuffin and Ravin Balakrishnan. Fitts' law and expanding targets: Experimental studies and designs for user interfaces. *ACM Transactions on Computer-Human Interactions, TOCHI*, 12(4):388–422, 2005.

[351] Leland McInnes, John Healy, Nathaniel Saul, and Lukas Großberger. UMAP: uniform manifold approximation and projection. *Journal of Open Source Software*, 3(29):861, 2018.

[352] John P. McIntire and Kristen K. Liggett. The (possible) utility of stereoscopic 3d displays for information visualization: The good, the bad, and the ugly. In *Proceedings of IEEE VIS International Workshop on 3DVis, 3DVis@IEEE VIS*, pages 1–9. IEEE, 2014.

[353] Raphael Menges, Chandan Kumar, and Steffen Staab. Improving user experience of eye tracking-based interaction: Introspecting and adapting interfaces. *ACM Transactions on Computer-Human Interactions*, 26(6):37:1–37:46, 2019.

[354] Corine S. Meppelink and Nadine Bol. Exploring the role of health literacy on attention to and recall of text-illustrated health information: An eye-tracking study. *Computers in Human Behavior*, 48:87–93, 2015.

[355] Slavko Milekic. Using eye- and gaze-tracking to interact with a visual display. In Stuart Dunn, Jonathan P. Bowen, and Kia Ng, editors, *Proceedings of Electronic Visualisation and the Arts, EVA*, Workshops in Computing. BCS, 2012.

[356] Kazuo Misue, Peter Eades, Wei Lai, and Kozo Sugiyama. Layout adjustment and the mental map. *Journal of Visual Languages and Computing*, 6(2):183–210, 1995.

[357] Parag K. Mital, Tim J Smith, Robin L. Hill, and John M. Henderson. Clustering of gaze during dynamic scene viewing is predicted by motion. *Cognitive Computation*, 3(1):5–24, 2011.

[358] Nadine Marie Moacdieh and Nadine Sarter. Using eye tracking to detect the effects of clutter on visual search in real time. *IEEE Transactions on Human-Machine Systems*, 47(6):896–902, 2017.

[359] Kenneth Moreland. A survey of visualization pipelines. *IEEE Transactions on Visualization and Computer Graphics*, 19(3):367–378, 2013.

[360] Carlos Hitoshi Morimoto and Marcio R. M. Mimica. Eye gaze tracking techniques for interactive applications. *Computer Vision and Image Understanding*, 98(1):4–24, 2005.

[361] Rajeev Motwani and Prabhakar Raghavan. *Randomized Algorithms*. Cambridge University Press, 1995.

[362] Haris Mumtaz, Shahid Latif, Fabian Beck, and Daniel Weiskopf. Exploranative code quality documents. *IEEE Transactions on Visualization and Computer Graphics*, 26(1):1129–1139, 2020.

[363] Tanja Munz, Michael Burch, Toon van Benthem, Yoeri Poels, Fabian Beck, and Daniel Weiskopf. Overlap-free drawing of generalized pythagoras trees for hierarchy visualization. In *Proceedings of 30th IEEE Visualization Conference, IEEE VIS - Short Papers*, pages 251–255. IEEE, 2019.

[364] Tamara Munzner. *Process and Pitfalls in Writing Information Visualization Research Papers*, pages 134–153. Springer Berlin Heidelberg, Berlin, Heidelberg, 2008.

[365] Minoru Nakayama and Yuko Hayashi. Estimation of viewer's response for contextual understanding of tasks using features of eye-movements. In Carlos Hitoshi Morimoto, Howell O. Istance, Aulikki Hyrskykari,

and Qiang Ji, editors, *Proceedings of the Symposium on Eye-Tracking Research & Applications, ETRA*, pages 53–56. ACM, 2010.

[366] Shah Nawaz, Alessandro Calefati, Nisar Ahmed, and Ignazio Gallo. Hand written characters recognition via deep metric learning. In *Proceedings of the 13th IAPR International Workshop on Document Analysis Systems, DAS*, pages 417–422. IEEE Computer Society, 2018.

[367] Saul B. Needleman and Christian D. Wunsch. A general method applicable to the search for similarities in the amino acid sequence of two proteins. *Journal of Molecular Biology*, 48(3):443 – 453, 1970.

[368] Keith V. Nesbitt and Carsten Friedrich. Applying gestalt principles to animated visualizations of network data. In *Proceedings of International Conference on Information Visualisation, IV*, pages 737–743. IEEE Computer Society, 2002.

[369] Rudolf Netzel, Michael Burch, and Daniel Weiskopf. Comparative eye tracking study on node-link visualizations of trajectories. *IEEE Transactions on Visualization and Computer Graphics*, 20(12):2221–2230, 2014.

[370] Rudolf Netzel, Michael Burch, and Daniel Weiskopf. Interactive scanpath-oriented annotation of fixations. In Pernilla Qvarfordt and Dan Witzner Hansen, editors, *Proceedings of the Ninth Biennial ACM Symposium on Eye Tracking Research & Applications, ETRA*, pages 183–187. ACM, 2016.

[371] Rudolf Netzel, Marcel Hlawatsch, Michael Burch, Sanjeev Balakrishnan, Hansjörg Schmauder, and Daniel Weiskopf. An evaluation of visual search support in maps. *IEEE Transactions on Visualization and Computer Graphics*, 23(1):421–430, 2017.

[372] Rudolf Netzel, Bettina Ohlhausen, Kuno Kurzhals, Robin Woods, Michael Burch, and Daniel Weiskopf. User performance and reading strategies for metro maps: An eye tracking study. *Spatial Cognition & Computation*, 17(1-2):39–64, 2017.

[373] Thanh-An Nguyen, Constantinos K. Coursaris, Pierre-Majorique Léger, Sylvain Sénécal, and Marc Fredette. Effectiveness of banner ads: An eye tracking and facial expression analysis. In Fiona Fui-Hoon Nah and Keng Siau, editors, *Proceedings of 7th International Conference in HCI in Business, Government and Organizations, HCIBGO*, volume 12204 of *Lecture Notes in Computer Science*, pages 445–455. Springer, 2020.

[374] Chris North and Ben Shneiderman. Snap-together visualization: can users construct and operate coordinated visualizations? *International Journal of Human Computer Studies*, 53(5):715–739, 2000.

[375] Dabid Noton and Lawrence Stark. Scanpaths in saccadic eye movements while viewing and recognizing patterns. *Vision Research*, 11(9):929–942, 1971.

[376] Lucy T. Nowell, Elizabeth G. Hetzler, and Ted Tanasse. Change blindness in information visualization: A case study. In Keith Andrews, Steven F. Roth, and Pak Chung Wong, editors, *Proceedings of IEEE Symposium on Information Visualization (INFOVIS)*, pages 15–22. IEEE Computer Society, 2001.

[377] Marcus Nyström, Ignace Hooge, and Kenneth Holmqvist. Post-saccadic oscillations in eye movement data recorded with pupil-based eye trackers reflect motion of the pupil inside the iris. *Vision Research*, 92:59–66, 2013.

[378] Jonathan O'Donovan, Jon Ward, Scott Hodgins, and Veronica Sundstedt. Rabbit run: Gaze and voice based game interaction. In *Proceedings of the 9th Irish Eurographics Workshop*, 2009.

[379] Mershack Okoe, Sayeed Safayet Alam, and Radu Jianu. A gaze-enabled graph visualization to improve graph reading tasks. *Computer Graphics Forum*, 33(3):251–260, 2014.

[380] Manuel M. Oliveira. Uncertainty visualization and color vision deficiency. In Charles D. Hansen, Min Chen, Christopher R. Johnson, Arie E. Kaufman, and Hans Hagen, editors, *Scientific Visualization*, Mathematics and Visualization, pages 29–33. Springer, 2014.

[381] Anneli Olsen, Albrecht Schmidt, Paul Marshall, and Veronica Sundstedt. Using eye tracking for interaction. In Desney S. Tan, Saleema Amershi, Bo Begole, Wendy A. Kellogg, and Manas Tungare, editors, *Proceedings of the International Conference on Human Factors in Computing Systems, CHI*, pages 741–744. ACM, 2011.

[382] Gustav Öquist and Kristin Lundin. Eye movement study of reading text on a mobile phone using paging, scrolling, leading, and RSVP. In Timo Ojala, editor, *Proceedings of the 6th International Conference on Mobile and Ubiquitous Multimedia, MUM*, volume 284 of *ACM International Conference Proceeding Series*, pages 176–183. ACM, 2007.

[383] Eliza O'Reilly, François Baccelli, Gustavo de Veciana, and Haris Vikalo. End-to-end optimization of high-throughput DNA sequencing. *Journal of Computational Biology*, 23(10):789–800, 2016.

[384] Oyebade Kayode Oyedotun. *Analyzing and Improving Very Deep Neural Networks: From Optimization, Generalization to Compression*. PhD thesis, University of Luxembourg, Luxembourg City, Luxembourg, 2020.

[385] Jeni Paay, Dimitrios Raptis, Jesper Kjeldskov, Bjarke M. Lauridsen, Ivan S. Penchev, Elias Ringhauge, and Eric V. Ruder. A comparison of techniques for cross-device interaction from mobile devices to large displays. *Journal of Mobile Multimedia*, 12(3&4):243–264, 2017.

[386] Lace M. K. Padilla, Spencer C. Castro, P. Samuel Quinan, Ian T. Ruginski, and Sarah H. Creem-Regehr. Toward objective evaluation of working memory in visualizations: A case study using pupillometry and a dual-task paradigm. *IEEE Transactions on Visualization and Computer Graphics*, 26(1):332–342, 2020.

[387] Oskar Palinko, Andrew L. Kun, Alexander Shyrokov, and Peter A. Heeman. Estimating cognitive load using remote eye tracking in a driving simulator. In Carlos Hitoshi Morimoto, Howell O. Istance, Aulikki Hyrskykari, and Qiang Ji, editors, *Proceedings of the 2010 Symposium on Eye-Tracking Research & Applications, ETRA*, pages 141–144. ACM, 2010.

[388] Richard Panchyk. *Charting the World: Geography and Maps from Cave Paintings to GPS with 21 Activities*. CHICAGO REVIEW PR; 1st edition, 2011.

[389] Aditeya Pandey, Uzma Haque Syeda, and Michelle A. Borkin. Towards identification and mitigation of task-based challenges in comparative visualization studies. In *Proceedings of IEEE Workshop on Evaluation and Beyond - Methodological Approaches to Visualization (BELIV)*, pages 55–64, 2020.

[390] Karen Panetta, Qianwen Wan, Srijith Rajeev, Aleksandra Kaszowska, Aaron L. Gardony, Kevin Naranjo, Holly A. Taylor, and Sos S. Agaian. ISeeColor: Method for advanced visual analytics of eye tracking data. *IEEE Access*, 8:52278–52287, 2020.

[391] Mohsen Parisay, Charalambos Poullis, and Marta Kersten. EyeTAP: A novel technique using voice inputs to address the midas touch problem for gaze-based interactions. *CoRR*, abs/2002.08455, 2020.

[392] Alvydas Paunksnis, Skaidra Kurapkiene, Audris Maciulis, Audris Kopustinskas, and Marija-Liucija Paunksniene. Ultrasound quantitative evaluation of human eye cataract. *Informatica*, 18(2):267–278, 2007.

[393] Fabio Pellacini, Lori Lorigo, and Geri Gay. Visualizing paths in context. In *Technical Report, Computer Science*, 2006.

[394] Thies Pfeiffer. Measuring and visualizing attention in space with 3d attention volumes. In Carlos Hitoshi Morimoto, Howell O. Istance, Stephen N. Spencer, Jeffrey B. Mulligan, and Pernilla Qvarfordt, editors, *Proceedings of the Symposium on Eye-Tracking Research and Applications, ETRA*, pages 29–36. ACM, 2012.

[395] Thies Pfeiffer, Patrick Renner, and Nadine Pfeiffer-Leßmann. Eyesee3d 2.0: model-based real-time analysis of mobile eye-tracking in static and dynamic three-dimensional scenes. In Pernilla Qvarfordt and Dan Witzner Hansen, editors, *Proceedings of the Ninth Biennial ACM Symposium on Eye Tracking Research & Applications, ETRA*, pages 189–196. ACM, 2016.

[396] William A. Pike, John T. Stasko, Remco Chang, and Theresa A. O'Connell. The science of interaction. *Information Visualization*, 8(4):263–274, 2009.

[397] Catherine Plaisant. The challenge of information visualization evaluation. In Maria Francesca Costabile, editor, *Proceedings of the working conference on Advanced visual interfaces, AVI*, pages 109–116. ACM Press, 2004.

[398] Catherine Plaisant, Jesse Grosjean, and Benjamin B. Bederson. Spacetree: Supporting exploration in large node link tree, design evolution and empirical evaluation. In Pak Chung Wong and Keith Andrews, editors, *Proceedings of IEEE Symposium on Information Visualization (InfoVis)*, pages 57–64. IEEE Computer Society, 2002.

[399] Christopher Plaue, Todd Miller, and John T. Stasko. Is a picture worth a thousand words? an evaluation of information awareness displays. In Wolfgang Heidrich and Ravin Balakrishnan, editors, *Proceedings of the Graphics Interface Conference*, volume 62 of *ACM International Conference Proceeding Series*, pages 117–126. Canadian Human-Computer Communications Society, 2004.

[400] Mathias Pohl, Markus Schmitt, and Stephan Diehl. Comparing the readability of graph layouts using eyetracking and task-oriented analysis. In *Proceedings of the Eurographics Workshop on Computational Aesthetics*, pages 49–56, 2009.

[401] Alex Poole and Linden J. Ball. Eye tracking in human-computer interaction and usability research: Current status and future prospects. In *Idea Group Inc., ch. E*, pages 211–219, 2006.

[402] Stanislav Popelka and Jitka Dolezalová. Non-photorealistic 3d visualization in city maps: An eye-tracking study. In Jan Brus, Alena Vondráková, and Vit Vozenilek, editors, *Modern Trends in Cartography - Selected Papers of CARTOCON*, Lecture Notes in Geoinformation and Cartography, pages 357–367. Springer, 2014.

[403] Marco Pretorius, André P. Calitz, and Darelle van Greunen. The value of eye tracking in information visualisation interface evaluations. *South African Computer Journal*, 36:115–123, 2006.

[404] Claudio M. Privitera and Lawrence W. Stark. Algorithms for defining visual regions-of-interest: Comparison with eye fixations. *IEEE Transactions on Pattern Analysis and Machine Intelligence*, 22(9):970–982, 2000.

[405] Helen C. Purchase. Effective information visualisation: a study of graph drawing aesthetics and algorithms. *Interacting with Computers*, 13(2):147–162, 2000.

[406] Helen C. Purchase. Metrics for graph drawing aesthetics. *Journal of Visual Languages and Computing*, 13(5):501–516, 2002.

[407] Helen C. Purchase, Robert F. Cohen, and Murray I. James. Validating graph drawing aesthetics. In *Proceedings of the Symposium on Graph Drawing*, pages 435–446, 1995.

[408] Helen C. Purchase, Eve Hoggan, and Carsten Görg. How important is the mental map? An empirical investigation of a dynamic graph layout algorithm. In *Proceedings of International Symposium on Graph Drawing*, pages 262–273, 2007.

[409] Helen C. Purchase, Christopher Pilcher, and Beryl Plimmer. Graph drawing aesthetics - created by users, not algorithms. *IEEE Transactions on Visualization and Computer Graphics*, 18(1):81–92, 2012.

[410] Wen Qi. How much information do you remember? - the effects of short-term memory on scientific visualization tasks. In Randall Shumaker, editor, *Proceedings of Virtual Reality, Second International Conference, ICVR*, volume 4563 of *Lecture Notes in Computer Science*, pages 338–347. Springer, 2007.

[411] Zhiming Qian and Dan Xu. Automatic eye detection using intensity filtering and k-means clustering. *Pattern Recognition Letters*, 31(12):1633–1640, 2010.

[412] Nordine Quadar, Abdellah Chehri, and Gwanggil Geon. Visual analytics methods for eye tracking data. In Alfred Zimmermann, Robert J. Howlett, and Lakhmi C. Jain, editors, *Proceedings of the*

Conference on Human Centred Intelligent Systems, KES-HCIS, volume 189 of *Smart Innovation, Systems and Technologies*, pages 3–12. Springer, 2020.

[413] Protiva Rahman, Lilong Jiang, and Arnab Nandi. Evaluating interactive data systems. *The VLDB Journal*, 29(1):119–146, 2020.

[414] Kari-Jouko Räihä, Anne Aula, Päivi Majaranta, Harri Rantala, and Kimmo Koivunen. Static visualization of temporal eye-tracking data. In Maria Francesca Costabile and Fabio Paternò, editors, *Proceedings of the International Conference on Human-Computer Interaction - INTERACT*, volume 3585 of *Lecture Notes in Computer Science*, pages 946–949. Springer, 2005.

[415] Rameshsharma Ramloll, Cheryl Trepagnier, Marc M. Sebrechts, and Jaishree Beedasy. Gaze data visualization tools: Opportunities and challenge. In *Proceedings of 8th International Conference on Information Visualisation, IV*, pages 173–180. IEEE Computer Society, 2004.

[416] Michael Raschke, Tanja Blascheck, and Thomas Ertl. Cognitive ergonomics in visualization. In Achim Ebert, Gerrit C. van der Veer, Gitta Domik, Nahum D. Gershon, and Inga Scheler, editors, *Proceedings of Building Bridges: HCI, Visualization, and Non-formal Modeling - IFIP WG 13.7 Workshops on Human-Computer Interaction and Visualization: 7th HCIV@ECCE and 8th HCIV@INTERACT, Revised Selected Papers*, volume 8345 of *Lecture Notes in Computer Science*, pages 80–94. Springer, 2011.

[417] Michael Raschke, Xuemei Chen, and Thomas Ertl. Parallel scan-path visualization. In Carlos Hitoshi Morimoto, Howell O. Istance, Stephen N. Spencer, Jeffrey B. Mulligan, and Pernilla Qvarfordt, editors, *Proceedings of the Symposium on Eye-Tracking Research and Applications, ETRA*, pages 165–168. ACM, 2012.

[418] Erik D. Reichle, Keith Rayner, and Alexander Pollatsek. The e-z reader model of eye-movement control in reading: comparisons to other models. *Behavioral and Brain Sciences*, 26(4):445–476, 2003.

[419] Edward M. Reingold and John S. Tilford. Tidier drawings of trees. *IEEE Transactions on Software Engineering*, 7(2):223–228, 1981.

[420] William Ribarsky, Brian D. Fisher, and William M. Pottenger. Science of analytical reasoning. *Information Visualization*, 8(4):254–262, 2009.

[421] Daniel C. Richardson and Rick Dale. Looking to understand: The coupling between speakers' and listeners' eye movements

and its relationship to discourse comprehension. *Cognitive Science*, 29(6):1045–1060, 2005.

[422] Patrick Riehmann, Manfred Hanfler, and Bernd Froehlich. Interactive sankey diagrams. In John T. Stasko and Matthew O. Ward, editors, *Proceedings of IEEE Symposium on Information Visualization (InfoVis)*, pages 233–240. IEEE Computer Society, 2005.

[423] Gordan Ristovski, Mathew Hunter, Bettina Olk, and Lars Linsen. Eyec: Coordinated views for interactive visual exploration of eye-tracking data. In Ebad Banissi et al., editor, *Proceedings of the 17th International Conference on Information Visualisation, IV*, pages 239–248. IEEE Computer Society, 2013.

[424] Jonathan C. Roberts. Guest editor's introduction: special issue on coordinated and multiple views in exploratory visualization. *Information Visualization*, 2(4):199–200, 2003.

[425] René Rosenbaum and Bernd Hamann. Evaluation of progressive treemaps to convey tree and node properties. In Pak Chung Wong, David L. Kao, Ming C. Hao, Chaomei Chen, Robert Kosara, Mark A. Livingston, Jinah Park, and Ian Roberts, editors, *Proceedings of Visualization and Data Analysis*, volume 8294 of *SPIE Proceedings*, page 82940F. SPIE, 2012.

[426] Ruth Rosenholtz, Yuanzhen Li, Jonathan Mansfield, and Zhenlan Jin. Feature congestion: a measure of display clutter. In Gerrit C. van der Veer and Carolyn Gale, editors, *Proceedings of the Conference on Human Factors in Computing Systems, CHI*, pages 761–770. ACM, 2005.

[427] Alessandro Rossi, Sara Ermini, Dario Bernabini, Dario Zanca, Marino Todisco, Alessandro Genovese, and Antonio Rizzo. End-to-end models for the analysis of system 1 and system 2 interactions based on eye-tracking data. In Stephanie Denison, Michael Mack, Yang Xu, and Blair C. Armstrong, editors, *Proceedings of the 42th Annual Meeting of the Cognitive Science Society - Developing a Mind: Learning in Humans, Animals, and Machines, CogSci*. cognitivesciencesociety.org, 2020.

[428] Manuel Rubio-Sánchez and Alberto Sánchez. Axis calibration for improving data attribute estimation in star coordinates plots. *IEEE Transactions on Visualization and Computer Graphics*, 20(12):2013–2022, 2014.

[429] Puripant Ruchikachorn and Pimmanee Rattanawicha. An eye-tracking study on sparklines within textual context. In Anna Puig and

Renata Georgia Raidou, editors, *Proceedings of 20th Eurographics Conference on Visualization, EuroVis - Posters*, pages 17–19. Eurographics Association, 2018.

[430] David Rudi, Peter Kiefer, Ioannis Giannopoulos, and Martin Raubal. Gaze-based interactions in the cockpit of the future: a survey. *Journal on Multimodal User Interfaces*, 14(1):25–48, 2020.

[431] David Rudi, Peter Kiefer, and Martin Raubal. Visualizing pilot eye movements for flight instructors. In Lewis L. Chuang, Michael Burch, and Kuno Kurzhals, editors, *Proceedings of the 3rd Workshop on Eye Tracking and Visualization, ETVIS@ETRA*, pages 7:1–7:5. ACM, 2018.

[432] David Rudi, Peter Kiefer, and Martin Raubal. The instructor assistant system (iASSYST) - utilizing eye tracking for commercial aviation training purposes. *Ergonomics*, 63(1):61–79, 2020.

[433] Dominik Sacha, Andreas Stoffel, Florian Stoffel, Bum Chul Kwon, Geoffrey P. Ellis, and Daniel A. Keim. Knowledge generation model for visual analytics. *IEEE Transactions on Visualization and Computer Graphics*, 20(12):1604–1613, 2014.

[434] Emil Saeed, Anna Bartocha, Piotr Wachulec, and Khalid Saeed. Influence of eye diseases on the retina pattern recognition. In Khalid Saeed and Václav Snásel, editors, *Proceedings of 13th IFIP TC8 International Conference on Computer Information Systems and Industrial Management*, volume 8838 of *Lecture Notes in Computer Science*, pages 130–140. Springer, 2014.

[435] Anthony Santella and Douglas DeCarlo. Robust clustering of eye movement recordings for quantification of visual interest. In Andrew T. Duchowski and Roel Vertegaal, editors, *Proceedings of the Eye Tracking Research & Application Symposium, ETRA*, pages 27–34. ACM, 2004.

[436] João Miguel Santos, Paulo Dias, and Beatriz Sousa Santos. Implementation and evaluation of an enhanced h-tree layout pedigree visualization. In Ebad Banissi et al., editor, *Proceedings of 16th International Conference on Information Visualisation, IV*, pages 24–29. IEEE Computer Society, 2012.

[437] Rubina Sarki, Khandakar Ahmed, and Yanchun Zhang. Early detection of diabetic eye disease through deep learning using fundus images. *EAI Endorsed Transactions on Pervasive Health Technologies*, 6(22), 2020.

[438] Mithileysh Sathiyanarayanan and Tobias Mulling. Wellformedness properties in euler diagrams: An eye tracking study for visualisation evaluation. *CoRR*, abs/1611.06587, 2016.

[439] Simon Schenk, Marc Dreiser, Gerhard Rigoll, and Michael Dorr. Gazeeverywhere: Enabling gaze-only user interaction on an unmodified desktop PC in everyday scenarios. In Gloria Mark, Susan R. Fussell, Cliff Lampe, M. C. Schraefel, Juan Pablo Hourcade, Caroline Appert, and Daniel Wigdor, editors, *Proceedings of the CHI Conference on Human Factors in Computing Systems*, pages 3034–3044. ACM, 2017.

[440] Karen B. Schloss, Connor Gramazio, Allison T. Silverman, Madeline L. Parker, and Audrey S. Wang. Mapping color to meaning in colormap data visualizations. *IEEE Transactions on Visualization and Computer Graphics*, 25(1):810–819, 2019.

[441] Hansjörg Schmauder, Michael Burch, Christoph Müller, and Daniel Weiskopf. Distributed visual analytics on large-scale high-resolution displays. In *Proceedings of the Symposium on Big Data Visual Analytics, BDVA*, pages 33–40. IEEE, 2015.

[442] Axel K. Schmitt, Martin Danišík, Erkan Aydar, Erdal Şen, İnan Ulusoy, and Oscar M. Lovera. Identifying the volcanic eruption depicted in a neolithic painting at Çatalhöyük, central anatolia, turkey. *PLOS ONE*, 9(1):1–10, 01 2014.

[443] Jean Scholtz. *User-Centered Evaluation of Visual Analytics*. Synthesis Lectures on Visualization. Morgan & Claypool Publishers, 2017.

[444] Christoph Schulz, Michael Burch, Fabian Beck, and Daniel Weiskopf. Visual data cleansing of low-level eye-tracking data. In Michael Burch, Lewis L. Chuang, Brian D. Fisher, Albrecht Schmidt, and Daniel Weiskopf, editors, *Proceedings of the Workshop on Eye Tracking and Visualization, ETVIS*, Mathematics and Visualization, pages 199–216. Springer, 2015.

[445] Hans-Jörg Schulz. Treevis.net: A tree visualization reference. *IEEE Computer Graphics and Applications*, 31(6):11–15, 2011.

[446] Hans-Jörg Schulz, Steffen Hadlak, and Heidrun Schumann. The design space of implicit hierarchy visualization: A survey. *IEEE Transactions on Visualization and Computer Graphics*, 17(4):393–411, 2011.

[447] Robert M. Schumacher and Kirsten E. Jerch. Measuring usability in healthcare it: It's a practice, not a competition. *Interactions*, 19(4):8–9, 2012.

[448] Leonard Scinto, Ramakrishna Pillalamarri, and Robert Karsh. Cognitive strategies for visual search. *Acta Psychologica*, 62(3):263–292, 1986.

[449] Michael Sedlmair, Petra Isenberg, Dominikus Baur, and Andreas Butz. Information visualization evaluation in large companies: Challenges, experiences and recommendations. *Information Visualization*, 10(3):248–266, 2011.

[450] Michael Sedlmair, Miriah D. Meyer, and Tamara Munzner. Design study methodology: Reflections from the trenches and the stacks. *IEEE Transactions on Visualization and Computer Graphics*, 18(12):2431–2440, 2012.

[451] Lin Shao, Nelson Silva, Eva Eggeling, and Tobias Schreck. Visual exploration of large scatter plot matrices by pattern recommendation based on eye tracking. In Dorota Glowacka, Evangelos E. Milios, Axel J. Soto, and Fernando Vieira Paulovich, editors, *Proceedings of the ACM Workshop on Exploratory Search and Interactive Data Analytics, ESIDA@IUI*, pages 9–16. ACM, 2017.

[452] Zohreh Sharafi, Timothy Shaffer, Bonita Sharif, and Yann-Gaël Guéhéneuc. Eye-tracking metrics in software engineering. In Jing Sun, Y. Raghu Reddy, Arun Bahulkar, and Anjaneyulu Pasala, editors, *Proceedings of the Asia-Pacific Software Engineering Conference, APSEC*, pages 96–103. IEEE Computer Society, 2015.

[453] Zohreh Sharafi, Bonita Sharif, Yann-Gaël Guéhéneuc, Andrew Begel, Roman Bednarik, and Martha E. Crosby. A practical guide on conducting eye tracking studies in software engineering. *Empirical Software Engineering*, 25(5):3128–3174, 2020.

[454] Bonita Sharif, Michael Falcone, and Jonathan I. Maletic. An eye-tracking study on the role of scan time in finding source code defects. In Carlos Hitoshi Morimoto, Howell O. Istance, Stephen N. Spencer, Jeffrey B. Mulligan, and Pernilla Qvarfordt, editors, *Proceedings of the Symposium on Eye-Tracking Research and Applications, ETRA*, pages 381–384. ACM, 2012.

[455] Kshitij Sharma, Michail N. Giannakos, and Pierre Dillenbourg. Eye-tracking and artificial intelligence to enhance motivation and learning. *Smart Learning Environments*, 7(1):13, 2020.

[456] Rishabh Sharma and Hubert Cecotti. Classification of graphical user interfaces through gaze-based features. In K. C. Santosh and Ravindra S. Hegadi, editors, *Proceedings of the International Conference on Recent Trends in Image Processing and Pattern*

Recognition, RTIP2R, volume 1035 of *Communications in Computer and Information Science*, pages 3–16. Springer, 2018.

[457] Chao Shi, Ayala Cohen, Ling Rothrock, and Tatiana Umansky. An investigation of placement of textual and graphical information using human performance and eye tracking data. In Sakae Yamamoto and Hirohiko Mori, editors, *Proceedings of 21st HCI International Conference on Human Interface and the Management of Information. Visual Information and Knowledge Management - Thematic Area, HIMI*, volume 11569 of *Lecture Notes in Computer Science*, pages 122–136. Springer, 2019.

[458] Ben Shneiderman. Tree visualization with tree-maps: 2-d space-filling approach. *ACM Transactions on Graphics*, 11(1):92–99, 1992.

[459] Ben Shneiderman. The eyes have it: A task by data type taxonomy for information visualizations. In *Proceedings of the IEEE Symposium on Visual Languages*, pages 336–343. IEEE Computer Society, 1996.

[460] Ben Shneiderman and Catherine Plaisant. Strategies for evaluating information visualization tools: multi-dimensional in-depth long-term case studies. In Enrico Bertini, Catherine Plaisant, and Giuseppe Santucci, editors, *Proceedings of the 2006 AVI Workshop on BEyond time and errors: novel evaluation methods for information visualization, BELIV*, pages 1–7. ACM Press, 2006.

[461] Ben Shneiderman, Catherine Plaisant, Maxine Cohen, Steven Jacobs, and Niklas Elmqvist. *Designing the User Interface - Strategies for Effective Human-Computer Interaction, 6th Edition*. Pearson, 2016.

[462] Mina Shojaeizadeh. Text simplification and eye tracking. In *Proceedings of the 22nd Americas Conference on Information Systems, AMCIS*. Association for Information Systems, 2016.

[463] Mina Shojaeizadeh, Soussan Djamasbi, and Andrew C. Trapp. Density of gaze points within a fixation and information processing behavior. In Margherita Antona and Constantine Stephanidis, editors, *Universal Access in Human-Computer Interaction. Methods, Techniques, and Best Practices*, pages 465–471. Springer International Publishing, 2016.

[464] Sav Shrestha and Kelsi Lenz. Eye gaze patterns while searching vs. browsing a website, 2007.

[465] Harri Siirtola, Tuuli Laivo, Tomi Heimonen, and Kari-Jouko Räihä. Visual perception of parallel coordinate visualizations. In Ebad Banissi et al., editor, *Proceedings of 13th International Conference on*

Information Visualisation, IV, pages 3–9. IEEE Computer Society, 2009.

[466] Nelson Silva, Tanja Blascheck, Radu Jianu, Nils Rodrigues, Daniel Weiskopf, Martin Raubal, and Tobias Schreck. Eye tracking support for visual analytics systems: foundations, current applications, and research challenges. In Krzysztof Krejtz and Bonita Sharif, editors, *Proceedings of the 11th ACM Symposium on Eye Tracking Research & Applications, ETRA*, pages 11:1–11:10. ACM, 2019.

[467] Nelson Silva, Tobias Schreck, Eduardo E. Veas, Vedran Sabol, Eva Eggeling, and Dieter W. Fellner. Leveraging eye-gaze and time-series features to predict user interests and build a recommendation model for visual analysis. In Bonita Sharif and Krzysztof Krejtz, editors, *Proceedings of the ACM Symposium on Eye Tracking Research & Applications, ETRA*, pages 13:1–13:9. ACM, 2018.

[468] Aniruddha Sinha, Rikayan Chaki, Bikram De Kumar, and Sanjoy Kumar Saha. Readability analysis of textual content using eye tracking. In Rituparna Chaki, Agostino Cortesi, Khalid Saeed, and Nabendu Chaki, editors, *Proceedings of 5th International Doctorial Symposium on Advanced Computing and Systems for Security and Applied Computation and Security Systems, ACSS*, volume 897 of *Advances in Intelligent Systems and Computing*, pages 73–88. Springer, 2018.

[469] Tim J. Smith and Parag K. Mital. Attentional synchrony and the influence of viewing task on gaze behavior in static and dynamic scenes. *Journal of Vision*, 13(8), 2013.

[470] Herman Snellen. Probebuchstaben zur bestimmung der sehschärfe, 1862.

[471] Hyunjoo Song, Bo Hyoung Kim, Bongshin Lee, and Jinwook Seo. A comparative evaluation on tree visualization methods for hierarchical structures with large fan-outs. In Elizabeth D. Mynatt, Don Schoner, Geraldine Fitzpatrick, Scott E. Hudson, W. Keith Edwards, and Tom Rodden, editors, *Proceedings of the 28th International Conference on Human Factors in Computing Systems, CHI*, pages 223–232. ACM, 2010.

[472] Hyunjoo Song, Jeongjin Lee, Tae Jung Kim, Kyoung Ho Lee, Bo Hyoung Kim, and Jinwook Seo. GazeDx: Interactive visual analytics framework for comparative gaze analysis with volumetric medical images. *IEEE Transactions on Visualization and Computer Graphics*, 23(1):311–320, 2017.

[473] Oleg Spakov and Darius Miniotas. Visualization of eye gaze data using heat maps. *Electronics and Electrical Engineering*, 74:55–58, 2007.

[474] Oleg Spakov and Kari-Jouko Räihä. KiEV: a tool for visualization of reading and writing processes in translation of text. In Kari-Jouko Räihä and Andrew T. Duchowski, editors, *Proceedings of the Eye Tracking Research & Application Symposium, ETRA*, pages 107–110. ACM, 2008.

[475] Robert Spence. Rapid, serial and visual: a presentation technique with potential. *Information Visualization*, 1(1):13–19, 2002.

[476] Robert Spence. *Information Visualization: Design for Interaction*. Pearson/Prentice Hall, 2 edition, 2007.

[477] Robert Spence and Mark Witkowski. *Rapid Serial Visual Presentation - Design for Cognition*. Springer Briefs in Computer Science. Springer, 2013.

[478] Bernd Carsten Stahl and David Wright. Ethics and privacy in AI and big data: Implementing responsible research and innovation. *IEEE Security & Privacy*, 16(3):26–33, 2018.

[479] John T. Stasko. Tango: A framework and system for algorithm animation. *Computer*, 23(9):27–39, 1990.

[480] John T. Stasko. Animating algorithms with XTANGO. *SIGACT News*, 23(2):67–71, 1992.

[481] Ben Steichen, Giuseppe Carenini, and Cristina Conati. User-adaptive information visualization: using eye gaze data to infer visualization tasks and user cognitive abilities. In Jihie Kim, Jeffrey Nichols, and Pedro A. Szekely, editors, *Proceedings of 18th International Conference on Intelligent User Interfaces, IUI*, pages 317–328. ACM, 2013.

[482] Ben Steichen, Oliver Schmid, Cristina Conati, and Giuseppe Carenini. Seeing how you're looking - using real-time eye gaze data for user-adaptive visualization. In Shlomo Berkovsky, Eelco Herder, Pasquale Lops, and Olga C. Santos, editors, *Proceedings of the 21st Conference on User Modeling, Adaptation, and Personalization*, volume 997. CEUR-WS.org, 2013.

[483] Sophie Stellmach, Lennart E. Nacke, and Raimund Dachselt. 3d attentional maps: aggregated gaze visualizations in three-dimensional virtual environments. In Giuseppe Santucci, editor, *Proceedings of the International Conference on Advanced Visual Interfaces, AVI*, pages 345–348. ACM Press, 2010.

[484] Sophie Stellmach, Lennart E. Nacke, and Raimund Dachselt. Advanced gaze visualizations for three-dimensional virtual environments. In Carlos Hitoshi Morimoto, Howell O. Istance, Aulikki Hyrskykari, and Qiang Ji, editors, *Proceedings of the Symposium on Eye-Tracking Research & Applications, ETRA*, pages 109–112. ACM, 2010.

[485] Robert J. Sternberg and Karin Sternberg. *Cognitive Psychology (6th ed.)*. Belmont, California: Cengage Learning, 2012.

[486] Judith B. Strother and Jan M. Ulijn. The challenge of information overload. In *Proceedings of IEEE International Professional Communication Conference*, pages 1–3. IEEE, 2012.

[487] Guodao Sun, Yingcai Wu, Ronghua Liang, and Shi-Xia Liu. A survey of visual analytics techniques and applications: State-of-the-art research and future challenges. *Journal of Computer Science and Technology*, 28(5):852–867, 2013.

[488] Jerzy M. Szymanski, Janusz Sobecki, Piotr Chynal, and Jedrzej Anisiewicz. Eye tracking in gesture based user interfaces usability testing. In Ngoc Thanh Nguyen, Bogdan Trawinski, and Raymond Kosala, editors, *Proceedings of the 7th Asian Conference on Intelligent Information and Database Systems, ACIIDS*, volume 9012 of *Lecture Notes in Computer Science*, pages 359–366. Springer, 2015.

[489] Kay Talmi and Jin Liu. Eye and gaze tracking for visually controlled interactive stereoscopic displays. *Signal Processing: Image Communication*, 14(10):799–810, 1999.

[490] Christiane Taras and Thomas Ertl. Interaction with colored graphical representations on braille devices. In Constantine Stephanidis, editor, *Proceedings of 5th International Conference on Universal Access in Human-Computer Interaction. Addressing Diversity, UAHCI*, volume 5614 of *Lecture Notes in Computer Science*, pages 164–173. Springer, 2009.

[491] Pawel Tarnowski, Marcin Kolodziej, Andrzej Majkowski, and Remigiusz Jan Rak. Eye-tracking analysis for emotion recognition. *Computational Intelligence and Neuroscience*, 2020:1–13, 2020.

[492] Benjamin W. Tatler, Nicholas J. Wade, Hoi Kwan, John M. Findlay, and Boris M. Velichkovsky. Yarbus, eye movements and vision. *Iperception*, 1(1):7–27, 2010.

[493] William R. Taylor. Multiple sequence alignment by a pairwise algorithm. *Computer Applications in the Biosciences*, 3(2):81–87, 1987.

[494] Soon Tee Teoh. A study on multiple views for tree visualization. In *Proceedings of the Conference on Visualization and Data Analysis*, volume 6495, 2007.

[495] James J. Thomas and Kristin A. Cook. *Illuminating the Path - The Research and Development Agenda for Visual Analytics*. National Visualization and Analytics Ctr, 2005.

[496] Christian Tominski. *Interaction for Visualization*. Synthesis Lectures on Visualization. Morgan & Claypool Publishers, 2015.

[497] Warren S. Torgerson. Multidimensional scaling I: Theory and method. *Psychometrika*, 17:401–419, 1952.

[498] Melanie Tory, M. Stella Atkins, Arthur E. Kirkpatrick, Marios Nicolaou, and Guang-Zhong Yang. Eyegaze analysis of displays with combined 2d and 3d views. In *Proceedings of the 16th IEEE Visualization Conference, IEEE*, pages 519–526. IEEE Computer Society, 2005.

[499] Michael Traoré and Christophe Hurter. Exploratory study with eye tracking devices to build interactive systems for air traffic controllers. In Guy A. Boy, editor, *Proceedings of the International Conference on Human-Computer Interaction in Aerospace, HCI-Aero*, pages 6:1–6:9. ACM, 2016.

[500] Anne Treisman. Preattentive processing in vision. *Computer Vision, Graphics, and Image Processing*, 31(2):156–177, 1985.

[501] Sandra Trösterer, Alexander Meschtscherjakov, David Wilfinger, and Manfred Tscheligi. Eye tracking in the car: Challenges in a dual-task scenario on a test track. In Linda Ng Boyle, Gary E. Burnett, Peter Fröhlich, Shamsi T. Iqbal, Erika Miller, and Yuqing Wu, editors, *Adjunct Proceedings of the 6th International Conference on Automotive User Interfaces and Interactive Vehicular Applications*, pages 12:1–12:6. ACM, 2014.

[502] Hoi Ying Tsang, Melanie Tory, and Colin Swindells. eSeeTrack - visualizing sequential fixation patterns. *IEEE Transactions on Visualization and Computer Graphics*, 16(6):953–962, 2010.

[503] Edward Rolf Tufte. *The visual display of quantitative information*. Graphics Press, 1992.

[504] Vilius Turenko, Simonas Baltulionis, Mindaugas Vasiljevas, and Robertas Damasevicius. Analysing program source code reading skills with eye tracking technology. In Robertas Damasevicius, Tomas Krilavicius, Audrius Lopata, Dawid Polap, and Marcin Wozniak, editors, *Proceedings of the International Conference on Information*

Technologies, IVUS, volume 2470 of *CEUR Workshop Proceedings*, pages 33–37. CEUR-WS.org, 2019.

[505] Barbara Tversky, Julie Bauer Morrison, and Mireille Bétrancourt. Animation: can it facilitate? *International Journal of Human Computer Studies*, 57(4):247–262, 2002.

[506] Timothy C. Urdan. *Statistics in Plain English, 5th Edition*. Routledge, 2021.

[507] André Calero Valdez, Martina Ziefle, and Michael Sedlmair. Priming and anchoring effects in visualization. *IEEE Transactions on Visualization and Computer Graphics*, 24(1):584–594, 2018.

[508] Niilo V. Valtakari, Ignace Th. C. Hooge, Charlotte Viktorsson, Pär Nyström, Terje Falck-Ytter, and Roy S. Hessels. Eye tracking in human interaction: Possibilities and limitations. In Khiet P. Truong, Dirk Heylen, Mary Czerwinski, Nadia Berthouze, Mohamed Chetouani, and Mikio Nakano, editors, *Proceedings of the International Conference on Multimodal Interaction, ICMI*, page 508. ACM, 2020.

[509] Huub van de Wetering, Nico Klaassen, and Michael Burch. Space-reclaiming icicle plots. In *Proceedings of the IEEE Pacific Visualization Symposium, PacificVis*, pages 121–130. IEEE, 2020.

[510] Jan van der Kamp and Veronica Sundstedt. Gaze and voice controlled drawing. In Veronica Sundstedt and Charlotte C. Sennersten, editors, *Proceedings of First Conference on Novel Gaze-Controlled Applications, NGCA*, page 9. ACM, 2011.

[511] Laurens van der Maaten and Geoffrey Hinton. Visualizing high-dimensional data using t-SNE. *Journal of Machine Learning Research*, 9:2579–2605, 2008.

[512] Tamara van Gog, Liesbeth Kester, Fleurie Nievelstein, Bas Giesbers, and Fred Paas. Uncovering cognitive processes: Different techniques that can contribute to cognitive load research and instruction. *Computers in Human Behavior*, 25(2):325–331, 2009.

[513] Jarke J. van Wijk. The value of visualization. In *Proceedings of 16th IEEE Visualization Conference, IEEE*, pages 79–86. IEEE Computer Society, 2005.

[514] Jarke J. van Wijk. Evaluation: A challenge for visual analytics. *Computer*, 46(7):56–60, 2013.

[515] Boris M. Velichkovsky and John Paulin Hansen. New technological windows into mind: There is more in eyes and brains for human-computer interaction. In Bonnie A. Nardi, Gerrit C. van der Veer, and Michael J. Tauber, editors, *Proceedings of the Conference on*

Human Factors in Computing Systems: Common Ground, CHI, pages 496–503. ACM, 1996.

[516] Boris M. Velichkovsky, Andreas Sprenger, and Pieter Unema. Towards gaze-mediated interaction: Collecting solutions of the "midas touch problem". In Steve Howard, Judy Hammond, and Gitte Lindgaard, editors, *Proceedings of the International Conference on Human-Computer Interaction, INTERACT*, volume 96, pages 509–516. Chapman & Hall, 1997.

[517] Anton Wachner, Janick Edinger, and Christian Becker. Towards gaze-based mobile device interaction for the disabled. In *Proceedings of the IEEE International Conference on Pervasive Computing and Communications Workshops, PerCom Workshops*, pages 397–402. IEEE Computer Society, 2018.

[518] Nicholas J. Wade. How were eye movements recorded before yarbus? *Perception*, 44(8–9):851–883, 2015.

[519] Nicholas J. Wade and Benjamin W. Tatler. Did javal measure eye movements during reading? *Journal of Eye Movement Research*, 2(5), 2009.

[520] Yue Wang, Soon Tee Teoh, and Kwan-Liu Ma. Evaluating the effectiveness of tree visualization systems for knowledge discovery. In Beatriz Sousa Santos, Thomas Ertl, and Kenneth I. Joy, editors, *Proceedings of 8th Joint Eurographics - IEEE VGTC Symposium on Visualization, EuroVis*, pages 67–74. Eurographics Association, 2006.

[521] Colin Ware. *Information Visualization: Perception for Design*. Morgan Kaufmann, 2004.

[522] Colin Ware. *Visual Thinking: for Design*. Morgan Kaufmann Series in Interactive Technologies, Paperback, 2008.

[523] Nadir Weibel, Adam Fouse, Colleen Emmenegger, Sara Kimmich, and Edwin Hutchins. Let's look at the cockpit: exploring mobile eye-tracking for observational research on the flight deck. In Carlos Hitoshi Morimoto, Howell O. Istance, Stephen N. Spencer, Jeffrey B. Mulligan, and Pernilla Qvarfordt, editors, *Proceedings of the Symposium on Eye-Tracking Research and Applications, ETRA*, pages 107–114. ACM, 2012.

[524] Dieter Weidlich, Sandra Scherer, and Markus Wabner. Analyses using VR/AR visualization. *IEEE Computer Graphics and Applications*, 28(5):84–86, 2008.

[525] Julia M. West, Anne R. Haake, Evelyn P. Rozanski, and Keith S. Karn. eyePatterns: software for identifying patterns and similarities across

fixation sequences. In Kari-Jouko Räihä and Andrew T. Duchowski, editors, *Proceedings of the Eye Tracking Research & Application Symposium, ETRA*, pages 149–154. ACM, 2006.

[526] Mark A. Whiting and Nick Cramer. Webtheme™: Understanding web information through visual analytics. In Ian Horrocks and James A. Hendler, editors, *Proceedings of the First International Semantic Web Conference - The Semantic Web - ISWC*, volume 2342 of *Lecture Notes in Computer Science*, pages 460–468. Springer, 2002.

[527] Andreas Wichert and Luis Sa-Couto. *Machine Learning - A Journey to Deep Learning - with Exercises and Answers*. WorldScientific, 2021.

[528] Pak Chung Wong, Han-Wei Shen, Christopher R. Johnson, Chaomei Chen, and Robert B. Ross. The top 10 challenges in extreme-scale visual analytics. *IEEE Computer Graphics and Applications*, 32(4):63–67, 2012.

[529] Pak Chung Wong and Jim Thomas. Visual analytics. *IEEE Computer Graphics and Applications*, 24(5):20–21, 2004.

[530] David S. Wooding. Fixation maps: quantifying eye-movement traces. In Andrew T. Duchowski, Roel Vertegaal, and John W. Senders, editors, *Proceedings of the Eye Tracking Research & Application Symposium, ETRA*, pages 31–36. ACM, 2002.

[531] Paul R. Woodward, David H. Porter, Michael R. Knox, Steven T. Andringa, Alex J. Larson, and Aaron Stender. A system for interactive volume visualization on the powerwall. In *Proceedings of 16th IEEE Visualization Conference, IEEE Vis*, page 110. IEEE Computer Society, 2005.

[532] Ping Wu, Jie Gu, and Tian Lu. Don't lie to me: Tracking eye movement and mouse trajectory to detect deception in sharing economy. In Constantine Stephanidis, editor, *Proceedings of 20th International Conference on Human-Computer Interaction, HCI*, volume 850 of *Communications in Computer and Information Science*, pages 377–381. Springer, 2018.

[533] Kai Xie, Jie Yang, and Yue Min Zhu. Real-time visualization of large volume datasets on standard PC hardware. *Computer Methods and Programs in Biomedicine*, 90(2):117–123, 2008.

[534] Kai Xu, Chris Rooney, Peter Passmore, Dong-Han Ham, and Phong H. Nguyen. A user study on curved edges in graph visualization. *IEEE Transactions on Visualization and Computer Graphics*, 18(12):2449–2456, 2012.

[535] Tohru Yagi, Yoshiaki Kuno, Kazuo Koga, and Toshiharu Mukai. Drifting and blinking compensation in electro-oculography (EOG) eye-gaze interface. In *Proceedings of the IEEE International Conference on Systems, Man and Cybernetics*, pages 3222–3226. IEEE, 2006.

[536] Mehmet Adil Yalçin, Niklas Elmqvist, and Benjamin B. Bederson. Raising the bars: Evaluating treemaps vs. wrapped bars for dense visualization of sorted numeric data. In Elmar Eisemann and Scott Bateman, editors, *Proceedings of the 43rd Graphics Interface Conference*, pages 41–49. Canadian Human-Computer Communications Society / ACM, 2017.

[537] Yasuko Yamagishi, Hiroshi Yanaka, Katsuhiko Suzuki, Seiji Tsuboi, Takehi Isse, Masayuki Obayashi, Hajimu Tamura, and Hiromichi Nagao. Visualization of geoscience data on google earth: Development of a data converter system for seismic tomographic models. *Computers & Geosciences*, 36(3):373–382, 2010.

[538] Chia-Kai Yang. Identifying reading patterns with eye-tracking visual analytics. In Andreas Bulling, Anke Huckauf, Eakta Jain, Ralph Radach, and Daniel Weiskopf, editors, *Proceedings of the Symposium on Eye Tracking Research and Applications, ETRA Adjunct*, pages 10:1–10:3. ACM, 2020.

[539] Alfred L. Yarbus. *Eye Movements and Vision*. Springer, 1967.

[540] Lora Yekhshatyan and John D. Lee. Changes in the correlation between eye and steering movements indicate driver distraction. *IEEE Transactions on Intelligent Transportation Systems*, 14(1):136–145, 2013.

[541] Chi-Hsien Yen, Aditya G. Parameswaran, and Wai-Tat Fu. An exploratory user study of visual causality analysis. *Computer Graphics Forum*, 38(3):173–184, 2019.

[542] Leelakrishna Yenigalla, Vinayak Sinha, Bonita Sharif, and Martha E. Crosby. How novices read source code in introductory courses on programming: An eye-tracking experiment. In Dylan D. Schmorrow and Cali M. Fidopiastis, editors, *Proceedings of 10th International Conference on Foundations of Augmented Cognition: Neuroergonomics and Operational Neuroscience, AC*, volume 9744 of *Lecture Notes in Computer Science*, pages 120–131. Springer, 2016.

[543] Yeliz Yesilada, Caroline Jay, Robert Stevens, and Simon Harper. Validating the use and role of visual elements of web pages in navigation with an eye-tracking study. In Jinpeng Huai, Robin Chen,

Hsiao-Wuen Hon, Yunhao Liu, Wei-Ying Ma, Andrew Tomkins, and Xiaodong Zhang, editors, *Proceedings of the 17th International Conference on World Wide Web, WWW*, pages 11–20. ACM, 2008.

[544] Ji Soo Yi, Youn ah Kang, John T. Stasko, and Julie A. Jacko. Toward a deeper understanding of the role of interaction in information visualization. *IEEE Transaction on Visualization and Computer Graphics*, 13(6):1224–1231, 2007.

[545] Nurul Hidayah Mat Zain, Fariza Hanis Abdul Razak, Azizah Jaafar, and Mohd Firdaus Zulkipli. Eye tracking in educational games environment: Evaluating user interface design through eye tracking patterns. In Halimah Badioze Zaman, Peter Robinson, Maria Petrou, Patrick Olivier, Timothy K. Shih, Sergio A. Velastin, and Ingela Nyström, editors, *Proceedings of Second International Visual Informatics Conference on Sustaining Research and Innovations, IVIC*, volume 7067 of *Lecture Notes in Computer Science*, pages 64–73. Springer, 2011.

[546] Jingyuan Zhang and Hao Shi. Geospatial visualization using google maps: A case study on conference presenters. In *Proceeding of the Second International Multi-Symposium of Computer and Computational Sciences IMSCCS*, pages 472–476. IEEE Computer Society, 2007.

[547] Li Zhang, Jianxin Sun, Cole S. Peterson, Bonita Sharif, and Hongfeng Yu. Exploring eye tracking data on source code via dual space analysis. In *Proceedings of the Working Conference on Software Visualization, VISSOFT*, pages 67–77. IEEE, 2019.

[548] Chenyang Zheng and Tsuyoshi Usagawa. A rapid webcam-based eye tracking method for human computer interaction. In *Proceedings of International Conference on Control, Automation and Information Sciences, ICCAIS*, pages 133–136. IEEE, 2018.

[549] Anjie Zhu, Shiwei Cheng, and Jing Fan. Eye tracking and gesture based interaction for target selection on large displays. In *Proceedings of the ACM International Joint Conference and International Symposium on Pervasive and Ubiquitous Computing and Wearable Computers, UbiComp/ISWC*, pages 319–322. ACM, 2018.

[550] Zhiwei Zhu and Qiang Ji. Eye and gaze tracking for interactive graphic display. *Machine Vision and Applications*, 15(3):139–148, 2004.

[551] Stephen T. Ziliak. Retrospectives: Guinnessometrics: The economic foundation of "student's" t. *Journal of Economic Perspectives*, 22(4):199–216, 2008.

[552] Torre Zuk, Lothar Schlesier, Petra Neumann, Mark S. Hancock, and Sheelagh Carpendale. Heuristics for information visualization evaluation. In Enrico Bertini, Catherine Plaisant, and Giuseppe Santucci, editors, *Proceedings of the AVI Workshop on BEyond time and errors: novel evaluation methods for information visualization, BELIV*, pages 1–6. ACM Press, 2006.

Index

2.5D treemap 169
3D mobile eye tracking 193

A

Abstract/Elaborate 56
Abstract data 27
Adaptive contextualization 174
Adjacency list 42
Adjacency matrix 42, 95
Aesthetics 65
Aesthetics graph drawing criteria 43, 67, 170
Afterimage 186
Age group 141
Aha effect 61
Albert Einstein 75
Alfred Yarbus 99, 153, 185
Algorithms 106, 110
Alternation 50
Alzheimer's disease 141
Amnestic disease 141
Analytical reasoning 75
Angle 23
Angular disparity 257
Animation 50, 68, 172
Anonymization 80, 81
Annotation 58, 80, 81
ANOVA 162
Anscombe's quartet 25, 158
Antecedent 96
AOI river 261
AOI transition graph 261
AOI-based visualization 261
Area chart 32
Area of interest 198, 204
Argus science 193
Arrangement 23
Artificial intelligence 253
Association rule 86, 96
Astigmtism 181, 184
Asymptotic runtime 112
Attribute 44
Autism 181
Autofocus mechanism 179
Augmented reality 37, 58
Aviation 189

B

Bar chart 24, 32, 98, 163, 256
Bar code 49
Bee swarm visualization 258
Bell-shaped distribution 164
Between-subjects design 129, 147
Bipartite layout 115
Bidirectional dialogue 53
Big data 38
Bimodal distribution 164
Binocular 193
Biovisualization 27
Bite bar 188
Bivariate data 24, 164
Blickshift 193
Blood pressure 207
Blood volume 207
Body movement 60
Box plot 165
Braille 57, 145
Branching factor 43
Browsing 51
Brushing and linking 57
Bubble hierarchy 168

C

Cafe wall illusion 72
Calibration detail 45

Calibration error 82
Car driving experiment 200
Cartesian coordinate system 24
Cartographic map 30
Cataract 178, 181
Catch-up saccade 181, 203
Categorical data 39, 40
Cave painting 29
Central vision 218
Change blindness 68, 101
Charles Joseph Minard 32
Chart junk 67
Checker shadow illusion 72
Chernoff faces 45
Chin rest 188
Classing 84
Classification 84
Closed-end question 156
Closure 70
Cluster 70
Clustering 83, 109
Cognition 27, 77
Cognitive overload 90
Cognitive psychology 19, 178
Collaborative interaction 37, 61
Collapse 47
Color blindness 57, 64, 145, 178, 181, 183
Color coding 57
Color deficiency 4, 100
Color vision 144
Combined treemap 169
Communication 20
Communicative visualization 22
Community detection task 102
Comparison task 52, 101
Comparative study 167
Complex data type 39
Computational pathology 118
Computer monitor 59, 60
Computing resources 85
Cones 179
Confidence 96
Conflicting data 75
Confounding variable 128, 153
Conjugate eye movement 180
Connect 56
Consensus matrix 243, 252
Consequent 96
Contact lense 178
Context 52
Continual eye movement 180
Continuous data 47, 80
Controlled study 130, 143, 157, 158, 172
Convex hull 259
Convolutional neural network 254
Cooperative eye hypothesis 29
Cornea 178
Corneal reflection 187, 191
Corrected-to-normal 145, 183
Correlation 43, 93, 95
Correlation analysis 251
Correlation task 101
Countertrend 50
Counting task 100, 108
Covert attention 190
Coxcomb 34
Crispness 23
Cross checking behavior 211
Cross eyes 178
Crowdsourcing study 135, 136
Cultural difference 142
Curvature 170
Cyber attack 120
Cyber security 120

D

Data 19
Data acquisition 190, 192, 232
Data aggregation 85, 248
Data annotation 81
Data anonymization 81
Data classification 247
Data classing 247
Data cleaning 106, 240
Data clustering 245
Data collapsing 85
Data collection 80
Data dimension 50
Data enhancement 82, 241
Data enrichment 241
Data ethics 38
Data handling 79

Data interpretation 81, 235
Data lineage 82
Data linking 81, 236
Data management 77, 79
Data migration 106
Data mining 77, 86, 167
Data normalization 85
Data ordering 244
Data parsing 82, 106
Data preparation 160
Data presentation 82
Data privacy 37, 82
Data projection 85, 249
Data provenance 38, 82
Data reading 82
Data sampling 106
Data screening 119
Data security 37, 82
Data sorting 244
Data storage 83, 238
Data stream 90
Data summarization 247
Data transformation 82, 242
Data type 38
Data validation 81, 82, 240
Data verification 81, 82, 240
Data volume 90
Data-to-ink ratio 67, 260
DBLP 26
Decision making 87
Deep eyelashes 145
Deep learning 89, 253
Degree of freedom 150
Dependent variable 39, 129, 153
Depth 43
Depth camera 193
Descriptive statistics 160
Design flaw 43, 61, 66
Design principle 62
Deterministic 107
Diaphragm 179
Digital camera 179
Digital camera lens 179
Dikablis glasses 193
Dimensionality reduction 70, 85, 109, 249
Disabled people 59
Discrete data 27
Disease 181
Disjunctive eye movement 180
Disorder 181
Dissemination 32, 87
Document data 45
Domain expert 125
Dot plot 40
Dry eyes 178
Dwell time 193
Dynamic AOI detection 201
Dynamic graph 117
Dynamic graph visualization 170
Dynamic rapid serial visual
presentation 116
Dynamic stability 68
Dynamic stimulus 150
Dynamic visual analytics 11, 123
D-lab 193

E

Ebbinghaus illusion 71
Eclipse software system 49
Edge 42
Edge bundling 43, 260
Edge representation style 43, 170
Edge splatting 117
EEG 199, 207
Electrocardiogram 51, 207
Electronic signal 179
Electro-oculography 188, 191
Emergence 69
EMG 207
Emotional state 144, 183
Empirical user evaluation 169
Encode 56
Ergoneers 193
Error rate 129
Error report 82
Estimation task 101
Ethics 125, 145
Ethics committee 146
ETRA 189
Euclidian distance 156
Evaluation 88, 125
Experience 21

Experimenter 157
Expert study 132
Explainanable artificial intelligence 89
Explicit link 43
Explore 56
Exponential runtime 112
Extraocular muscle 180
Eye 177
Eye anatomy 178
Eye cataract 141
Eye lashes 178
Eye movement 179
Eye movement data 44, 122
Eye movement direction plot 140
Eye muscle 179
Eye rotation 180
Eye tracking 175
Eye tracking study 33, 138
Eye tracking visual analytics 263
Eye Tribe 214
Eye Tribe SDK 225
Eye twitching 178
EyeLink 1000 eye tracker 219
EyeLink 2 193
EyeSee 193
EyeTech 193
EyeVido 193
EyeWare 193
Eyestrain 178
Eyes-off-road detection 196
Eye-controllable monitor 194
Eye-mind hypothesis 178, 190

F

F distribution 162
Face expressions 45
Factorial ANOVA 162
False negative 156
False positive 156
Faraday's law 191
Farsightedness 184
Fatigue effect 128
Feasibility study 128
Feedback loop 105
Field study 136
Film camera 188
Filter 56, 57
Filtering 26, 66, 252
Fitt's law 172
Fixation 187, 198, 202
Fixation deviation 257
Fixation duration 44, 156, 163
Fixation map 258
Fixation radius 202, 203
Flickering stimulus 146
Florence Nightingale 32
Focal plane 183
Focus-and-context 57
Fovea 178
Francis Galton 32
Friedman test 162
Fuzzy clustering 246
F-shape pattern 216
F-test 162

G

Galvanic skin response 199, 207
Gaze 59, 198
Gaze depth 201
Gaze Intelligence 193
Gaze plot 45, 98, 230, 258
Gaze point 198, 202
Gaze point cluster 202
GazeHawk 193
Gazepoint 193
Gaze-assisted interaction 54, 59, 176, 215
Geodesic-path tendency 214
Geographic map 32
Gestalt law 68
Gestalt psychology 68
Gesture 59, 60
GGobi 158
Glaucoma 178, 181
Glasses 178
Glyph 44
Goal-oriented activity 54
Google Earth 31
Google Maps 31
Graph visualization 27, 169
Graph 42
Graphical primitive 26
Graying out 57

Grouping 69, 70
Guiding line 64
Gypsum cap 187

H
Hand control 59
Harry Beck 32
Head mounted eye tracker 132, 189, 191
Head movement 60
Head rest 188
Head tracking 193
Heat map 258
Herman grid illusion 72
Heterogeneous data 78, 80
Heuristic 107
Hierarchical clustering 246
Hierarchical data 168
Hierarchical granularity 49, 168
Hierarchy 43
Hierarchy visualization 37, 168
Highlighting 56, 64
Histogram 163, 164, 256
History of visualization 28
Hot spot 48, 258
HTC Vive Pro Eye 195
Hue 23
Human eye 144
Human user 138
Human-computer interaction 36, 54, 77
Human-in-the-loop 77
Human-is-the-loop 77
Hybrid output 60
Hyperbolic browser 169
Hyperopia 181, 184
Hypothesis 27, 97, 148
H-tree 169

I
Iconic graphics 41, 63
Illuminating the path 78
Immersive analytics 37, 58
iMotions 193
Indentation 43
Indented tree 169
Independent variable 39, 129, 153
Industrialization 32
Inferential statistics 161
Infographic 32
Information monitor 59
Information overflow 64
Information pyramid 169
Information visualization 19, 27, 36
Infrared oculography 191
Inner eye component 178
Input device 53, 58
Input modality 59
Input parameter 110
Insight 104
Instrumentation 150
Interaction effect 162
Interaction graph 201
Interaction history 55, 57
Interactive responsiveness 6
Interaction technique 53, 171
Interactive timeslicing 172
Interactive visual interface 75
Interpolation 164, 241
Interview 147
Intrusive 190
Invariance 69
Invasive 190
Invasiveness 187
Iris 178
ISCAN 193
ISCAN RK726/RK520 eye tracker 217
Ishihara color perception test 144
iViewX eye tracker 214

J
Jacques Bertin 41
JASP 158
Jock Mackinlay 41
John Snow 32
Joint attention 185
Joystick 59
Judgement task 100
J-shaped distribution 164

K
Keyboard 59
Knowledge 104
Knowledge discovery in databases 77, 86

Knowledge state 201
Kruskal-Wallis test 162
Kymograph 186
k-means clustering 246

L

Lab study 136
Label 31, 64
Laboratory 136
Landau symbol 112
Large-scale interaction 60
Las Vegas 107
Law of common fate 70
Law of continuity 70
Law of good form 70
Layout 43
Layout algorithm 43
Lazy eye 178
LC Technologies 193
Learning effect 128
Least common ancestor 43
Left-skewed distribution 164
Left-to-right reading direction 64
Legend 63
Lens 178
Leo Cherne 75
Leonard Euler 42
Level of abstraction 56
Level of expertise 139
Level of granularity 56
Levenshtein distance 252
Lie factor 67
Likert scale 156
Limited-number population study 135
Line chart 164
Line graph 32, 256
Linear runtime 112
Line-based diagrams 45, 67, 102
Logarithmic runtime 112
London Underground Tube map 34
Longitudinal study 134
Low-level task 99

M

Machine learning 77
Macular edema 182
Magnetic field 191
Malware analysis 119
Map 30
MATLAB 159
Mean 161
Mechanical Turk 135
Median 161
Medium-scale interaction 60
Memory consumption 114
Mental map 55, 68, 103
Menu 58
Metadata 39, 52, 82
Microsaccade 202
Microsoft Hololens 195
Midas touch problem 59, 176, 193, 269
Minard map 33
Minimal linear arrangement 107
Minimalistic diagram 67
Mirametrix 193
Mobile eye tracking 189, 191
Mobile phone 59
Monte Carlo 107
Motion-compensated attention map 259
Mouse 59
Müller-Lyer illusion 72
Multidimensional scaling 108
Multiple coordinated views 14, 18, 35, 56, 65
Multiple sequence alignment 252
Multistability 69
Multivariate data 44, 66, 83, 95
Multi-camera eye tracker 191
Multi-dimensional data 43
Multi-layered neural network 254
Multi-modal interaction 215
Myopia 181, 184

N

Natural language processing 109
Navigation system 102
Navigation task 172
NCBI taxonomy 43
Nearsightedness 184
Needleman-Wunsch algorithm 252
Negative correlation 24, 95, 96, 139, 165, 251
Nerve impulse 179
Nesting 43

Neural network 254
Neuron 177
Night blindness 178
Node-link diagram 42
Node-link tree 169
Nominal data 39, 40
Non-expert study 132
Non-invasive 188
Normal color vision 145
Normal distribution 164
Null hypothesis 162

O

Occlusion effect 48, 106
Oculometer 189
Oculomotor system 180
Offline 107
Off-screen fixation 204
Off-screen target 172
One-way ANOVA 162
Online 107
Online experiment 131
Online eye tracking 132, 268
On-screen fixation 204
Open-ended question 156
Ophtalmologist 184
Optic nerve 179
Optical eye tracker 187
Optical flow 259
Optical illusions 71
Optimal 107
Optimal linear arrangement problem 107
Optometrist 184
Order of visual attention 153
Ordering 83, 109
Ordinal data 40
Orientation 23
Oscillation 50
Outer eye component 178
Output device 53, 58
Overt attention 190
Overview-and-detail 57

P

Pairwise sequence alignment 252
Panning 31, 56
Parallel coordinates plots 44, 95, 165
Parameter space 156
Parent-child relationship 43
Parkinson's disease 181
Partial link 154
Participants 77, 128
Part-to-a-whole relationship 163
Pattern 93, 94
Pattern identification task 101
Pattern recognition 90
Pedigree tree 113
Pen 59
Perception 77
Perceptual ability 177
Performance 112
Peripheral vision 100, 218
Permutation 140
Perspective distortion 58
Photographic 187
Photographic tape 188
Photoreceptor 179
Photosensitive 187
Physical device 58
Physical disablement 194
Physiological measure 45, 76, 199, 206
Pie chart 24, 32, 163
Pilot study 127, 128
Pixel-based representation 66
Pointing device 172
Point-based visualization 257
Polar area diagram 23, 34
Polyline 44, 96
Polynomial runtime 112
Ponzo illusion 72
Population 128
Portable eye tracker 191, 193
Position 23
Position in a common scale 23
Positive correlation 43, 95, 165
Post process analysis 12
Post-saccadic oscillation 202
Powerwall high resolution display 60
Practice runthrough 139
Prefix tag cloud 46
Pre-attentive 64
Primary color 185

Primary visual analytics system 115
Primitive data types 39
Principal component analysis 108
PRISM 159
Privacy 125
Probabilistic model 102
Problem-driven study 147
Progressive treemap 169
Projection 70, 109
Proximity 70
Pseudo-isochromatic plate 144
PSPP 158
Public transport map 213
Pupil 178
Pupil dilation 199, 207
Pupil Labs 193
Pythagoras tree 103
p-value 162

Q

Quadratic runtime 112
Qualitative study 129
Quantitative data 39
Quantitative study 129
Quantitative values 24
Quartile 165

R

R 159
Radial tree 168
Radial visualization 66
Randomization 140
Randomized algorithm 102, 107
Rapid eye movement 180
Rapid serial visual presentation 51, 68
RapidMiner 159
Reading direction 142
Reading task 100, 185
Real-time analysis 12
Real-time data 35, 107
Real-time eye tracking 193
Reconfigure 56
RED500 193
Redo 54, 57
Reference graphic 33
Reflex eye movement action 180
Refractive error 181, 184
Region of interest 261
Reification 69
Reinforcement learning 254
Relational data 43, 84
Remote eye tracking 130, 182, 191
Replication 140
Research question 148
Resolution 23
Respiration activity 207
Response latency 257
Response time 129
Retina 178
Retinal disorder 181
Retinal variables 23
Return sweep 187
Right-skewed distribution 164
RINGS 168
Rods 179
Root node 43
Route finding task 101, 109, 213
Rubber suction cap 189
Rule 93
Runtime 91, 102, 114
Runtime complexity 111

S

Saccade 156, 181, 187, 198, 202
Saccade length 156, 163
Saccade orientation 156
Saliency 254
Sampling frequency 202
Sankey diagram 32, 261
SAS 159
Scalar field 47
Scale 52, 64
Scanpath 45, 156, 202
Scanpath length 156
Scarf plot 262
Scatterplot 25, 44, 52, 53, 95, 118, 250, 257
Scatterplot matrix 45, 293
Scientific visualization 27, 36
Sclera 178
Scleral search coil 189
Search coil 191
Search task 99

Secondary visual analytics system 115
Select 56
Self-explanatory diagram 39
Semantic segmentation 264
Semi-supervised learning 254
Sense 29, 38, 78, 134
Sensemaking 173
Senso-motoric 62, 141
Sequence comparison 109
Sequence rule 86, 96
Set visualization 47
Seven Bridges of Königsberg 42
Shape 23, 41
Short-term memory 258
Short-term study 134
Significance level 162
Similarity 70
Single-camera eye tracker 191
Size 23
Skin conductance 207
Skin potential 191
Sliding time window 116, 260
Small multiples 172
Small-scale interaction 60
Small-scale study 139
Smart Eye 193
Smart Eye eye tracker 217
Smartphone 60
SMI 193
SMI EyeLink I 219
SMI RED250 eye tracker 214
SMI RED500 eye tracker 219
SMI REDn eye tracker 222
Smooth animation 68
Smooth pursuit 179, 181, 203
Smooth transition 56
Snellen chart 144
Software visualization 27, 49
Sorting 83, 109
Space-reclaiming icicle plot 169
Space-time cube 257
Sparklines 66
Spatial data 27
Spatial fixation coverage 156
Spatio-temporal data 44
Speech 60
Spinning dancer illusion 72
Sport map 34
SPSS 159
Squinting 183
SR Research 193
Stacking 43
Standard deviation 161
Standard error of the means 166
Star plot 257
Stata 159
Static stimulus 150
STATISTICA 159
Statistical evaluation 158
Statistical plot 256
Statistical test 160, 161
Statistics 25, 159
Stemming 109
Stereoscopic display 60
Stimulus 45, 149, 198
Stochastic model 155
Strabismus 181
Student's t-test 162
Study design 147
Study type 127
Suction cap 186
Summarization 84
Sunburst 168
Supervised learning 254
Support 96
Swarm behavior 71, 102
Symmetry 70
Synchronization task 198
Synergy effect 7
Synthetic data 127

T

Tag cloud 46
Talk-aloud study 130
Tangible visualization 59
Tapered edge 154
Task 97, 151
Technique-driven study 147
Temporal cluster 117
Tensor field 47
Text data 45
Text reading task 51, 64

Texture 23, 41
The Eye Tribe 193
The Unexpected Visitor 153
ThemeRiver 50
Think-aloud study 130
Three-way ANOVA 162
Time to first fixation 199
Timeline plot 257
Time-dependent data 39
Time-series plot 52
Time-to-space mapping 50, 68
Time-to-time mapping 50, 68
Tobii 193
Tobii 1720 eye tracker 219
Tobii 1750 eye tracker 216
Tobii 2150 eye tracker 215
Tobii EyeX eye tracker 222
Tobii Pro Glasses 3 190
Tobii REX eye tracker 216
Tobii T120 eye tracker 214
Tobii T60 XL eye tracker 211
Tobii X120 eye tracker 214
Tobii X2-60 eye tracker 215
Tobii x5 eye tracker 222
Tokenization 109
Top-to-bottom layout 49
Touch 59
Touch device 59
Tracker Pro 193
Tracking rate 202
Trajectory 44
Transition 198, 204
Transition probability 55
Transparency 23
Transparent overlay 258
Treemap 168
Trend 50
Trend analysis 250
Trial 147
Trivariate data 164
Two-way ANOVA 162
Type I error 156, 162
Type II error 156
t-distributed stochastic neighbor embedding 108
t-test 162

U

Uncontrolled study 130
Undo 54, 57
Uniform distribution 164
Uniform manifold approximation and projection 108
Univariate data 164
Unsupervised learning 254
Usability 59
User evaluation 37, 125
User friendliness 62
User interface eye tracking 221
Users-in-the-loop 54, 61, 139

V

Value 23
Variance 161
Vector field 48
Verbal feedback 45, 60
Vertex 42
Video data 121
Video oculography 191
Video stimulus 151
Video surveillance 121
Virtual reality 59, 171
Vision deficiency 144
Visual acuity 64, 144, 183
Visual analytics 36, 75, 172, 254
Visual analytics eye tracking 223
Visual analytics pipeline 91, 92
Visual annotation 32
Visual attention 32
Visual attention contour map 140
Visual attention map 48, 230, 258
Visual attention strategy 202
Visual clutter 43, 67, 169, 258
Visual complexity 155
Visual cortex 179
Visual decoration 63
Visual deficiency 57
Visual encoding 39
Visual illusion 71
Visual information seeking mantra 66

Visual metaphor 26
Visual pattern 21
Visual search pattern 212
Visual stimulus 149, 199
Visual system 177
Visual task solution strategy 213
Visual transformation 20
Visual variable 19, 22
Visualization 35
Visualization framework 38
Visualization library 38
Visualization pipeline 54
Visualization technique 254
Voice 59
Voice recognition 194
Volumetric data 47
Voronoi region 206

W

Walk-and-interact 216
Walk-then-interact 216
Wearable eye tracker 132, 189, 191
Web-based environment 38
Web-base visualization 66
Wedge chart 34
Weighted browsing 51, 121
Welch's test 162
William Playfair 32
Within-subjects design 129, 147
Word cloud 46
World wide web 38
Wrapped bar 169

Z

Zooming 31

About the Author

Michael Burch studied computer science and mathematics at the Saarland University in Saarbrücken, Germany. He received his PhD from the University of Trier in 2010 in the fields of information visualization and visual analytics. After 8 years of having been a PostDoc in the Visualization Research Center (VISUS) in Stuttgart, he moved to the Eindhoven University of Technology (TU/e) as an assistant professor for visual analytics. Michael Burch is in many international program committees and published more than 160 conference papers and journal articles in the field of visualization. His main interests are in information visualization, visual analytics, eye tracking, data science, and software engineering. Currently, he works as an associate professor at the University of Applied Sciences in Chur, Switzerland in the center for data analytics, visualization, and simulation (DAViS).